IET MATERIALS, CIRCUITS AND DEVICES SERIES 76

Secured Hardware Accelerators for DSP and Image Processing Applications

Other volumes in this series:

Volume 2 **Analogue IC Design: The current-mode approach** C. Toumazou, F.J. Lidgey and D.G. Haigh (Editors)
Volume 3 **Analogue–Digital ASICs: Circuit techniques, design tools and applications** R.S. Soin, F. Maloberti and J. France (Editors)
Volume 4 **Algorithmic and Knowledge-Based CAD for VLSI** G.E. Taylor and G. Russell (Editors)
Volume 5 **Switched Currents: An analogue technique for digital technology** C. Toumazou, J.B.C. Hughes and N.C. Battersby (Editors)
Volume 6 **High-Frequency Circuit Engineering** F. Nibler *et al.*
Volume 8 **Low-Power High-Frequency Microelectronics: A unified approach** G. Machado (Editor)
Volume 9 **VLSI Testing: Digital and mixed analogue/digital techniques** S.L. Hurst
Volume 10 **Distributed Feedback Semiconductor Lasers** J.E. Carroll, J.E.A. Whiteaway and R.G.S. Plumb
Volume 11 **Selected Topics in Advanced Solid State and Fibre Optic Sensors** S.M. Vaezi-Nejad (Editor)
Volume 12 **Strained Silicon Heterostructures: Materials and devices** C.K. Maiti, N.B. Chakrabarti and S.K. Ray
Volume 13 **RFIC and MMIC Design and Technology** I.D. Robertson and S. Lucyzyn (Editors)
Volume 14 **Design of High Frequency Integrated Analogue Filters** Y. Sun (Editor)
Volume 15 **Foundations of Digital Signal Processing: Theory, algorithms and hardware design** P. Gaydecki
Volume 16 **Wireless Communications Circuits and Systems** Y. Sun (Editor)
Volume 17 **The Switching Function: Analysis of power electronic circuits** C. Marouchos
Volume 18 **System on Chip: Next generation electronics** B. Al-Hashimi (Editor)
Volume 19 **Test and Diagnosis of Analogue, Mixed-Signal and RF Integrated Circuits: The system on chip approach** Y. Sun (Editor)
Volume 20 **Low Power and Low Voltage Circuit Design with the FGMOS Transistor** E. Rodriguez-Villegas
Volume 21 **Technology Computer Aided Design for Si, SiGe and GaAs Integrated Circuits** C.K. Maiti and G.A. Armstrong
Volume 22 **Nanotechnologies** M. Wautelet *et al.*
Volume 23 **Understandable Electric Circuits** M. Wang
Volume 24 **Fundamentals of Electromagnetic Levitation: Engineering sustainability through efficiency** A.J. Sangster
Volume 25 **Optical MEMS for Chemical Analysis and Biomedicine** H. Jiang (Editor)
Volume 26 **High Speed Data Converters** A.M.A. Ali
Volume 27 **Nano-Scaled Semiconductor Devices** E.A. Gutiérrez-D (Editor)
Volume 28 **Security and Privacy for Big Data, Cloud Computing and Applications** L. Wang, W. Ren, K.R. Choo and F. Xhafa (Editors)
Volume 29 **Nano-CMOS and Post-CMOS Electronics: Devices and modelling** S.P. Mohanty and A. Srivastava
Volume 30 **Nano-CMOS and Post-CMOS Electronics: Circuits and design** S.P. Mohanty and A. Srivastava
Volume 32 **Oscillator Circuits: Frontiers in design, analysis and applications** Y. Nishio (Editor)
Volume 33 **High Frequency MOSFET Gate Drivers** Z. Zhang and Y. Liu
Volume 34 **RF and Microwave Module Level Design and Integration** M. Almalkawi
Volume 35 **Design of Terahertz CMOS Integrated Circuits for High-Speed Wireless Communication** M. Fujishima and S. Amakawa
Volume 38 **System Design with Memristor Technologies** L. Guckert and E.E. Swartzlander Jr.
Volume 39 **Functionality-Enhanced Devices: An alternative to Moore's law** P.-E. Gaillardon (Editor)
Volume 40 **Digitally Enhanced Mixed Signal Systems** C. Jabbour, P. Desgreys and D. Dallett (Editors)

Volume 43	**Negative Group Delay Devices: From concepts to applications** B. Ravelo (Editor)
Volume 45	**Characterisation and Control of Defects in Semiconductors** F. Tuomisto (Editor)
Volume 47	**Understandable Electric Circuits: Key concepts, 2nd Edition** M. Wang
Volume 48	**Gyrators, Simulated Inductors and Related Immittances: Realizations and applications** R. Senani, D.R. Bhaskar, V.K. Singh and A.K. Singh
Volume 49	**Advanced Technologies for Next Generation Integrated Circuits** A. Srivastava and S. Mohanty (Editors)
Volume 51	**Modelling Methodologies in Analogue Integrated Circuit Design** G. Dundar and M.B. Yelten (Editors)
Volume 53	**VLSI Architectures for Future Video Coding** M. Martina (Editor)
Volume 54	**Advances in High-Power Fiber and Diode Laser Engineering** I. Divliansky (Editor)
Volume 55	**Hardware Architectures for Deep Learning** M. Daneshtalab and M. Modarressi
Volume 57	**Cross-Layer Reliability of Computing Systems** G. Di Natale, A. Bosio, R. Canal, S. Di Carlo and D. Gizopoulos (Editors)
Volume 58	**Magnetorheological Materials and Their Applications** S. Choi and W. Li (Editors)
Volume 59	**Analysis and Design of CMOS Clocking Circuits for Low Phase Noise** W. Bae and D.K. Jeong
Volume 60	**IP Core Protection and Hardware-Assisted Security for Consumer Electronics** A. Sengupta and S. Mohanty
Volume 64	**Phase-Locked Frequency Generation and Clocking: Architectures and circuits for modem wireless and wireline systems** W. Rhee (Editor)
Volume 65	**MEMS Resonator Filters** R.M. Patrikar (Editor)
Volume 66	**Frontiers in Hardware Security and Trust: Theory, design and practice** C.H. Chang and Y. Cao (Editors)
Volume 67	**Frontiers in Securing IP Cores; Forensic Detective Control and Obfuscation Techniques** A. Sengupta
Volume 68	**High Quality Liquid Crystal Displays and Smart Devices: Vol. 1 and Vol. 2** S. Ishihara, S. Kobayashi and Y. Ukai (Editors)
Volume 69	**Fibre Bragg Gratings in Harsh and Space Environments: Principles and applications** B. Aïssa, E.I. Haddad, R.V. Kruzelecky and W.R. Jamroz
Volume 70	**Self-Healing Materials: From fundamental concepts to advanced space and electronics applications, 2nd Edition** B. Aïssa, E.I. Haddad, R.V. Kruzelecky and W.R. Jamroz
Volume 71	**Radio Frequency and Microwave Power Amplifiers: Vol. 1 and Vol. 2** A. Grebennikov (Editor)
Volume 72	**Tensorial Analysis of Networks (TAN) Modelling for PCB Signal Integrity and EMC Analysis** B. Ravelo and Z. Xu (Editors)
Volume 73	**VLSI and Post-CMOS Electronics Vol. 1: VLSI and post-CMOS electronics and Vol. 2: Materials, devices and interconnects** R. Dhiman and R. Chandel (Editors)
Volume 77	**Integrated Optics Vol. 1: Modeling, material platforms and fabrication techniques and Vol. 2: Characterization, devices, and applications** G. Righini and M. Ferrari (Editors)

Secured Hardware Accelerators for DSP and Image Processing Applications

Anirban Sengupta

The Institution of Engineering and Technology

Published by The Institution of Engineering and Technology, London, United Kingdom

The Institution of Engineering and Technology is registered as a Charity in England & Wales (no. 211014) and Scotland (no. SC038698).

First published 2020

The Institution of Engineering and Technology
Michael Faraday House
Six Hills Way, Stevenage
Herts, SG1 2AY, United Kingdom

www.theiet.org

British Library Cataloguing in Publication Data
A catalogue record for this product is available from the British Library

ISBN 978-1-83953-306-8 (hardback)
ISBN 978-1-83953-307-5 (PDF)

Typeset in India by MPS Limited

Contents

Preface

The book *Secured Hardware Accelerators for DSP and Image Processing Applications* presents state-of-the-art technological solutions for securing and protecting hardware accelerators of digital signal processing (DSP) and image processing applications against major cyberthreats. Hardware accelerators such as image processing filters (blurring filter, sharpening filter, embossing filter, etc.), discrete Fourier transform, finite impulse response filters and JPEG compression hardware are widely used in several consumer, medical, military and space applications. They are an integral component of these sophisticated electronics systems and are responsible for computationally intensive, data crunching and control-intensive applications. All modern electronics gadgets having complex system-on-chips (SoCs) rely heavily on these data-intensive hardware accelerators from DSP and image processing applications. Thus, security/protection of these hardware accelerators against standard threats, IP abuse/misuse, etc. becomes highly essential. This book presents state-of-the art security solutions and optimization algorithms employed for designing secured hardware accelerators for DSP, multimedia and image processing applications.

Broadly, the theme of this book includes the following:

1. Introduction: Secured and optimized hardware accelerators for DSP and image processing applications
2. Cryptography-driven IP steganography for DSP hardware accelerators
3. Double line of defence to secure JPEG codec hardware for medical imaging systems
4. Integrating multi-key-based structural obfuscation and low-level watermarking for double line of defence of DSP hardware accelerators
5. Multimodal hardware accelerators for image processing filters – secured and optimized designs
6. Fingerprint biometric for securing hardware accelerators
7. Key-triggered hash-chaining-based encoded hardware steganography for securing DSP hardware accelerators
8. Designing *N*-point DFT hardware accelerator using obfuscation and steganography
9. Structural transformation and obfuscation frameworks for data-intensive IPs

Chapter 1 presents an *Introduction: secured and optimized hardware accelerators for DSP and image processing applications*. The significant features of this chapter include providing a background on hardware accelerators, why these are important for modern electronics systems, its applications to several domains such as consumer electronics, medical followed by discussion on modern hardware security threats and its solutions. The hardware security tools developed by the author, corresponding to the security solutions presented in the book, are also discussed. The four tools are available for free download publicly at http://www.anirban-sengupta.com/Hardware_Security_Tools.php.

Chapter 2 presents *Cryptography-driven IP steganography for DSP hardware accelerators*. The significant features of this chapter include defining hardware steganography, basics of crypto-steganography, advantages of crypto-steganography over traditional techniques and methodology for securing DSP hardware accelerators using crypto-steganography.

Chapter 3 presents *Double line of defence to secure JPEG codec hardware for medical imaging systems*. The significant features of this chapter include defining JPEG compression/decompression hardware accelerator, discussion on the motivation of using JPEG compression in medical imaging systems, security methodology using double line of defence and analysis on case studies.

Chapter 4 presents *Integrating multi-key-based structural obfuscation and low-level watermarking for double line of defence of DSP hardware accelerators*. The significant features of this chapter include the impact of multi-key-based structural obfuscation, enhanced security possible by integrating key-based structural obfuscation and physical level watermarking and methodology of double line of defence for securing DSP hardware accelerators.

Chapter 5 presents *Multimodal hardware accelerators for image processing filters*. The significant features of this chapter include defining multimodal hardware accelerators for image processing filters, discussion on different image processing filters, design of application specific circuits for several major convolution filters, design of functionally reconfigurable processors for convolution filter types and analysis on case studies.

Chapter 6 presents *Fingerprint biometric for securing hardware accelerators*. The significant features of this chapter include an overview on fingerprint biometric, major steps involved in extracting digital template from a biometric fingerprint, how to embed the corresponding digital template of a biometric fingerprint into a complex hardware accelerator, its design and finally the impact of several fingerprint biometrics on security of hardware accelerators.

Chapter 7 presents *Key-triggered hash-chaining-based encoded hardware steganography for securing DSP hardware accelerators*. The significant features of this chapter include an introduction to hash-chaining-based hardware steganography, multi-level encoding, how to exploit hash-chaining for deriving secret hardware security constraints for hardware accelerators, analysis on case studies and finally discussion on some security metrics.

Chapter 8 presents *Designing a secured N-point DFT hardware accelerator using obfuscation and steganography*. The significant features of this chapter include a background on *N*-point DFT, its mathematical function, design process of secured *N*-point DFT using structural obfuscation and steganography, and finally analysis on case studies.

Chapter 9 presents *Structural transformation-based obfuscation using pseudo-operation mixing for securing data-intensive IP cores*. The significant features of this chapter include discussion on some other structural transformation and functional obfuscation techniques used for securing DSP hardware accelerators as well as the impact of these approaches on hardware security and design overhead.

Authors believe that there is no book that presents details of secured hardware accelerators for DSP and image processing applications, under one canopy. Electronics/CAD/EDA/hardware/VLSI community comprises people from diverse backgrounds. Electronics design industry in future is a heading for a paradigm shift towards secured and low-cost hardware accelerators from the conventional ones. By covering chapters under this special topic, it will enable readers to push their boundaries of knowledge to dive into some emerging security and design aspects of modern hardware accelerators, especially from image processing, multimedia and DSP. This book aims to present novel solutions for secured hardware accelerators for DSP, multimedia and image processing applications. The theme is important to researchers in different areas of specialization, as it encompasses overlapping contents of hardware design security, VLSI and finally hardware accelerator design for various health cares, consumers and medical applications. All the aforesaid topics are encapsulated in the proposed theme where researchers, practitioners and industry experts are expected to show interest in reading. The book is prepared keeping in mind that it can be easily integrated to any graduate level course. Furthermore, it also serves as a designer's handbook, who is eager to learn designing secured hardware accelerators for DSP and image processing applications.

Dr Anirban Sengupta, Ph.D., Assoc. Professor
Fellow of IET, Fellow of British Computer Society (BCS), Senior Member of IEEE
IEEE Distinguished Lecturer (IEEE Consumer Electronics Society)
IEEE Distinguished Visitor (IEEE Computer Society)
Former Ex-Officio Member, IEEE Consumer Electronics Society Board of Governors
Former Chair, IEEE Computer Society Technical Committee on VLSI
Founder and Chair, IEEE Consumer Electronics Society Bombay Chapter
Deputy Editor-in-Chief, IET Computers and Digital Techniques,
Former Editor-in-Chief, IEEE VLSI Circuits and Systems Letter
Featured in Researcher Spotlight, ACM Special Interest Group on Design Automation (SIGDA) Newsletter
Awardee, IEEE Chester Sall Memorial Consumer Electronics Award (IEEE CE Society)
Associate Editor IEEE Transactions on VLSI Systems, IEEE Transactions on Aerospace and Electronic Systems, IEEE Transactions on Consumer Electronics, IEEE Letters of the Computer Society, IEEE Canadian Journal of Electrical and Computer Engineering
Former Editorial Board Member IEEE Access, IEEE Consumer Electronics Magazine, IET Computers and Digital Techniques, Elsevier Microelectronics Journal
General Chair, 37th IEEE International Conference on Consumer Electronics (ICCE), Las Vegas
General Chair, 23rd International Symposium on VLSI Design and Test (VDAT-2019), India

Executive Committee, IEEE International Conference on
Consumer Electronics (ICCE) – Berlin and Las Vegas
IEEE Distinguished Lecturer Nominations Committee, IEEE CE Society
Computer Science and Engineering
Indian Institute of Technology Indore
Email: asengupt@iiti.ac.in
Web: http://www.anirban-sengupta.com

Acknowledgements

I would like to thank my family and friends for the support and encouragement throughout the execution of the book project. I would also like to thank Indian Institute of Technology (IIT) Indore for the support in executing this work.

About the author

Anirban Sengupta is an Associate Professor in Computer Science and Engineering at Indian Institute of Technology (IIT) Indore, where he directs the research lab on 'CAD for Consumer Electronics Hardware Device Security and Reliability'. He is an elected *Fellow of IET and Fellow of British Computer Society (FBCS), United Kingdom*. He holds a Ph.D. and an M.A.Sc. in Electrical and Computer Engineering from Ryerson University, Toronto (Canada) and is a registered Professional Engineer of Ontario (P.Eng.). He has been an active researcher in the emerging areas of 'Hardware Security', 'IP Core Protection' and 'Digital Rights Management for Electronics Devices'. He has been awarded prestigious *IEEE Distinguished Lecturer* by IEEE Consumer Electronics Society in 2017 and *IEEE Distinguished Visitor* by IEEE Computer Society in 2019. He was an *Ex-Officio Member of Board of Governors* of IEEE Consumer Electronics Society. He has featured in *Researcher Spotlight* of prestigious ACM Special Interest Group on Design Automation (SIGDA) Newsletter for his contributions on hardware security. He is the awardee of IEEE Chester Sall Memorial Consumer Electronics Award in 2020.

He has *230 publications and patents*. He is the author of two books from IET—*IP Core Protection and Hardware-Assisted Security for Consumer Electronics* and *"Frontiers in Securing IP Cores – Forensic Detective Control and Obfuscation Techniques* published in 2019 and 2020, respectively, from the United Kingdom. He is also the author of an edited book from Springer on '*VLSI Design and Test'* published in 2020. He is currently the *Deputy Editor-in-Chief* of *IET Computers and Digital Techniques* journal that has a publishing history of over 40 years and the *Editor-in-Chief* of IEEE VLSI Circuits and Systems Letter of IEEE Computer Society TCVLSI. He is also currently the *Chairman* of IEEE Computer Society TCVLSI. He currently serves/served in several editorial positions as Senior Editor, Associate Editor, Editor and Guest Editor of several IEEE Transactions/Journals, IET and Elsevier Journals, including *IEEE Transactions on Aerospace and Electronic Systems (TAES), IEEE Transactions on VLSI Systems, IEEE Transactions on Consumer Electronics, IEEE Access Journal, IEEE Letters of Computer Society, IET Journal on Computer and Digital Techniques, IEEE Consumer Electronics Magazine, IEEE Canadian Journal of Electrical and Computer Engineering, IEEE VLSI Circuits and Systems Letter and Elsevier Microelectronics Journal*. He further serves as a Guest Editor of IEEE Transactions on VLSI Systems, IEEE Access and IET Computers and Digital Techniques. He was the *General/Conference Chair* of 37th IEEE International Symposium on Consumer Electronics (ICCE) 2019, Las Vegas, *General/Conference*

Chair of 23rd International Symposium on VLSI Design and Test – VDAT and *Technical Program Chairs* of 36th IEEE International Conference on Consumer Electronics (ICCE) 2018 in Las Vegas, 10th IEEE International Conference on Consumer Electronics (ICCE) – Berlin 2020, 9th IEEE International Conference on Consumer Electronics (ICCE) – Berlin 2019, 15th IEEE International Conference on Information Technology (ICIT) 2016 and 3rd IEEE International Symposium on Nanoelectronic and Information Systems (iNIS) 2017. Furthermore, he has served in *Executive Committee* of IEEE International Conference on Consumer Electronics (ICCE) – Berlin, IEEE International Conference on Consumer Electronics (ICCE) – Las Vegas as well as *International Advisor* of IEEE International Conference on Consumer Electronics (ICCE) – Las Vegas.

More than a dozen of his IEEE publications have appeared in 'Top 50 Most Popular Articles' with few in 'Top 5 Most Popular Articles' from IEEE Periodicals. His patents have been cited in industry patents of *IBM Corporation, Siemens Corporation, Qualcomm, Amazon Technologies, Siemens (Germany), Mathworks Inc., Ryerson University and STC University of Mexico* multiple times. His professional works have received wide media coverage nationally and internationally such as in *IET International News (UK), Times of India, Central Chronicle, DBPOST News, Free Press Journal, Dainik Bhaskar*. He has supervised more than 35 candidates, including several graduated Ph.D. candidates, Research Assistants, Associates and B. Techs, all of whom are/were placed in academia and industry. He has successfully commissioned special issues in IEEE TVLSI, IEEE TCAD, IET CDT, IEEE Access as well as IEEE CEM. He has been awarded the highest rating 'Excellent' by the Department of Science and Technology (DST) based on the performance in funded project in 2017. His ideas have been awarded funding from the Department of Science and Technology (DST), the Council of Scientific and Industrial Research (CSIR) and the Department of Electronics and IT (DEITY). Complete details available at http//www.anirban-sengupta.com/index.php.

List of acronyms

Chapter 1

CPU	central processing unit
GPU	graphics processing unit
FPGA	field-programmable gate array
ASIC	application-specific integrated circuit
DSP	digital signal processing
AI	artificial intelligence
IoT	Internet of Things
HD	high definition
HDL	hardware description language
IC	integrated circuit
IP	intellectual property
SoC	system-on-chip
HLS	high-level synthesis
RTL	register transfer level
DSE	design space exploration
VLSI	very large scale integration
EDA	electronic design automation
CE	consumer electronics
ESL	electronic system level
CDFG	control data flow graph
FU	functional unit
DCT	discrete cosine transform
FIR	finite impulse response
MAC	multiply accumulate
DFT	discrete Fourier transform
JPEG	joint photographic experts group
codec	compression–decompression
KHC	key-based hash chaining

KSO-PW key-based structural obfuscation–physical level watermarking
POM-SO pseudo-operation mixing–structural obfuscation

Chapter 2

CPU central processing unit
GPU graphics processing unit
FPGA field-programmable gate array
ASIC application-specific integrated circuit
NRE non-recurring engineering
DSP digital signal processing
AI artificial intelligence
IC integrated circuit
IP intellectual property
SoC system-on-chip
HLS high-level synthesis
RTL register transfer level
DSE design space exploration
CDFG control data flow graph
CIG coloured interval graph
FU functional unit
AES advanced encryption standard
MDS maximum distance separable
GUI graphical user interface
DCT discrete cosine transform
FIR finite impulse response
IIR infinite impulse response
DWT discrete wavelet transform
ARF auto-regression filter
JPEG joint photographic experts group
MPEG moving picture experts group
EWF elliptic wave filter
IDCT inverse discrete cosine transform

Chapter 3

CT computed tomography
MRI magnetic resonance imaging

ROI	region of interest
JPEG	joint photographic experts group
codec	compression–decompression
DSP	digital signal processing
RE	reverse engineering
IP	intellectual property
HLS	high-level synthesis
THT	tree-height transformation
RTL	register transfer level
CDFG	control data flow graph
CIG	coloured interval graph
FU	functional unit
GF	Galois field
MDS	maximum distance separable
GUI	graphical user interface
DCT	discrete cosine transform
1D	one dimensional
2D	two dimensional
IDCT	inverse discrete cosine transform
PSNR	peak signal-to-noise ratio
MSE	mean square error

Chapter 4

HD	high definition
ASIC	application specific integrated circuit
FPGA	field-programmable gate array
DSP	digital signal processing
DFS	design for security
3PIP	third-party intellectual property
SoC	system-on-chip
VLSI	very large scale integration
HLS	high-level synthesis
RTL	register transfer level
PSO	particle swarm optimization
DSE	design space exploration
RE	reverse engineering

THT	tree-height transformation
LU	loop unrolling
LT	logic transformation
ROE	redundant operation elimination
LICM	loop invariant code motion
UF	unrolling factor
CDFG	control data flow graph
CIG	coloured interval graph
FU	functional unit
GUI	graphical user interface
DCT	discrete cosine transform
FIR	finite impulse response
IIR	infinite impulse response
DWT	discrete wavelet transform
ARF	auto-regression filter
DWT	discrete wavelet transform
FFT	fast Fourier transform
JPEG	joint photographic experts group
DE	differential equation
IDCT	inverse discrete cosine transform
SO	structural obfuscation
KSO-PW	key-based structural obfuscation–physical level watermarking

Chapter 5

CE	consumer electronics
2D	two dimensional
FPGA	field-programmable gate array
HLS	high-level synthesis
RTL	register transfer level
DSE	design space exploration
RE	reverse engineering
THT	tree-height transformation
UF	unrolling factor
DFG	data flow graph
FU	functional unit
VE	vertical embossment

HE	horizontal embossment
ED	edge detection

Chapter 6

VLSI	very large scale integration
SoC	system-on-chip
DSP	digital signal processing
CE	consumer electronics
IC	integrated circuit
IP	intellectual property
3PIP	third-party intellectual property
HLS	high-level synthesis
RTL	register transfer level
CDFG	control data flow graph
CIG	coloured interval graph
SHA	secure hash algorithm
FU	functional unit
DCT	discrete cosine transform
FFT	fast Fourier transform
JPEG	joint photographic experts group
Codec	compression–decompression
2D	two dimensional
CN	crossing number
PMDF	point matching difference function

Chapter 7

DSP	digital signal processing
IC	integrated circuit
IP	intellectual property
HLS	high-level synthesis
DFG	data flow graph
CIG	coloured interval graph
FU	functional unit
SB	switch block
HU	hash unit
RFC	round function computation

GUI graphical user interface
FIR finite impulse response

Chapter 8

DSP digital signal processing
DFT discrete Fourier transform
VLSI very large scale integration
HLS high-level synthesis
RTL register transfer level
RE reverse engineering
DFG data flow graph
CIG coloured interval graph
FU functional unit
THT tree-height transformation
MDS maximum distance separable

Chapter 9

DSP digital signal processing
RE reverse engineering
IP intellectual property
HLS high-level synthesis
RTL register transfer level
DFG data flow graph
FU functional unit
GUI graphical user interface
DWT discrete wavelet transform
VLSI very large scale integration
SoC system-on-chip
POM-SO pseudo-operation mixing – structural obfuscation

List of notations

Chapter 1

$X[n]$	input to an FIR filter
$Y[n]$	output of an FIR filter
N	order of an FIR filter
h	FIR filter coefficients
$w[n]$	input sequence to DFT
$W[k]$	output sequence from DFT
$d[0]$ to $d[7]$	inputs to 1D-DCT (8-point)
$b1$–$b8$	generic values of DCT coefficients
$D[n]$	nth output value of 1D-DCT where n varies from 0 to 7
O_V	Vth output value of 2D convolution (image processing filters)
I	input pixels matrix
K	kernel matrix
O	output matrix of convolution filters
N and M	dimensions of input matrix $[I]_{M\times N}$
Y	output matrix of image brightness and contrast hardware accelerator
B_R	coefficient of brightness
C_N	coefficient of contrast
n and m	dimensions of a generic filter/kernel matrix $[K]_{n\times m}$
P	an 8×8 block of image pixels
B	2D-DCT coefficient matrix
B'	transpose of 2D-DCT coefficient matrix B
W	DCT transformed 8×8 block of image pixels
W_{ij}	pixel valuc at position ij after DCT transformation
W'_{ij}	first pixel of compressed JPEG image after quantization
t_{ij}	quantization coefficient in quantization matrix at position ij

Chapter 2

*E*th	entropy threshold value
S	storage variable
(*Si*, *Sj*)	a node/storage variable pair in CIG
V1 and V2	vendor type 1 and Vendor type 2
A	set representing secret design data
A_j^i	*j*th instance of an adder resource unit from vendor type *i*
M_j^i	*j*th instance of a multiplier resource unit from vendor type *i*
I	a digit in set *A*
Q	control step
M_S	state matrix
M_B	matrix after bit manipulation
M_{Rd}	matrix post row diffusion
abc	encrypted output of an alphabet post Trifid cipher
M_{AS}	matrix post alphabet substitution
M_T	transposed matrix
M_{Cd}	matrix post-performing mix column diffusion
Pc	probability of coincidence
h	number of colours used in the CIG before implanting steganography
*k*1	number of effective constraints embedded into the CIG/register allocation phase
*k*2	number of stego-constraints embedded during the FU vendor allocation phase
$N(U_j)$	number of resources of FU type U_j
m	total types of FU resources used in design
$C_d(U_i)$	the design cost post-embedding steganography with resource constraints U_i
L_d	design latency
A_d	design area
L_m	maximum execution latency
A_m	maximum hardware area
ρ_1	user-specified weight for design latency
ρ_2	user-specified weight for design area

Chapter 3

S	storage variable
(*Si*, *Sj*)	a node (storage variable) pair in CIG
V1 and V2	vendor type 1 and vendor type 2

A	set representing secret design data
A_j^i	*j*th instance of an adder resource unit from vendor type *i*
M_j^i	*j*th instance of a multiplier resource unit from vendor type *i*
P, I, V, G, Y, O, R, B	eight distinct colours representing eight distinct registers
Q	control step
M_S	state matrix
M_B	matrix after bit manipulation
M_{Rd}	matrix post row diffusion
abc	encrypted output of an alphabet post Trifid cipher
M_{AS}	matrix post alphabet substitution
M_T	transposed matrix
M_{Cd}	matrix post performing mix column diffusion
P	an 8×8 block of image pixels
B	2D-DCT coefficient matrix
B'	transpose of 2D-DCT coefficient matrix B
W	DCT transformed 8×8 block of image pixels
g	elements in the matrix $[B*P]$
b	DCT coefficients in the matrix B
p	pixel values in the matrix P
$W11'$	first pixel of compressed JPEG image
t	quantization coefficient in the quantization matrix T
R	register
O	operation
Pc	probability of coincidence
h	number of colours used in the CIG before implanting Steganography
$k1$	number of effective constraints embedded into the CIG/register allocation phase
$k2$	number of stego-constraints embedded during the FU vendor allocation phase
$N(U_j)$	number of resources of FU type U_j
m	total types of FU resources used in design
$C_d(U_i)$	the design cost with resource constraints U_i
L_d	design latency
A_d	design area
L_m	maximum execution latency
A_m	maximum hardware area

ρ_1 user-specified weight for design latency

ρ_2 user-specified weight for design area

Chapter 4

α, β and γ	signature variables
S	intermediate signal variable
SO-key	structural obfuscation key
K	maximum number of iterations in a loop of DSP algorithm
K1–K8	generic values of DCT coefficients
$x[0]$–$x[7]$	inputs to 1D-DCT
$X[0]$	first output sample of 1D-DCT
C	number of cuts applied on CDFG
$C+1$	number of partitions
P	a partition of CDFG
R	set of RTL components
$R1$	subset of R containing only FU components (multipliers, adders, etc.)
$R2$	subset of R containing only Mux components
$R3$	subset of R containing only Demux components
M	multiplier
A	adder
C	comparator
x	multiplexer
d	demultiplexer
Q	control step
KS_{total}	total SO-key size
V_{new}	current velocity of a particle
V_{old}	old velocity of a particle
ω	inertia weight
b_1 and b_2	acceleration coefficients
r_1 and r_2	random values between 0 and 1
R_{curr}, R_{lb}	current and best position of the current particle
R_{gb}	the best position of a particle
$R_{\max}$	maximum resource constraints
$R_{\min}$	minimum resource constraints
R_{new}	new position of a particle

R_{old}	old position of a particle
$V^i_{\max}$	maximum velocity of a particle in *i*th dimension
Pc	probability of coincidence
$k1$	denotes total number of FU resource components of type F_p
p	total types of FU resources used in design
$k2$	number of multiplexers of size X_q
q	different sizes of multiplexers in the design
$k3$	number of demultiplexers of size D_r
r	different sizes of demultiplexers in the design
T^P	tamper tolerance capability
Z	number of signature variables used in the watermark
Q	size of the author's signature
S^B	probability of finding correct signature using brute force analysis
$C_d(U_i)$	the design cost with resource constraints U_i
L_d	design latency
A_d	design area
L_m	maximum execution latency
A_m	maximum hardware area
ρ_1	user specified weight for design latency
ρ_2	user specified weight for design area

Chapter 5

I	input matrix corresponding to an input image
X_{ij}	elements in input matrix
$A \times B$	dimensions of input matrix
K	kernel matrix
K_{rs}	elements in kernel matrix
$n \times m$	dimensions of kernel matrix
w	size of filter kernel
L	factor for zero padding in input matrix
$N \times M$	modified dimensions of input matrix post zero padding
Y_{pq}	elements in modified input matrix post zero padding
O	output matrix
O_{ij}	elements in output matrix
$(N-n+1)\times (M-m+1)$	dimensions of output matrix

O_V	an output pixel value, where V varies from 0 to $[(N-n+1)\times(M-m+1)-1]$
$E1$–$E5$	intermediate signal variables
P1–P6	six partitions of DFG
Q	control step
K^B	kernel matrix of a blur filter
K^S	kernel matrix of a sharpening filter
K^{VE}	kernel matrix of a vertical embossment filter
K^{HE}	kernel matrix of a horizontal embossment filter
K^{ED}	kernel matrix of Laplace edge detection filter
$C_d(U_i)$	the design cost with resource constraints U_i
L_d	design latency
A_d	design area
L_m	maximum execution latency
A_m	maximum hardware area
ρ_1	user-specified weight for design latency
ρ_2	user-specified weight for design area

Chapter 6

X_m	x-coordinate of the location of a minutia point
Y_m	y-coordinate of the location of a minutia point
M_t	minutia type
R_a	ridge direction or angle
Di	digital template of ith minutiae point
X_m^b	binary representation of the minutiae attribute X_m
Y_m^b	binary representation of the minutiae attribute Y_m
M_t^b	binary representation of the minutiae attribute M_t
R_a^b	binary representations of the minutiae attribute R_a
n	total number of minutiae points extracted from a fingerprint image
DT	final digital template
P	an 8×8 block of image pixels
B	2D-DCT coefficient matrix
B'	transpose of 2D-DCT coefficient matrix B
W	DCT transformed 8×8 block of image pixels
g	elements in the matrix $[B*P]$
b	DCT coefficients in the matrix B

p	pixel values in the matrix P
W'_{11}	first pixel of compressed JPEG image
t	quantization coefficient in the quantization matrix T
R	registers
O	operation
S	storage variable
(Si, Sj)	a node/storage variable pair in CIG
A	adder
M	multiplier
Q	control step
Pc	probability of coincidence
h	number of colours used in the CIG before
$k1$	number of effective constraints embedded into the CIG/register allocation phase
$k2$	number of stego-constraints embedded during the FU vendor allocation phase
$N(U_j)$	number of resources of FU type U_j
m	total types of FU resources used in design
G	total constraints size ($k1+k2$)
$C_d(U_i)$	the design cost with resource constraints U_i
L_d	design latency
A_d	design area
L_m	maximum execution latency
A_m	maximum hardware area
ρ_1	user-specified weight for design latency
ρ_2	user-specified weight for design area
MP_i	decimal representation of a minutia point of the implanted fingerprint
MP_j	decimal representation of a minutia point of the IP vendor's fingerprint

Chapter 7

S	storage variable
$\langle Si, Sj \rangle$	a node/storage variable pair in CIG
V1 and V2	vendor type 1 and vendor type 2
A_j^i	jth instance of an adder resource unit from vendor type i
M_j^i	jth instance of a multiplier resource unit from vendor type i
Q	control step
n	total number of operations in a DSP application

k	number of encoded bitstreams chosen by the designer
Z	encoding number
e_{sk}^{m}	attacker's maximum effort of finding the stego-key
e_{H}^{eb}	attacker effort of finding the encoded bits through brute-force
O	operation
Pc	probability of coincidence
h	number of colours used in the CIG before implanting steganography
$k1$	number of effective constraints embedded into the CIG/register allocation phase
$k2$	number of stego-constraints embedded during the FU vendor allocation phase
$N(U_j)$	number of resources of FU type U_j
m	total types of FU resources used in design
W	total constraints size
$C_d(U_i)$	the design cost post-embedding steganography with resource constraints U_i
L_d	design latency
A_d	design area
L_m	maximum execution latency
A_m	maximum hardware area
ρ_1	user-specified weight for design latency
ρ_2	user-specified weight for design area

Chapter 8

$w[n]$	input sequence to DFT
$W[k]$	output sequence from DFT
S	storage variable
(Si, Sj)	a node/storage variable pair in CIG
V1 and V2	vendor type 1 and vendor type 2
A	set representing secret design data
A_j^i	jth instance of an adder resource unit from vendor type i
M_j^i	jth instance of a multiplier resource unit from vendor type i
Q	control step
M_S	state matrix
M_B	matrix after bit manipulation
M_{Rd}	matrix post row diffusion
abc	encrypted output of an alphabet post Trifid cipher

M_{AS}	matrix post alphabet substitution
M_T	transposed matrix
M_{Cd}	matrix post-performing mix column diffusion
NGx	difference in gate count pre- and post-structural obfuscation
NGy	number of gates modified post-structural obfuscation
Pc	probability of coincidence
h	number of colours used in the CIG before implanting steganography
$k1$	number of effective constraints embedded into the CIG/register allocation phase
$k2$	number of stego-constraints embedded during the FU vendor allocation phase
$N(U_j)$	number of resources of FU type U_j
m	total types of FU resources used in design
$C_d(U_i)$	the design cost with resource constraints U_i
L_d	design latency
A_d	design area
L_m	maximum execution latency
A_m	maximum hardware area
ρ_1	user-specified weight for design latency
ρ_2	user-specified weight for design area

Chapter 9

M^c	multiplier resource constraints
A^c	adder resource constraints
M_i	number of multiplier instances in ith control step
A_i	number of adder instances in ith control step
W	list of pseudo-operations
Q	control step
REG	register
SOB	strength of structural obfuscation
AG	total affected gate count (with respect to baseline) post structural obfuscation
BG	total gate count of baseline (un-obfuscated) design
G^{AR}	gate count of affected resources post obfuscation
G^C	change in gate count post obfuscation
$C_d(U_i)$	design cost with resource constraints U_i
L_d	design latency

A_d	design area
L_m	maximum execution latency
A_m	maximum hardware area
ρ_1	user-specified weight for design latency
ρ_2	user-specified weight for design area

Chapter 1

Introduction: secured and optimized hardware accelerators for DSP and image processing applications

Anirban Sengupta[1]

This chapter provides a background introduction on hardware accelerators, followed by its relevance in today's digital world as well as the security modules/algorithms being used to secure a hardware accelerator and finally ending with the paradigm shift needed for the future.

The chapter is organized as follows: Section 1.1 discusses about the definition, significance and application of hardware accelerators, followed by the role of electronic system level (ESL) synthesis in hardware accelerator design in Section 1.2; Section 1.3 provides significant details on the popular hardware accelerators for digital signal processing (DSP) and image processing applications by including details of its mathematical function/algorithm. Section 1.4 presents a background summary of important security algorithm/modules used for securing hardware accelerators by especially giving reference to the chapters where it is discussed; Section 1.5 explains the new paradigm shift expected in future for hardware and very large scale integration (VLSI) communities; Section 1.6 concludes the chapter, while Section 1.7 provides questions and exercise for the readers.

1.1 Hardware accelerators: an introduction, definition, significance and applications

Now we are living in an era wherein internet speed has reached 5G, 8D audio songs are mesmerising listeners, high-definitional videos and graphics are enthralling today's generation. Moreover, the rise of Internet of Things network and artificial intelligence (AI) has led in making our life very sophisticated, faster and comfortable. However, with rapid growth in modern technology, the demands of security of digital information and authorized access are also prevailing. Therefore, cryptography, biometric fingerprinting, ear biometric and face recognition biometric,

[1]Computer Science and Engineering, Indian Institute of Technology Indore, Indore, India

etc. are also playing a pivotal role in the advancement of technology. Towards aforementioned achievements and advancement, the role of hardware accelerators cannot be overlooked. For example, cryptographic applications are facilitated by cryptographic accelerators; fingerprint, ear and face recognition biometric require DSP and image processing hardware accelerators; AI requires AI accelerators and so forth. This chapter highlights the significance of hardware accelerators towards the modern technology advancement. Further, this book focuses on security of DSP and image processing filter hardware accelerators using some useful hardware security techniques.

Hardware acceleration is a hardware-facilitated process of performing data or computational intensive tasks in order to achieve high performance and increased throughput of a system. The underlying hardware responsible for performing hardware acceleration is referred to as a hardware accelerator. Generically, hardware accelerators are designed as following variations: application-specific integrated circuits (ICs) or application-specific processor, graphics processing units and field-programmable gate arrays (FPGAs). Application-specific processors are customized particularly towards a specific task or application, hence enhancing overall performance as it emphasises solely on the execution of one function. FPGAs are hardware-description-language-driven ICs which are designed as a reconfigurable circuit, capable of being reconfigured according to desired logic functionality. Using FPGAs in systems, parts of an algorithm/process can be accelerated, or sharing of different portions of the computation between a general-purpose processor and the FPGA can be enabled. GPU hardware accelerators handle the motion of images, data-intensive calculations and acceleration of a part of an application (reaming part continues execution on the central processing unit). Thereby, hardware accelerators lead to the following advantages: (i) high-speed computation, (ii) high parallelism and (iii) less power consumption.

Following are some popular hardware accelerators which are designed to execute dedicated tasks: (i) DSP hardware accelerators, (ii) image processing filter hardware accelerators, (iii) AI accelerators, (iv) network interface controller, (v) GPU hardware accelerator, (vi) sound card and (vii) crypto-processor or cryptographic accelerator and so forth. These hardware accelerators are employed to facilitate following applications respectively in order to enhance the system performance: (i) DSP applications, (ii) image processing applications, (iii) AI applications, (iv) computer networking applications, (v) computer graphics applications, (vi) sound processing applications, (vii) cryptographic applications and so forth.

1.2 Role of ESL synthesis in hardware accelerator design

Hardware accelerators are employed to perform computational intensive part of complex applications or algorithms such as DSP algorithms. Because of higher complexity and larger size, it is not easy to design hardware of such application starting from lower level of VLSI design process such as gate level. This is because of huge gate count (thousands of gates) in the gate-level structure. Further, designing from one level above in the design process, i.e. register transfer level (RTL), is also a tedious task because of complex architecture involving a number of functional unit (FU) resources such as multipliers and adders/subtractors, interconnect-hardware such as multiplexers and demultiplexers and storage hardware such as registers and latches. Additionally, designing from RTL does not enable the exploration of design space in order to achieve an optimal architecture (Sengupta, 2020).

In order to avoid the limitations of lower phases of VLSI design process, it is efficient to start the design process of hardware accelerators from high-level synthesis (HLS) or ESL synthesis (McFarland *et al.*, 1988). The ESL synthesis plays a significant role in hardware accelerator design process because of the following reasons:

1. Complexity of a hardware accelerator design is lesser when it is in the form of high level or algorithmic or system-level description, compared to RTL or gate-level description of design. Hence, it is an efficient practice to automatically obtain the RTL structural from high-level description using ESL synthesis process, at less effort and shorter time (McFarland *et al.*, 1988).
2. High-level or ESL synthesis process offers the flexibility of exploring design space in order to obtain an optimal architecture. More explicitly, possible design solutions of a hardware accelerator architecture can iteratively be explored using a design space exploration process to reach such a design solution which satisfies given design constraints such as time, area and power constraints.
3. ESL synthesis offers more flexibility of employing security mechanisms such as hardware watermarking, hardware steganography and hardware obfuscation. More explicitly, there are different phases of ESL synthesis process such as high-level transformation, scheduling, allocation, binding, data path synthesis and controller synthesis which can be leveraged to perform security algorithms to secure hardware accelerator design against prevalent hardware threats (Koushanfar *et al.*, 2005; Le Gal and Bossuet, 2012; Sengupta, 2017).
4. The integration of security mechanism with ESL synthesis process also ensures the security of hardware accelerator designs at subsequent lower abstraction levels (such as RTL, gate level and layout level) of VLSI design process.
5. Employing security and optimization during ESL synthesis process of hardware accelerator design offers the flexibility of performing security and design cost trade-off. It helps in obtaining a low-cost secured hardware accelerator design.

1.3 Hardware accelerators for popular DSP and image processing applications

This section discusses hardware accelerators for popular DSP and image processing applications in terms of their transfer function or input–output relationship or computation function and their basic functionality. Figure 1.1 depicts the taxonomy

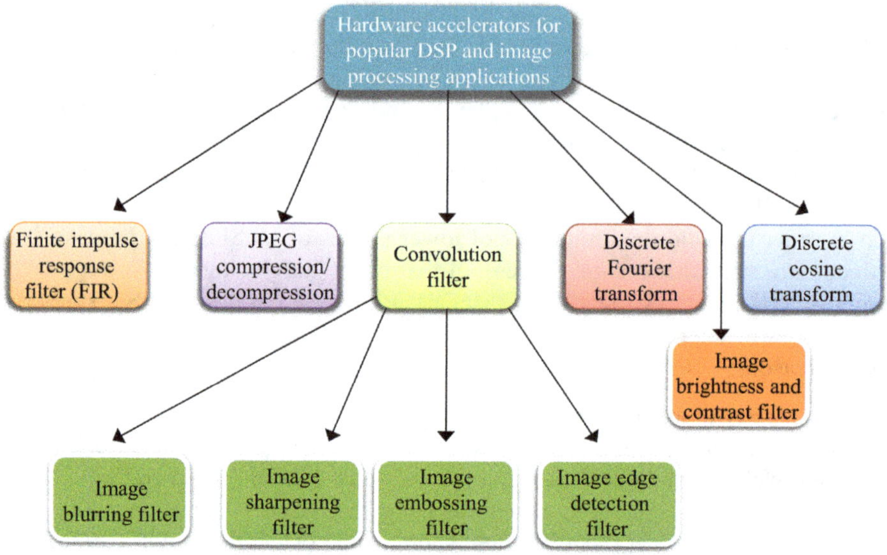

Figure 1.1 Hardware accelerators for popular DSP and image processing applications

of popular hardware accelerators that have been targeted to employ various security algorithms/mechanisms discussed in this book. Let us start discussion with finite impulse response (FIR) filter followed by discrete cosine transform (DCT) core, joint photographic experts group (JPEG) compression–decompression (codec), discrete Fourier transform (DFT) and convolution filters such as image blurring filter, image sharpening filter, image embossing filter and edge detection filter used in image processing.

1.3.1 Finite impulse response (FIR) filter

An FIR filter is a DSP algorithm/application whose impulse response is of finite length because of the absence of feedback path. More explicitly, the computation of the current output sample depends only on input samples (no dependency on previous output samples). This makes it inherently stable and a linear phase filter. The FIR filters can be leveraged as a low-pass filter, a high-pass filter, a notch filter and a band-pass filter in DSP. The FIR filter equation or computation function is given as follows:

$$Y[n] = h_0 * X[n] + h_1 * X[n-1] + h_2 * X[n-2] + \cdots + h_N * X[n-N] \tag{1.1}$$

where $X[n]$ and $Y[n]$ indicate the input and output of the FIR filter, whereas $X[n-1]$, $X[n\text{–}2]$ and $X[n\text{–}3]$ indicate the previous values of inputs, and h_0, h_1, h_2 and h_N indicate the input coefficients of the FIR filter. This filter can be represented as a loop-based/iterative DSP application which can be unrolled on the basis of the chosen value of unrolling factor, while designing the filter.

The function of the loop-based FIR filter is given as follows:

$$Y[n] = \sum_{i=0}^{N} h[i] * X[n-i] \tag{1.2}$$

where N indicates the order of filter. An Nth-order FIR filter has $N+1$ taps (pairs of coefficient-delayed input). An FIR filter performs one multiply–accumulate operation per tap. The detailed information of filter transfer function and its conversion into corresponding control data flow graph (CDFG) representation is available in Sengupta and Mohanty (2019). The DFG/CDFG representation of the computation function (shown in (1.1) or (1.2)) of the FIR filter is fed as inputs to the ESL synthesis process in order to obtain its hardware accelerator design.

1.3.2 Discrete cosine transform (DCT) core

DCT application coverts input signal/sequence from time or spatial representation to frequency representation. The generic equation or computation function of 1D-DCT core (8-point) is given as follows (Sengupta and Rathor, 2019a):

$$\begin{aligned} D[n] = {} & b1 * d[0] + b2 * d[1] + b3 * d[2] + b4 * d[3] + b5 * d[4] + b6 * d[5] \\ & + b7 * d[6] + b8 * d[7] \end{aligned} \tag{1.3}$$

where $d[0]$–$d[7]$ indicate input values and $b1$–$b8$ indicate generic values of DCT coefficients. In addition, $D[n]$ indicates the nth output value where n varies from 0 to 7. More details on derivation of generic equation of 1D-DCT, coefficient matrix of DCT and conversion of DCT function into corresponding DFG representation are available in Sengupta and Mohanty (2019). The DFG representation of computation function (shown in (1.3)) of 1D-DCT is fed as inputs to the ESL synthesis process in order to obtain its hardware accelerator design.

1.3.3 JPEG codec

The JPEG image compression process performs the compression by first converting the input images from spatial representation to frequency representation followed by quantization. The conversion of spatial representation (two dimensional (2D) discrete data) of an input image to the frequency representation is performed using 2D-DCT function. The computation function of 2D-DCT transformation of an 8×8 block of input pixel matrix is as follows (Sengupta and Rathor, 2020a):

$$W = (B * P) * B' \tag{1.4}$$

where W indicates a DCT-transformed 8×8 block of image pixels, P indicates an 8×8 block of input image pixels. Further, B and B' represent the 2D-DCT coefficient matrix and its transpose, respectively. Here, the matrix $[B*P]$ generates the transformation of matrix P (8×8 block of image pixels) in one dimension, which is further multiplied to matrix B' to produce 2D transformed matrix. Post DCT transformation, the entire image is segregated into portions of distinct frequencies. Further, the actual compression phase is performed by quantization process which discards less important frequency components and keeps only most important frequency components. The quantization process is performed on DCT-transformed blocks of image pixels. The computation function to perform quantization on each DCT transformed pixel value is as follows:

$$W_{ij}^{'} = W_{ij} * \frac{1}{t_{ij}} \tag{1.5}$$

where $W_{ij}^{'}$ indicates a pixel value of compressed image after quantization, W_{ij} indicates corresponding pixel value after DCT transformation, and t_{ij} indicates a coefficient in the quantization matrix at the respective position ij. More details on the derivation of computation functions of DCT transformation and quantization have been discussed in Chapter 3. Further, the formation of corresponding DFG representation of the computation part of JPEG compression application has also been discussed in Chapter 3. The DFG representation of the computation function of JPEG-compression processor, constructed using (1.4) and (1.5), is fed as inputs to the ESL synthesis process in order to obtain its hardware accelerator design.

1.3.4 Discrete Fourier transform (DFT) core

DFT is a transformation of a discrete signal from its discrete-time representation to a discrete-frequency representation. A generic equation or computation function of N-point DFT is given as follows (Rathor and Sengupta, 2020):

$$W[k] = \sum_{n=0}^{N-1} w[n]e^{-j2\pi nk/N}, \; k = 0, \; 1, \; 2, \; 3, \; \ldots, \; N-1 \tag{1.6}$$

where input discrete-data sequence is represented by $w[n]$ and output discrete-data sequence is represented by $W[k]$. In the case of 4-point DFT, each discrete value of output sequence is computed as follows:

$$W[0] = w[0] * 1 + w[1] * 1 + w[2] * 1 + w[3] * 1 \tag{1.7}$$

$$W[1] = w[0] * 1 + w[1]e^{-j\pi/2} + w[2]e^{-j\pi} + w[3]e^{-j3\pi/2} \tag{1.8}$$

$$W[2] = w[0] * 1 + w[1]e^{-j\pi} + w[2]e^{-j2\pi} + w[3]e^{-j3\pi} \tag{1.9}$$

$$W[3] = w[0] * 1 + w[1]e^{-j3\pi/2} + w[2]e^{-j3\pi} + w[3]e^{-j9\pi/2} \tag{1.10}$$

The formation of the corresponding DFG representation of the computation function of the 4-point DFT application has been discussed in detail in Chapter 8. The DFG representation of computation functions of an N-point DFT processor is fed as input to the ESL synthesis process in order to obtain its hardware accelerator design.

1.3.5 Convolution filters used in image processing

An FIR filter is also termed a convolution filter as it performs convolution between input data sequence and impulse response of the filter. Convolution filters used in image processing are the 2D FIR filters which are applied to images to perform blurring, sharpening, embossment, etc. The 2D FIR filters are implemented using 2D-convolution. Output pixels of a filtered image (output of 2D convolution) are computed using following pixel computation function (Sengupta and Rathor, 2020e):

$$\text{for } (V = 0; V < (N - n + 1) \times (M - m + 1); V + +)$$

$$\left\{ O_V = \sum_{p,r=\text{min value}}^{p,r=\text{max value}} \left(\sum_{q,s=\text{min value}}^{q,s=\text{max value}} I_{pq} \times K_{rs} \right) \right\} \tag{1.11}$$

Output values of 2D convolution are indicated by O_V, where V varies from 0 to $[(N-n+1)\times(M-m+1)-1]$. N and M are the dimensions of input matrix $[I]_{M\times N}$ post modifying using zero padding to perform same convolution. The same convolution results into the same size of an output image (or matrix) as an input image (or matrix). Further, m and n are the dimensions of a generic filter/kernel matrix $[K]_{n\times m}$. The values in the kernel matrix have generically been denoted by K_{rs}, where r and s vary from 0 to $n-1$ and $m-1$, respectively. Generally, a kernel matrix $[K]$ is a square matrix, where $m=m$. The popular kernel sizes of convolution filters are 3×3 and 5×5. Further, entries in the input matrix $[I]$ have generically been denoted by I_{pq}, where p and q vary from 0 to $N-1$ and $M-1$, respectively.

Let us see output pixel computation function of various kinds of convolution filters used in image processing (Sengupta and Rathor, 2020e). The pixel computation functions are based on a 3×3 kernel matrix of the respective filter type. The kernel matrices are given in Chapter 5.

1. *Image blurring filter:*

   The computation function for first pixel output of the image blurring filter is as follows (derived from (1.11)):

$$O_0 = \left[(I_{00} + I_{01} + I_{02} + I_{10} + I_{11} + I_{12} + I_{20} + I_{21} + I_{22}) \times \left(\frac{1}{9}\right) \right] \tag{1.12}$$

2. *Image sharpening filter:*

   The computation function for first pixel output of a sharpening filter is as follows (derived from (1.11)):

$$O_0 = [(I_{00} + I_{01} + I_{02} + I_{10} + I_{12} + I_{20} + I_{21} + I_{22}) \times (-1)] + (I_{11} \times 9) \tag{1.13}$$

3. *Image embossment filter:*

   The computation function for the first pixel output of a vertical embossment filter is as follows (derived from (1.11)):

$$\text{O}_0 = [(I_{12})] + [(I_{10} \times (-1))] \tag{1.14}$$

The computation function for the first pixel output of a horizontal emboss-ment filter is as follows (derived from (1.11)):

$$O_0 = [(I_{21})] + [(I_{01} \times (-1))] \tag{1.15}$$

4. *Edge detection filter:*

The computation function for the first pixel output of a Laplace edge detection filter is as follows (derived from (1.11)):

$$O_0 = [(I_{01} + I_{10} + I_{12} + I_{21}) \times (-1)] + (I_{11} \times 4) \tag{1.16}$$

Further, the derivation of computation function of image processing filters and the formation of corresponding DFG representations have been discussed in Chapter 5. The DFG representations of computation functions of image processing filters are fed as inputs to the ESL synthesis process in order to obtain their hardware accelerator designs.

Additionally, image brightness and contrast is another image processing application. However, it is not a convolution filter, hence not derived from (1.11). The output pixel-computation function of image brightness and contrast application is given as follows:

$$[Y] = [I] * B_R + C_N \tag{1.17}$$

where I indicates the input pixel matrix and Y indicates the output pixel matrix. Further, B_R and C_N indicate the coefficient of brightness and contrast, respectively. By varying the values of B_R and C_N, brightness and contrast of the images can be adjusted. The computation function given in (1.17) is converted into corresponding DFG representation which is fed to ESL synthesis process to generate hardware accelerator design.

1.4 Security techniques/algorithms/modules for securing hardware accelerators

The useful security techniques/algorithms that have been discussed in this book, for securing hardware accelerators, are highlighted in Figure 1.2. This subsection highlights the basic functionality and the goal of useful security techniques, viz. (i) crypto-steganography, (ii) integrated crypto-steganography and structural obfuscation, (iii) integrated watermarking and key-based structural obfuscation, (iv) biometric-fingerprinting-based hardware security and (v) key-based hash-chaining-driven steganography, which are the key contributions of this book. These security techniques have been discussed in detail in this book.

1.4.1 Crypto-steganography (Sengupta and Rathor, 2019b)

This is a kind of hardware steganography approach which generates a robust stego-mark (or stego-constraints) and implants into the hardware accelerator design during two distinct phases of HLS process, viz. register allocation phase and FU

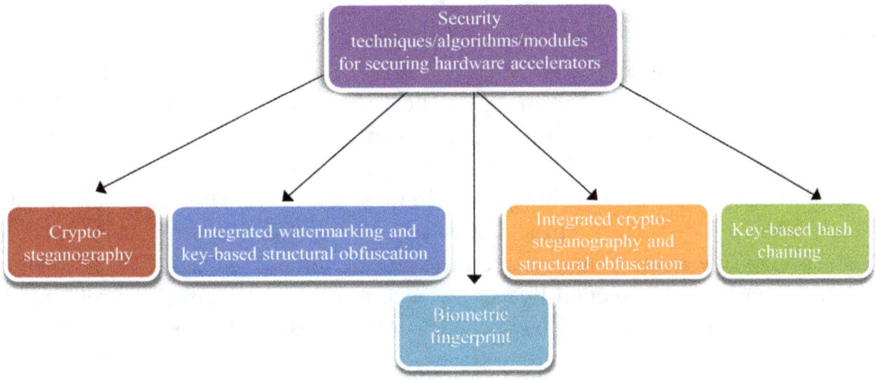

Figure 1.2 Useful hardware security techniques for securing hardware accelerators

vendor allocation phase. The stego-encoder of crypto-steganography approach generates stego-constraints by performing multiple key-driven steps, which includes some crypto-graphic modules such as byte substitution using S-box, row and column diffusion and Trifid-cipher-based encryption. The goal of crypto-steganography approach is to secure DSP hardware accelerators against piracy (resulting into counterfeiting or cloning) and false claim of ownership threats. This security technique has been discussed in detail in Chapter 2.

1.4.2 Integrated crypto-steganography and structural obfuscation (Sengupta and Rathor, 2020a)

This hardware security technique integrates crypto-steganography and structural obfuscation techniques to enhance the security of multimedia hardware accelerators such as JPEG compression processor. Structural obfuscation is performed during high-level transformation phase of HLS. Further, crypto-steganography is performed during scheduling and allocation phases of HLS to obtain a stego-embedded structurally obfuscated design. The goal of this approach is to provide double line of defence against popular hardware threats such as Trojan insertion, counterfeiting and cloning, to secure JPEG compression processors used in medical imaging systems. The structural obfuscation acts as a first line of defence to ensure preventive control against aforementioned threats, whereas crypto-steganography acts as a second line of defence to enable detective control against piracy. This security technique has been discussed in detail in Chapter 3.

1.4.3 Integrated watermarking and key-based structural obfuscation (Sengupta and Rathor, 2020b)

This hardware security technique integrates watermarking and key-based structural obfuscation techniques to enhance the security of DSP hardware accelerators. Multiple key-driven techniques of structural obfuscation such as key-driven loop unrolling, key-driven partitioning, key-driven redundant node elimination, key-driven tree height transformation and key-driven folding are performed during high-level transformation phase of HLS. Further, watermarking is performed on structurally obfuscated design during physical-level synthesis. The physical-level watermarking is performed during the floorplanning phase. The goal of this approach is to offer a double line of defence against popular hardware threats such as Trojan insertion, counterfeiting and cloning, to secure DSP hardware accelerators. The key-driven structural obfuscation acts as a first line of defence to ensure preventive control against aforementioned threats, whereas physical-level watermarking acts as a second line of defence to enable detective control against piracy. This security technique has been discussed in detail in Chapter 4.

1.4.4 Biometric-fingerprinting-based hardware security (Sengupta and Rathor, 2020c)

This hardware security technique secures DSP and multimedia hardware accelerators using biometric fingerprint of a vendor or designer. The unique features of a biometric fingerprint of a person are minutiae points (ridge ending and bifurcation). The minutiae points are extracted from the fingerprint of the vendor and converted into the digital template. This unique digital template of the biometric fingerprint is embedded into the design during HLS process. The goal of this approach is to offer security against false claim of ownership and piracy threats. This security technique has been discussed in detail in Chapter 6.

1.4.5 Key-based hash-chaining-driven steganography (Sengupta and Rathor, 2020d)

This is another kind of hardware steganography approach which generates a highly robust stego-mark and implants into the hardware accelerator design during the HLS process. The stego-encoder of key-based hash-chaining-driven steganography approach generates stego-constraints by performing multiple encodings of scheduled DSP hardware accelerator, followed by a hash-chaining process which comprises a number of key-driven hash units. The goal of key-based hash-chaining-driven steganography approach is to secure DSP hardware accelerators against false claim of ownership, counterfeiting and cloning threats. The generated stego-mark is so robust that an attacker fails to regenerate or extract it. This disables the attacker from escaping counterfeit detection, by copying the genuine owner's stego-mark in the counterfeited designs. This security technique has been discussed in detail in Chapter 7.

1.5 A new paradigm in future ahead for EDA/VLSI/CE communities

This book suggests new paradigm shifts to electronic design automation (EDA)/VLSI/consumer electronics (CE) communities towards the following.

1.5.1 Security-aware integrated circuit (IC)/hardware accelerator design tools

Looking through the reality of security risks, EDA/VLSI/CE communities need to adapt to security-aware IC or hardware accelerator design tool. If design automation tools are aware of security at high abstraction level of design phases, then the security of design is also ensured at subsequently lower design phases also. Therefore, a security-aware HLS tool has a paramount importance in ensuring security of ICs or hardware accelerators. This book introduces four security-aware HLS tools, shown in Figure 1.3, which generate secured hardware accelerator designs. Highlights of security-aware design automation tools and their goals are presented here as follows.

1.5.1.1 Security tool 1: crypto-stego tool

Crypto-stego tool is a security-aware design tool which integrates crypto-steganography security mechanism with the scheduling, allocation and binding phases of HLS process and generates a secured scheduled and resource-allocated design. The objective of this tool is to generate steganography-embedded DSP hardware accelerator designs to secure them against piracy and false claim of ownership threats. This tool takes input DSP application in the form of DFG representation. The DFG is fed post converting it into a textual representation. Other inputs are resources constraints, module library file, stego-keys and size of stego-constraints. The module library file contains information about area, delay and power consumption of RTL modules such as multipliers, adders, multiplexers, demultiplexers and registers. The output of this tool is in the form of stego-embedded scheduled and resource-allocated design. Besides, intermediate outputs of the approach such as initial scheduling and register allocation, secret design data, initial state matrix, matrix post byte substitution, matrix post row diffusion, output

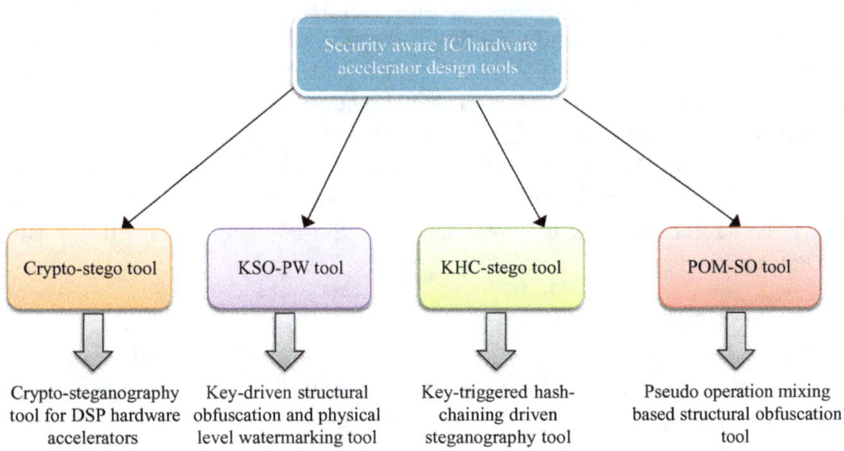

Figure 1.3 Useful hardware security tools for designing security-aware hardware accelerators

of Trifid cipher, output of column diffusion and finally generated stego-constraints can also be seen onto the tool. Moreover, scheduling and registering allocation post embedding stego-constraints, security metric and design cost can be seen onto the tool. The details of the tool with demonstration have been given in Chapter 2. The tool is publicly available. The link to download the crypto-stego tool is as follows: http://www.anirban-sengupta.com/Hardware_Security_Tools.php.

1.5.1.2 Security tool 2: KHC-stego tool

KHC-stego tool is a security-aware design tool which integrates key-driven hash-chaining-based steganography mechanism with the scheduling, allocation and binding phases of HLS process and generates a secure, scheduled and resource-allocated design. The objective of this tool is to generate steganography-embedded DSP hardware accelerator designs to secure them against piracy and false claim of ownership threats. Input DSP application, resource constraints and module library file fed to the KHC-stego tool are the same as crypto-stego tool. Besides, other inputs to the tool are the number of encodings, chosen encoding number, stego-keys, number of rounds of hashes and constraints size. The output is in the form of stego-embedded scheduled and resource-allocated design. The tool also shows the intermediate outputs, value of security metric and design cost. The details of the tool with demonstration have been given in Chapter 7. The tool is publicly available. The link to download the KHC-stego tool is as follows: http://www.anirban-sengupta.com/Hardware_Security_Tools.php.

1.5.1.3 Security tool 3: KSO-PW tool

KSO-PW tool is a security-aware design tool which integrates key-driven structural obfuscation mechanism with the HLS process and performs watermarking on early floorplan of structurally obfuscated RTL design. The objective of this tool is to generate structurally obfuscated RTL design and watermarked floorplan of RTL modules to secure the design against piracy and Trojan threats. Input DSP application, resource constraints and module library file fed to the KSO-PW tool are the same as crypto-stego tool. Besides, other inputs to the tool are the decimal value of structural obfuscation keys and author's signature for watermarking. The tool generates structurally obfuscated RTL of partitioned DFG and final watermarked floorplan at output. The tool also shows the intermediate outputs and design cost. The details of the tool with demonstration have been given in Chapter 4. The tool is publicly available. The link to download the KSO-PW tool is as follows: http://www.anirban-sengupta.com/Hardware_Security_Tools.php.

1.5.1.4 Security tool 4: POM-SO tool

POM-SO tool is a security-aware design tool which integrates pseudo-operations mixing-based structural obfuscation mechanism with the HLS process to generate secure designs. The objective of this tool is to generate a structurally obfuscated DSP hardware accelerator design to ensure security against Trojan threats. Input DSP application, resource constraints and module library file fed to the POM-SO tool are the same as previously discussed security tools. The tool generates structurally obfuscated scheduled and allocated design at output. The tool also shows the intermediate outputs, strength of obfuscation at RTL and design cost pre and post obfuscation. The details of the tool with demonstration have been given in Chapter 9. The tool is publicly available. The link to download the POM-SO tool is as follows: http://www.anirban-sengupta.com/Hardware_Security_Tools.php.

1.5.2 Using natural uniqueness such as biometric info as digital evidence in an intellectual property (IP)/IC

Towards ensuring security of IPs/ICs/hardware accelerators against false claim of ownership and piracy, EDA/VLSI/CE communities can make a paradigm shift from tradition hardware security approaches such as hardware watermarking, hardware steganography and non-biometric fingerprint-based security. This book introduces a new paradigm shift using natural uniqueness such as biometric info as digital evidence in an IP/IC. Sengupta and Rathor (2020c) proposed hardware security based on natural biometric fingerprint of an IP/IC vendor/designer. Because of uniqueness of biometric fingerprint of each person, the corresponding digital evidence embedded into the design cannot be claimed by an adversary. It overcomes the limitations of hardware watermarking, steganography and non-biometric fingerprint-based approaches as the security constraints can be compromised/theft/regenerated/extracted and claimed/reutilized by the adversaries for personal benefits. Hence, the biometric fingerprinting security approach (Sengupta and Rathor, 2020c) for securing IPs/ICs/ hardware accelerators is an important milestone towards ending of false claim of IP ownership threat. More details of this security technique are available in Chapter 6. Database of acquired fingerprints is available for download at http://www.anirban-sengupta.com/Our_Biometric_Fingerprint_Database.php.

1.5.3 Designing application-specific processors/hardware accelerators and functionally reconfigurable processors for image processing filters

Image processing applications have been executed using general purpose processors since their advent. However, increasing application execution load on general purpose processors in today's systems leads to poor performance/latency with high power consumption. Further, FPGAs have been used to facilitate image processing applications. Again, it is hard to customize area, power and delay requirements using FPGAs. Therefore, in order to enable execution of image processing applications within desired/custom area, power and performance requirements, application-specific processor or hardware accelerator design of image processing filters is a convincing solution. This book introduces such designs and its methodologies to readers. Designing process of application-specific processors or hardware accelerators for image processing filters (Sengupta and Rathor, 2020e) has been discussed in detail in Chapter 5. Further, Chapter 5 also discusses a design process of a functionally reconfigurable processor for image processing filters, where various convolution filters of different image processing applications can be executed using the same processor only by reconfiguring it through select lines.

1.5.4 Design flow that incorporates a double line of defence to secure IPs/ICs/hardware accelerators

For enhancing hardware security, there is need of paradigm shift towards the double line of defence-based security mechanisms. This is because advancement in technology has also offered means such as sophisticated tools to adversaries/attackers to nullify the security mechanism employed in the designs. Therefore, the integration of preventive-control-based security with detective control mechanism needs to be adapted to enhance the security of IPs/ICs/hardware accelerators against Trojan insertion and piracy threats. This book introduces the double line of defence-based hardware security techniques to readers. Rathor and Sengupta (2020) and Sengupta and Rathor (2020a, 2020b) have developed design flow that incorporates a double line of defence to secure IPs/ICs/hardware accelerators. Sengupta and Rathor (2020a) and Rathor and Sengupta (2020) integrated crypto-steganography with structural obfuscation to offer a double line of defence to DSP and multimedia hardware accelerators. More details of this technique are available in Chapters 3 and 8. Further, Sengupta and Rathor (2020b) integrated watermarking with key-based structural obfuscation to offer a double line of defence. More details of this double line of defence technique are available in Chapter 4.

1.6 Conclusion

This chapter presented the relevance of hardware accelerators in today's world digital electronics as well as highlighted the importance of hardware accelerators for DSP and image processing applications. Besides these, some of the other salient features of this chapter were as follows:

1. The role of ESL synthesis in hardware accelerator design.
2. Discussion on some popular DSP and image processing applications.
3. Summary of some well-known security algorithms used for securing hardware accelerators, along with reference to the respective chapter where those are elaborately explained.
4. Discussion on the future paradigm shift for EDA/VLSI/CE communities.
5. Introduction to the four security tools developed by the author and his team that can handle various hardware threats in the context of hardware accelerators.

1.7 Questions and exercise

1. What are hardware accelerators and their generic classification?
2. What is the role of hardware accelerators in modern technology advancement?
3. List out any five applications and their corresponding hardware accelerators.
4. What role does ESL synthesis play in designing of hardware accelerators?

5. What is the computation function of an FIR filter hardware accelerator and salient features of an FIR filter?
6. What is the computation function of a JPEG codec hardware accelerator and what is its major functionality?
7. What are 2D FIR filters and where are they used?
8. What are different types of convolution filters used in image processing applications?
9. What are the target threats of crypto-steganography and key-driven hash-chaining-based steganography security techniques?
10. What is the need of integration of watermarking or steganography approach with structural obfuscation approach?
11. What is the double line of defence-based security? What threats are targeted by employing a double line of defence?
12. What is biometric-fingerprinting-based hardware security approach?
13. Why is the need of security-aware design automation tools for EDA/VLSI/CE communities?
14. Why is paradigm shift to biometric fingerprint approach from traditional security approaches such as hardware watermarking and steganography, required?
15. Why is the need of designing application-specific processors/hardware accelerators and functionally reconfigurable processors for image processing filters?

References

F. Koushanfar, I. Hong, and M. Potkonjak (2005), 'Behavioral synthesis techniques for intellectual property protection,' *ACM Trans. Des. Autom. Electron. Syst.*, vol. 10(3), pp. 523–545.

B. Le Gal and L. Bossuet (2012), 'Automatic low-cost IP watermarking technique based on output mark insertions,' *Des. Autom. Embedded Syst.*, vol. 16(2), pp. 71–92.

M. C. McFarland, A. C. Parker and R. Camposano (1988), 'Tutorial on high-level synthesis,' *DAC '88 Proceedings of the 25th ACM/IEEE Design Automation*, vol. 27(1), pp. 330–336.

M. Rathor and A. Sengupta (2020), 'Design flow of secured N-point DFT application specific processor using obfuscation and steganography,' *Lett. IEEE Comput. Soc.*, vol. 3(1), pp. 13–16.

A. Sengupta (2017), 'Hardware security of CE devices [hardware matters],' *IEEE Consum. Electron. Mag.*, vol. 6(1), pp. 130–133.

A. Sengupta (2020), 'Frontiers in securing IP cores – forensic detective control and obfuscation techniques,' *The Institute of Engineering and Technology (IET)*, ISBN-10: 1-83953-031-6, ISBN-13: 978-1-83953-031-9.

A. Sengupta and S. P. Mohanty (2019), 'Trojan security aware DSP IP core and integrated circuits,' *IP Core Protection and Hardware-Assisted Security for*

Consumer Electronics, e-ISBN: 9781785618000, Chapter doi: 10.1049/PBCS060E_ch.

A. Sengupta and M. Rathor (2019a), 'Protecting DSP kernels using robust hologram-based obfuscation,' *IEEE Trans. Consum. Electron.*, vol. 65(1), pp. 99–108.

A. Sengupta and M. Rathor (2019b), 'Crypto-based dual-phase hardware steganography for securing IP cores,' *Lett. IEEE Comput. Soc.*, vol. 2(4), pp. 32–35.

A. Sengupta and M. Rathor (2020a), 'Structural obfuscation and crypto-steganography-based secured JPEG compression hardware for medical imaging systems,' *IEEE Access*, vol. 8, pp. 6543–6565.

A. Sengupta and M. Rathor (2020b), 'Enhanced security of DSP circuits using multi-key based structural obfuscation and physical-level watermarking for consumer electronics systems,' *IEEE Trans. Consum. Electron.*, doi: 10.1109/TCE.2020.2972808.

A. Sengupta and M. Rathor (2020c), 'Securing hardware accelerators for CE systems using biometric fingerprinting,' *IEEE Transactions on Very Large Scale Integration (VLSI) Systems*, vol. 28(9), pp. 1979–1992.

A. Sengupta and M. Rathor (2020d), 'IP core steganography using switch based key-driven hash-chaining and encoding for securing DSP kernels used in CE systems,' *IEEE Transactions on Consumer Electronics*, vol. 66(3), pp. 251–260.

A. Sengupta and M. Rathor (2020e), 'Obfuscated hardware accelerators for image processing filters - application specific and functionally reconfigurable processors,' *IEEE Transactions on Consumer Electronics*, accepted, doi: 10.1109/TCE.2020.3027760.

Chapter 2

Cryptography-driven IP steganography for DSP hardware accelerators

Anirban Sengupta[1]

The chapter describes a cryptography-driven intellectual property (IP) steganography process for securing hardware accelerators. The chapter focusses on hardware accelerators that are used popularly in digital signal processing (DSP) applications for modern electronics systems/products. A detailed elaboration on the salient features of cryptography-driven IP steganography process, its differences from DSP watermarking approaches, other hardware steganography approaches, details of secret steganography constraint generation process, embedding process, detection process and details on case studies have been provided. The chapter is organized as follows: Section 2.1 discusses the background of this topic; Section 2.2 presents the contemporary approaches for securing hardware accelerators. Section 2.3 presents the crypto-based steganography process for securing hardware accelerators; Section 2.4 introduces a new crypto-stego tool for securing hardware accelerators; Section 2.5 presents the case studies on DSP hardware accelerators; Section 2.6 concludes the chapter; Section 2.7 provides some exercise for the readers.

2.1 Introduction

Hardware acceleration is a mechanism of realizing computationally intensive tasks using hardware, in order to boost up system performance and throughput. In other words, general-purpose processors, such as central processing units, and custom hardware work together to enhance overall performance and throughput of an electronic system. Some popular custom hardware used for hardware acceleration are field programmable gate array, application-specific integrated circuits (IC) and graphics processing units (GPUs). The goal of making hardware and software work together is to simultaneously leverage the advantages of both. Software part of the system leads to advantages such as (i) faster system development, i.e. lesser time to market, (ii) less complications in updating features, (iii) easiness in locating and

[1]Computer Science and Engineering, Indian Institute of Technology Indore, Indore, India

patching bugs and (iv) reduced non-recurring engineering costs. However, the software part pays off in terms of poor performance when it comes to performing highly data-intensive tasks. This performance lagging can be managed with the aid of hardware accelerators, which leads to the following advantages: (i) high-speed computation, (ii) high parallelism and (iii) less power consumption. Following are some applications and the corresponding hardware accelerators employed to enhance performance: (i) DSP applications using DSP hardware accelerators, (ii) artificial intelligence (AI) applications using AI accelerators, (iii) computer networking applications using network processor and network interface controller, (iv) computer graphics using GPU hardware accelerator, (v) sound processing using sound card and (vi) cryptography applications using crypto-processor or cryptographic accelerator and so forth. The focus of this chapter is mainly on DSP hardware accelerators.

Hardware accelerators have paramount importance for DSP or image processing applications because of following reasons: (i) computational intensiveness (i.e. a large number of operations are required to be computed in a predefined time constraint) and (ii) vast utilization of DSP applications in low power portable devices such as mobile phones, digital camera, laptop, tablet, etc. Considering the aforementioned facts, it is highly efficient to realize DSP algorithms using hardware accelerators for achieving high performance at low power (Mahdiany *et al.*, 2001; Schneiderman, 2010).

Apart from design objectives such as low latency, low power and low area, one more objective is grasping the attention of hardware accelerator IC developers. Here, we are highlighting about the objective of *'security' or 'protection' of the hardware accelerators* against hardware counterfeiting, cloning, false claim of IP ownership and Trojan insertion threats. The security standpoint is highly relevant today because the entire process of IC design/development involves various multivendor third parties. The entire design process mainly involves following entities: IP vendors, system-on-chip (SoC) integrators and IC fabrication unit (foundry). For an SoC integrator, IP vendors and fabrication unit are considered third parties. For a product integrator/designer/manufacturer, an SoC integrator is considered a third party. In conclusion, IP vendors, SoC integrators and foundries all play a role of third party somewhere in the entire process of designing of an end electronic system product. These third parties are situated globally and may have their own personal or national interests. This leads to malfunctioning or infringement of hardware designs at different phases of IC development. Therefore, third parties involved in the hardware accelerator IC development process are considered to be unreliable. Thereby untrustworthiness of offshore third parties poses security concerns for hardware accelerator designs, and hence they are required to be secured, during their design process, against ownership abuse, counterfeiting, cloning and hardware Trojan threats (Castillo *et al.*, 2007; Plaza and Markov, 2015; Sengupta, 2016, 2017; Roy and Sengupta, 2019). If we discuss especially about DSP hardware accelerators, their design process is initiated with the high-level synthesis (HLS) phase (McFarland *et al.*, 1988) of IC design. This is due to the fact that the DSP algorithms are highly complex and large in size; hence it is challenging to be implemented directly at lower abstraction levels such as register transfer level (RTL) or gate level.

In addition, embedding security during higher abstraction level is relatively easier and also enables exploration of low-cost security solution using design space exploration process (Sengupta *et al.*, 2010; Mishra and Sengupta, 2014).

The security of DSP hardware accelerators (Pilato *et al.*, 2018) against counterfeiting, cloning and false claim of IP ownership threats can be enabled using detective control mechanisms such as hardware watermarking and hardware steganography. Hardware watermarking (Koushanfar *et al.* 2005; Ziener and Teich, 2008; Le Gal and Bossuet, 2012; Sengupta and Bhadauria, 2016; Sengupta *et al.*, 2019; Sengupta and Mohanty, 2019a, 2019b) inserts vendor's secret signature into the design to make the hardware accelerators authorized and authenticated. By detecting the vendor's secret signature into the design, fake hardware accelerators can be separated from authenticated ones. Let us come to the hardware steganography (Sengupta and Rathor, 2019a, 2019b) which is a newer approach of securing DSP hardware accelerators than watermarking approach. This is a signature-free approach of securing designs against aforementioned threats. The steganography approach uses stego-encoder mechanism to produce stego-constraints to be embedded into the design (Sengupta and Rathor, 2019a). *Hardware steganography differs from watermarking in terms of the following, in the context of HLS for DSP hardware accelerators* (Sengupta and Rathor, 2019a, 2019b):

1. *Kind of secret constraints to be embedded*: Watermarking embeds vendor's signature comprising two or multiple variables, where each variable is encoded as security constraint to be embedded. However, hardware steganography embeds secret information in the form of stego-constraints which are mapped to the hardware security constraints using designer's specified mapping rules.
2. *Secret (or security) constraints generation process*: In watermarking approaches, first a desired signature is chosen which is then converted to security constraints using designer's encoding rules. In hardware steganography, stego-constraints are generated by a stego-encoder process which employs a more scientific or mathematical algorithm driven through a controlling parameter. This controlling parameter is referred to as stego-key or hardware entropy depending on the type of hardware steganography to be employed.
3. *Controllability of amount of security employed*: In watermarking, designer has less control over the amount of secret constraints embedded because one cannot predetermine a particular size and combination of a signature that would correspond to maximum security constraints. Sometimes a larger size signature can result into lesser security constraints embedded. This is because some security constraints corresponding to a large size signature may not be implanted because of their default existence in the design. Moreover, the vendor's signature is vulnerable to theft by an adversary. Once the vendor's signature is compromised, she/he fails to prove it, and the goal of watermarking is defeated. However, steganography approach offers flexibility of controlling the amount of security employed using a controlling parameter (entropy threshold or stego-key). By increasing the value of entropy threshold, more security constraints can be embedded which leads to higher security.

Similarly, by increasing the stego-key size, more security against the theft of security constraints can be achieved. Even if the attacker somehow gets the access of security constraints, she/he cannot prove them without the knowledge of secret entropy threshold or secret stego-key value. In conclusion, a steganography approach offers more controllability over the security employed than watermarking approaches as well as stronger digital evidence.

4. *Controllability of design overhead incurred due to embedding secret constraints*: In watermarking approaches, it is challenging to estimate in advance the impact of vendor's signature on design overhead. Various signature combinations may pose different impact on design overhead. However, for entropy-based steganography approach, design cost overhead is controllable by threshold entropy value. Design overhead may increase with the increase in entropy threshold value. Further, in the case of stego-key-based steganography, the design cost overhead can be controlled using designer's chosen size of stego-constraints. Design overhead may increase with the increase in stego-constraints size.

The previous discussion also highlights the advantages of hardware steganography over watermarking approaches.

The focus of this chapter is on the discussion of a cryptography-driven hardware stenography approach (Sengupta and Rathor, 2019b) to secure DSP hardware accelerators against piracy and false claim of ownership threats. This approach is capable of providing higher security than other steganography and watermarking approaches. The robustness of the approach lies in the fact that the stego-constraints generation process is highly intricate to be back engineered by an adversary. This is because various complex crypto-mechanisms are incorporated in the stego-constraints generation process. In addition, a very large size key drives the stego-constraints generation process. Therefore, it is infeasible for an adversary to regenerate/extract and prove stego-constraints during forensic detection (Sengupta and Rathor, 2019b). A discussion on contemporary steganography and watermarking approaches of securing DSP hardware accelerators is briefed in the next section.

2.2 Contemporary approaches for securing hardware accelerators

Some major contemporary approaches of securing hardware accelerators using hardware steganography and hardware watermarking are discussed in the following subsection.

2.2.1 Entropy-threshold-based hardware steganography (Sengupta and Rathor, 2019a)

Sengupta and Rathor (2019a) proposed first hardware steganography for securing DSP hardware accelerators against counterfeiting and cloning threats. This hardware steganography enables counterfeiting and cloning detection by embedding stego-constraints in the register allocation phase of HLS process. Control data flow graph (CDFG) representation (a high-level representation) of DSP hardware accelerator application is fed as inputs to this approach. Upon embedding security constraints during HLS, a stego-embedded DSP hardware accelerator is generated as outputs. The main steps of the threshold-entropy-based steganography approach are highlighted in Figure 2.1. Stego-constraints generation process leverages a coloured interval graph (CIG) representation of register allocation (to the storage variables of the design) to list out all the possible constraints to be embedded. Further, the set of final constraints to be embedded are shortlisted using a controlling parameter called entropy threshold. The secret constraints are embedded in the form of additional artificial edges in the CIG. Added edge constraints in the CIG are reflected in the scheduled and hardware allocated design in the form of enforced register allocation to the storage variables of the design. Amount of security constraints to be embedded can be controlled by the designer by appropriately choosing the desired entropy threshold value. This approach being signature free eliminates the possibility of leaking of signature to an adversary, unlike watermarking approaches. Hence, this steganography approach emerged as more secure solution against the targeted hardware threats, in contrast to DSP watermarking approaches. However, this approach has some limitations such as the non-involvement of stego-key in the stego-constraints generation process and embedding stego-constraints only in the single phase (i.e. register allocation phase) of HLS process. This weakens the secrecy of stego-constraints and renders the regeneration/extraction easier for an attacker.

2.2.2 Cryptography-driven hardware steganography approach (Sengupta and Rathor, 2019b)

Accounting the limitations of entropy-based hardware steganography and watermarking approaches, Sengupta and Rathor (2019b) proposed a highly robust steganography mechanism which is driven through a very large size stego-key. Moreover, stego-constraints are embedded during two different phases of HLS, viz. register allocation phase and functional unit (FU) vendor allocation phase. This leads to embedding of higher and distributed digital evidence into the design as well as deeper embedding of stego-constraints (because of the level of embedding constraints is enhanced to two phases). In addition, the stego-encoder employed to generate stego-constraints is highly complex to be back engineered or cracked by an adversary. This is because the stego-constraints generation process employs a number of cryptographic mechanisms such as *byte substitution using S-Box, row diffusion, column diffusion, Trifid cipher*. The overview diagram capturing inputs, outputs and basic process of the cryptography-driven steganography approach for hardware accelerators is shown in Figure 2.2. As shown in the figure, primary inputs required to be fed to the cryptography-driven hardware steganography approach are as follows:

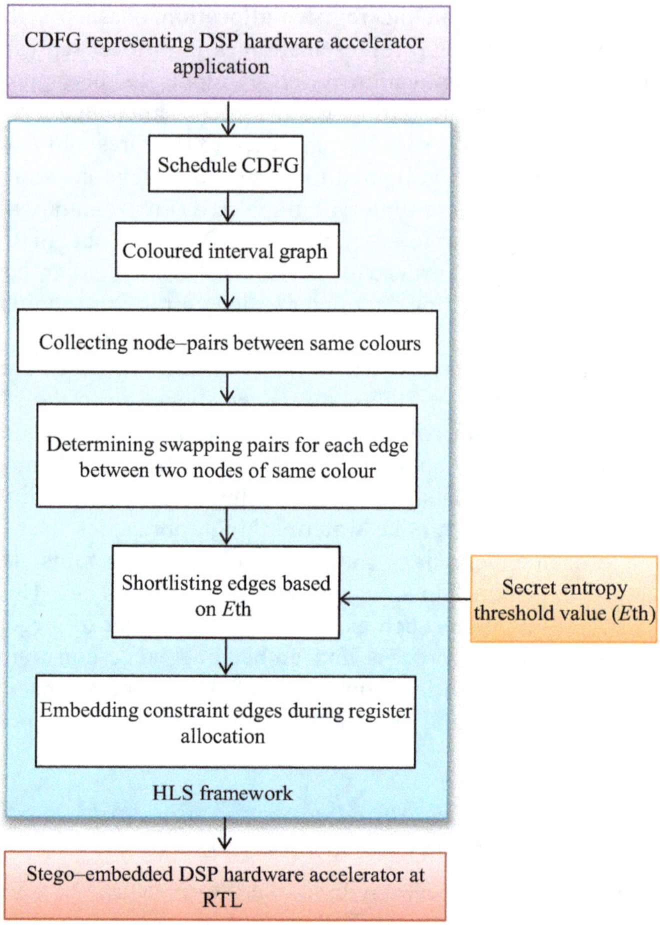

Figure 2.1 Hardware steganography approach based on entropy threshold (Sengupta and Rathor, 2019a)

1. High-level description of DSP hardware accelerator application. The high-level description can be in the form of C/C++ code, transfer function or CDFG (intermediate high-level representation)
2. Resource/hardware constraints
3. Module library
4. Stego-keys

The stego-encoder accepts secret design data, cover design data and stego-keys as inputs to generate stego-constraints. The secret design data and cover design data are generated from an intermediate step of HLS process (the details of formation of secret design data and cover design data and generation of stego-constraints are

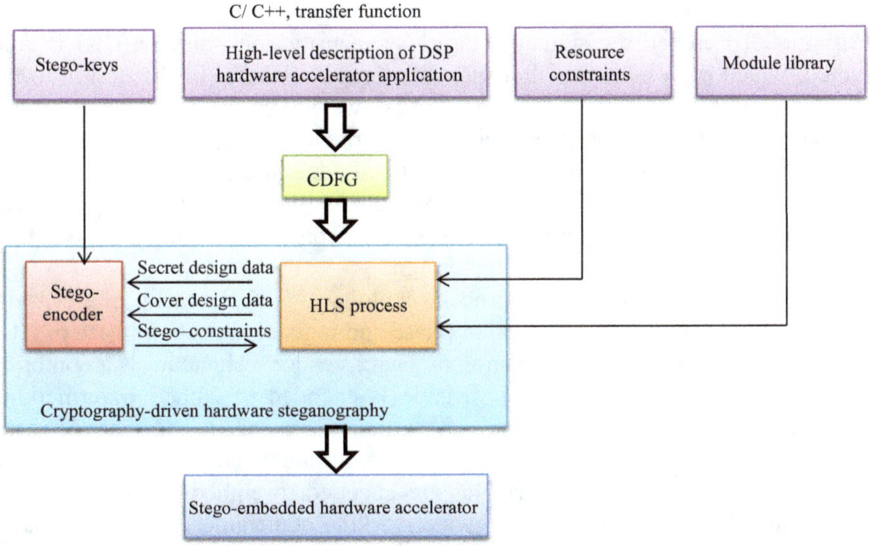

Figure 2.2 High-level view of cryptography-driven hardware steganography approach (Sengupta and Rathor, 2019b)

discussed in subsequent sections of this chapter). The stego-encoder process comprises several key-based cryptographic processes that execute in a sequence to generate stego-constraints. Thus, obtained stego-constraints are in the form of a bitstream, which is truncated to the designer chosen secret size. Further, stego-constraints are embedded into the design during register allocation and FU vendor allocation phase of HLS process. Post embedding stego-constraints, the output is a stego-embedded DSP hardware accelerator design.

2.2.3 Watermarking approaches (Sengupta and Bhadauria, 2016; Sengupta and Roy, 2017, 2018; Sengupta et al., 2018)

Some watermarking approaches for securing DSP hardware accelerators against piracy and false claim of ownership have been employed during higher phase of design process, i.e. architectural or behavioural level. This subsection discusses two different kinds of watermarking approaches employed during HLS, viz. single-phase watermarking (Sengupta and Bhadauria, 2016; Sengupta and Roy, 2017) and multiphase watermarking (Sengupta and Roy, 2018; Sengupta *et al.*, 2018). Single-phase watermarking approach exploits only single phase, i.e. register allocation phase of HLS to embed secret watermarking constraints. In the single-phase watermarking approach, vendor's signature is a combination of four variables where each variable can be utilized multiple times in order to upsurge the magnitude of digits in the signature. The vendor or designer associates an encoding rule with each signature variable in order to covert the signature into respective security constraints. To embed the signature digits as security constraints, a CIG is constructed, which represents the allocation of storage variables of the design to the minimum possible register. In the single-phase watermarking approach, Sengupta and Bhadauria (2016) encoded each signature variable in such a way that each digit of the signature is embedded as an extra artificial edge in the CIG. However, some edge constraints corresponding to signature digits may exist by default in the CIG. This causes diminution in number of effective constraints, corresponding to a vendor's signature, embedded into the design.

Multiphase watermarking approach exploits three divergent phases, viz. scheduling phase, FU vendor allocation phase and register allocation phase of HLS to embed secret watermarking constraints. Here, vendor's signature is a combination of seven variables where one variable is encoded to embed watermarking constraints in the scheduling phase, two variables are encoded to embed watermarking constraints in the FU (e.g. adders, multipliers, etc.) vendor allocation phase and the remaining four variables are encoded to embed constraints in the register allocation phase. Embedding watermarking constraints at multiple phases of HLS process enables embedding of deeper and higher amount digital evidence into the design. This leads to a stronger proof of authorship to nullify a false claim of authorship threat by an adversary or providing detective control of piracy. Figure 2.3 captures a very basic difference of multiphase and single-phase watermarking, highlighting the broader coverage of HLS design phases for embedding constraints using multiphase watermarking.

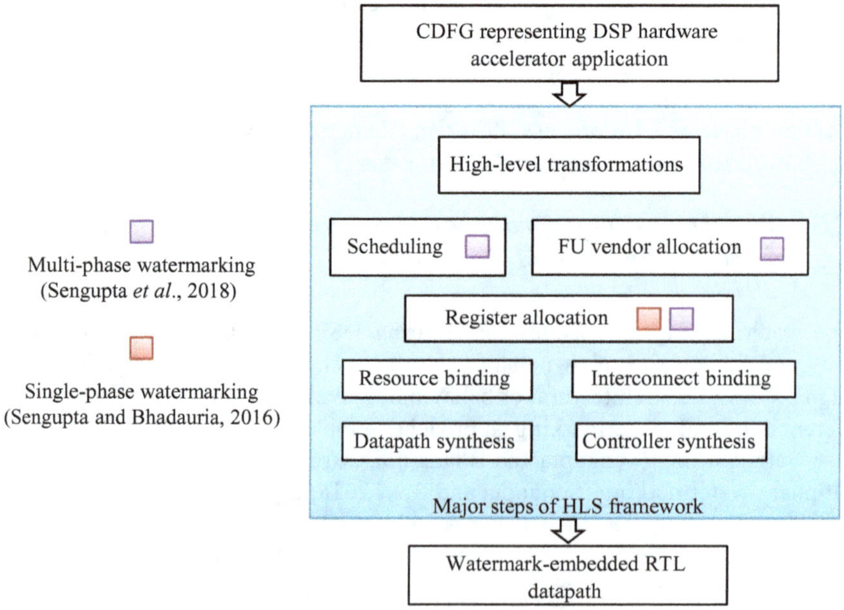

Figure 2.3 Basic difference between single-phase and multiphase watermarking during HLS

2.3 Crypto-based steganography for securing hardware accelerators (Sengupta and Rathor, 2019b)

Sengupta and Rathor (2019b) proposed a crypto-based hardware steganography approach that also leverages HLS framework to embed secret stego-constraints like related steganography approach (Sengupta and Rathor, 2019a). However, unlike related steganography approach, the generation process of stego-constraints is stego-key driven and exploits a number of security properties/mechanisms in sequence to generate secret stego-constraints. The exploited security properties/mechanisms are of two types, viz. cryptographic and non-cryptographic. Following are the cryptographic and non-cryptographic security properties incorporated in the stego-constraints generation process: (i) bit manipulation or byte substitution using cryptographic S-box, (ii) cryptographic row diffusion, (iii) multilayered Trifid-cipher-based encryption, (iv) alphabet substitution, (v) matrix transposition, (vi) cryptographic mix column diffusion, (vii) byte concatenation, (viii) bitstream truncation and (ix) bitmapping. The stego-encoder system of crypto-based hardware steganography approach performs aforementioned security algorithms to generate stego-constraints. It takes secret design data and stego-keys as inputs to generate stego-constraints. The stego-key is a combination of five sub-keys where each sub-key controls an intermediate step of stego-constraints generation process.

The detailed flow of the crypto-based hardware steganography approach is shown in Figure 2.4. As shown in the figure, a high-level description of DSP application is first converted into an intermediate representation in the form of CDFG. Further, designer's specified resource constraints and module library are used to generate a scheduled CDFG. Thereafter, a CIG is created using the information of allocation of storage variables of the design to the registers. Nodes in the CIG represent storage variables (*S*) of the design and the colour of a node represents its assignment to a register. Therefore, the total number of distinct colours used in the CIG is equal to the minimum number of registers required to accommodate all storage variables of the design. Further, overlapping of lifetime of storage variables is indicated by drawing edges between nodes in the CIG. Thus, obtained CIG is leveraged to generate secret design data which is fed to stego-encoder system. In addition, cover design data and stego-keys are also fed to the stego-encoder. The stego-encoder system uses secret design data and secret stego-keys to generate stego-constraints through following steps: (i) state-matrix formation, (ii) bit manipulation, (iii) row diffusion, (iv) Trifid-cipher-based encryption, (v) alphabet substitution, (vi) matrix transposition, (vii) mix column diffusion, (viii) byte concatenation, (ix) bitstream truncation and (x) bitmapping (these steps are discussed in detail in the later part of this section). Here, steps (i)–(viii) generate an encrypted bitstream. This encrypted bitstream is truncated based on designer's chosen secret constraints size as shown in Figure 2.4. Further, each bit in the bitstream is converted to the hardware security constraints (or stego-constraints) by using designer's specified mapping rules for bit '0' and bit '1'. The mapping rules (Sengupta and Rathor, 2019b) are shown in Figure 2.5. Thus, generated secret

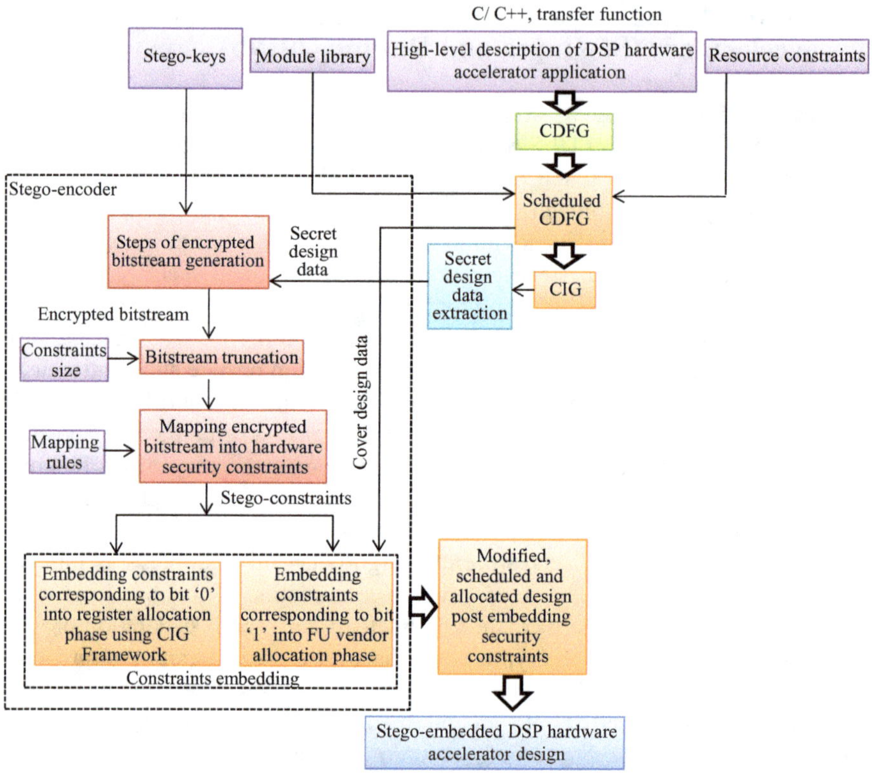

Figure 2.4 Flow of cryptography-driven hardware steganography approach (Sengupta and Rathor, 2019b)

stego-constraints are finally embedded into the cover design data during HLS process. This modifies the scheduled and hardware allocated design. Thus, incurred modifications reflect the stego-constraints embedded into the design. Thus, a steganography-embedded hardware accelerator design is generated.

The brief description of the three inputs to the stego-encoder is as follows (Sengupta and Rathor, 2019b):

1. *Secret design data*: The secret design data is obtained from CIG and used to generate stego-constraints. To obtain secret design data from the CIG, all possible pairs of the nodes of the same colours are extracted. The set or collection of indices (i, j) of all node-pair (S_i,S_j) of the same colours in the CIG represents the secret design data. The secret design data depends on the DSP application and register allocation scheme.

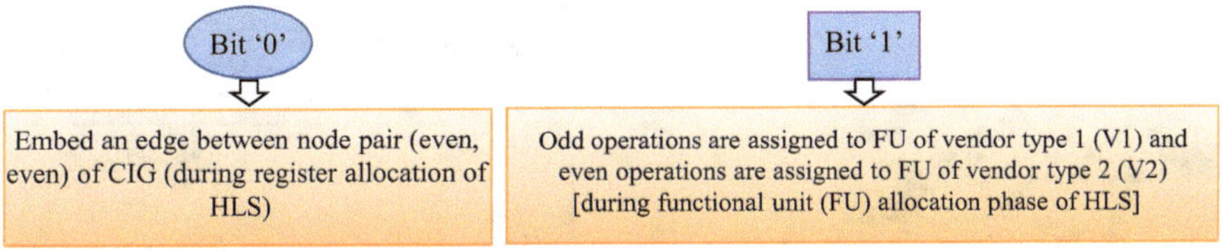

Figure 2.5 Mapping of bits of encrypted bitstream into stego-constraints (Sengupta and Rathor, 2019b)

2. *Cover design data*: The generated stego-constraints are embedded into the cover design data. The scheduled and allocated CDFG is exploited as cover design data to embed stego-constraints. Stego-constraints are embedded into the scheduled and allocated CDFG by performing register reallocation and FU vendor reallocation in HLS. Thereby, two different phases of HLS are utilized to embed stego-constraints into the cover data. Hence, this approach (Sengupta and Rathor, 2019b) is referred to as **dual-phase crypto-based steganography**.
3. *Stego-key*: The stego-key used in cryptography-based steganography is a combination of five sub-keys, viz. stego-key1, stego-key2, stego-key3, stego-key4 and stego-key5. The stego-key1 to stego-key5 control the following steps of stego-constraints generation process, respectively: *state-matrix formation, row diffusion, Trifid-cipher-based encryption, alphabet substitution and byte concatenation*. The role of each sub-key and its different modes of usage are highlighted in Figure 2.6. The stego key regulates the amount of stego-information embedded into the design and the security employed. The larger the size of stego-key, the higher (stronger) the security of generated stego-constraints. This is because as the key size increases, the difficulty level for an attacker in extracting/regenerating secret stego-constraints escalates.

2.3.1 Process of designing stego-embedded hardware accelerator for DCT core

So far, we have discussed the basic flow of crypto-based steganography approach, where we noted that the stego-encoder generates stego-constraints using secret design data and stego-keys. The generated stego-constraints are embedded into the scheduled and allocated CDFG that acts as cover design data. Further, we have discussed the basic definition of secret design data, cover design data and roles/different modes of stego-keys. However, the different steps used in the stego-encoder to generate the secret stego-constraints are yet to be discussed in detail. All the steps of secret stego-constraints generation and the basic function of each step are highlighted in Figure 2.7. Their details have been discussed in this subsection.

Let us discuss each step of crypto-based steganography approach in more detail with the aid of demonstration on 4-point discrete cosine transform (DCT) core. A DCT core is a DSP hardware accelerator used to accelerate the process of image compression. Therefore, it finds wide utility in such consumer electronics systems where image compression is required. Since the crypto-based steganography

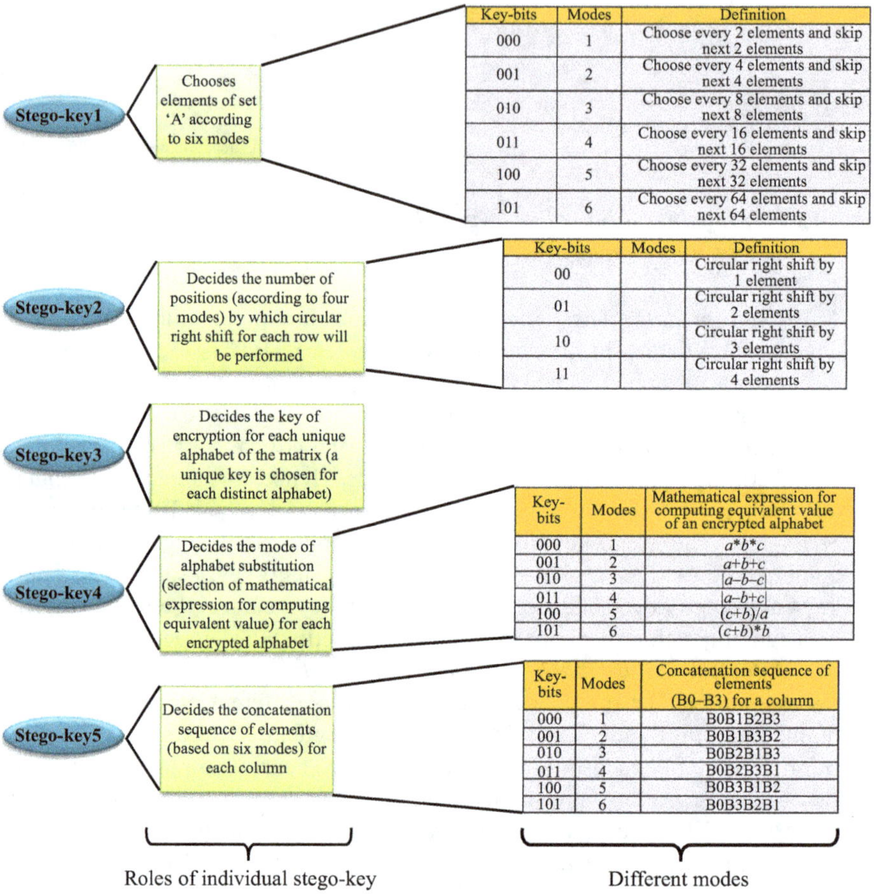

Figure 2.6 Roles of five stego-keys and definition of their different modes

approach requires HLS framework to embed steganography information, a high-level description of 4-point DCT core is fed as input to the HLS process. The high-level description of the 4-point DCT core is first converted into a DFG representation. Further, the DFG is scheduled using LIST scheduling based on designer's chosen resource constraints of two multipliers (M) and one adder (A). This results in a scheduled DFG as shown in Figure 2.8. There are total seven operations (four multiplications and three additions) that are executing in four control steps Q1–Q4. A two-vendor allocation scheme has been used to allocate hardware resources (FUs) to the operations. In this scheme, two or more operations of the same type in the same control step are assigned to the respective FU from two different vendors (V1 and V2). FU allocations to the operations using two vendors are shown in Figure 2.8. For an FU A_j^i or M_j^i, superscript i indicates vendor type, and subscript j indicates an instance number. Further, in the scheduled DFG, primary/internal inputs and outputs

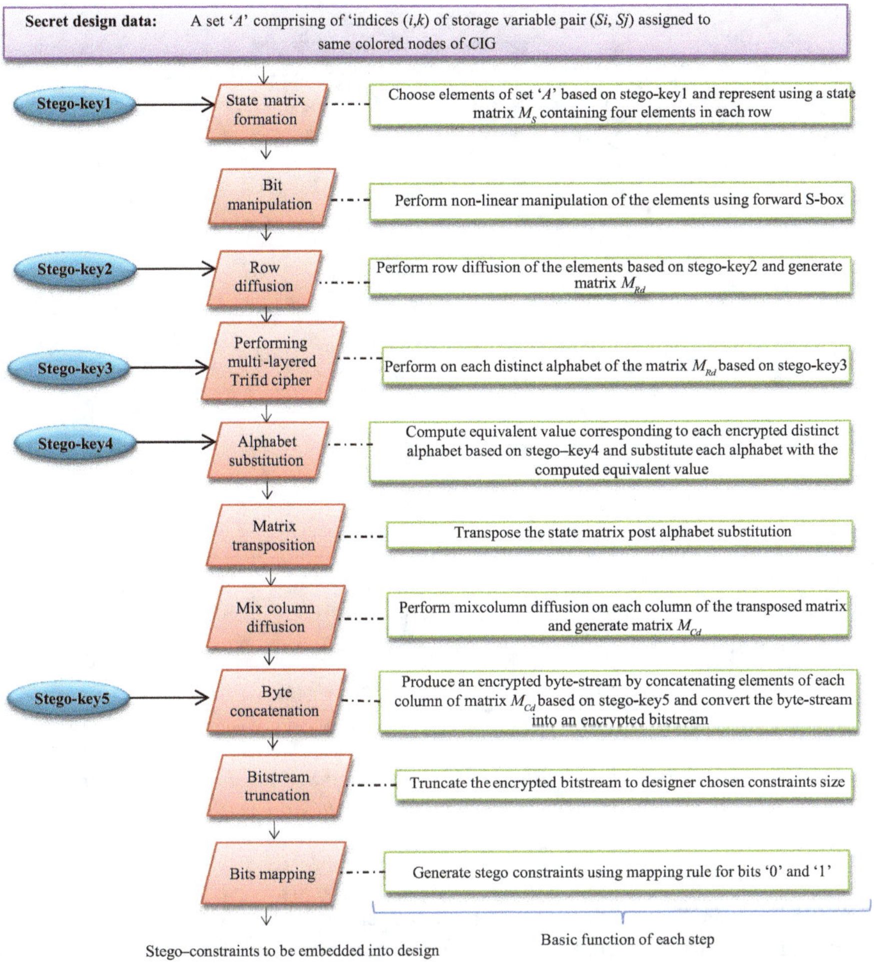

Figure 2.7 Steps of generating stego-constraints through a stego-encoder in the crypto-based dual-phase steganography

of the design have been assigned to 11 storage variables *S*0–*S*10. These eleven storage variables are executing through four distinct registers which have been represented by four distinct colours, viz. red, indigo, green and orange. Two or more storage variables executing in the same control steps are essentially assigned to distinct registers/colours in order to avoid overlapping of lifetime of storage variables. Further, the information of register assignment of storage variables and their lifetime is extracted from scheduled DFG to create a CIG. The CIG of 4-point DCT is shown in Figure 2.9(a), and the corresponding register allocation is shown in Table 2.1. Nodes in the CIG represent storage variables (*S*) of the design, and the colour of a node represents its assignment to a register. And, overlapping of lifetime of storage

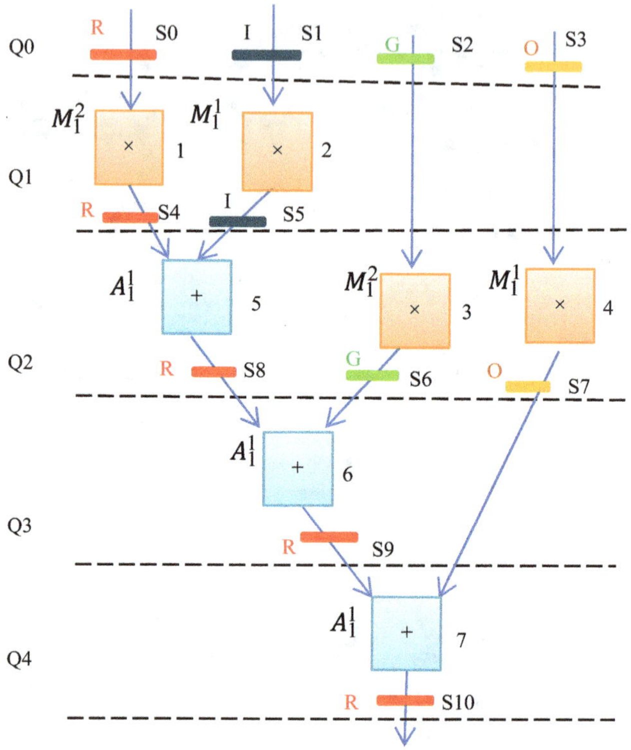

Figure 2.8 Scheduled and hardware allocated 4-point DCT using resource constraints of 1 (+) and 2(*) (Sengupta and Rathor, 2019b)

variables are indicated by drawing edges between nodes in the CIG. Further, secret design data is extracted from the CIG. This secret design data is fed as inputs to stego-encoder. For the 4-point DCT core, the secret design data is represented in terms of a set A as follows:

$$A = \{(0,4),(0,8),(0,9),(0,10),(4,8),(4,9),(4,10),(8,9),(8,10),(9,10),\\(1,5),(2,6),(3,7)\}$$

where each element of the set indicates the indices of node pairs of the same colours in the CIG. If any digit I in the set is greater than 15, then it is reduced using the following expression: $I'=I$ mod 15. Thereafter, the set A is updated after applying modulo-15 reduction. This reduction is performed because each digit in the set is further represented in hexadecimal notation. Hence, the updated secret design data, post-conversion into hexadecimal notation, is given as follows:

$$A = \{(0,4),(0,8),(0,9),(0,A),(4,8),(4,9),(4,A),(8,9),(8,A),(9,A),\\(1,5),(2,6),(3,7)\}$$

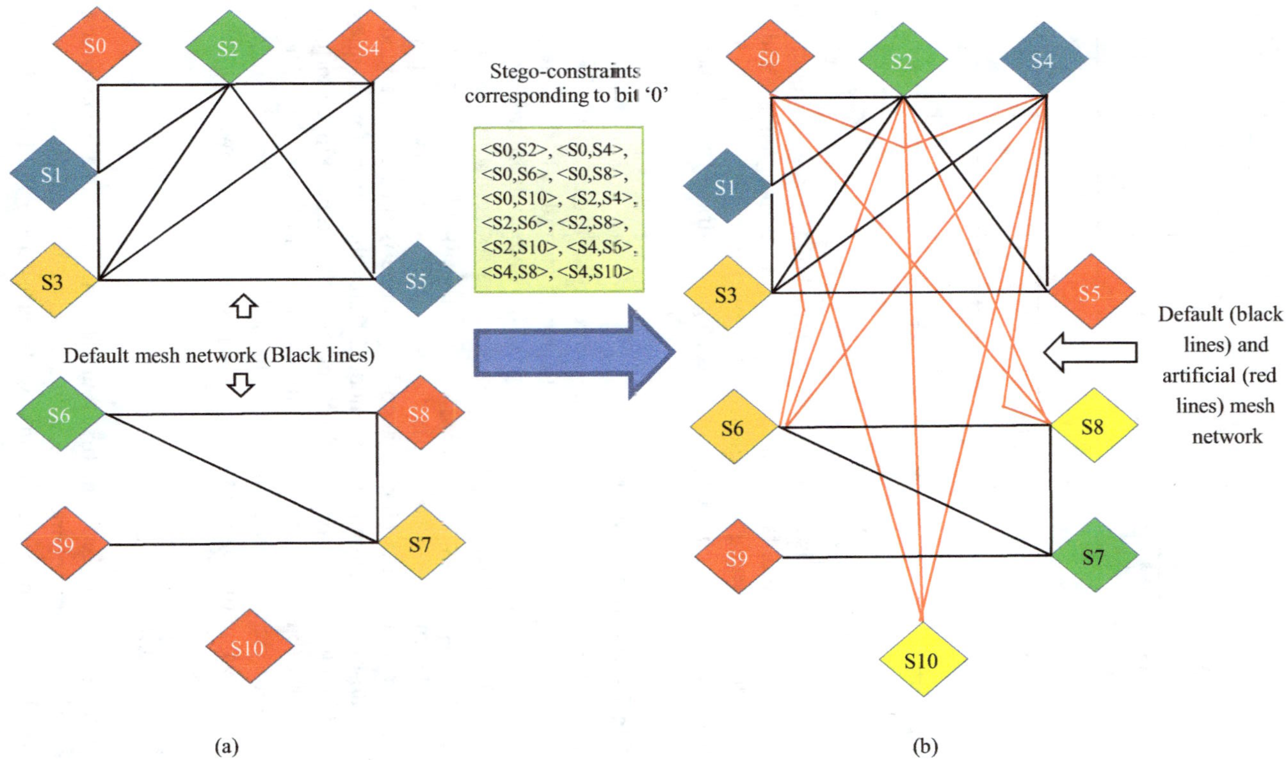

Figure 2.9 (a) CIG pre-embedding stego-constraints and (b) CIG post embedding stego-constraints (Sengupta and Rathor, 2019b)

Table 2.1 Register allocation of 4-point DCT before implanting stego-constraints corresponding to bit 0

Control Steps	R	I	G	O
Q0	S0	S1	S2	S3
Q1	S4	S5	S2	S3
Q2	S8	–	S6	S7
Q3	S9	–	–	S7
Q4	S10	–	–	–

Once the secret design data is obtained, it is exploited by the stego-encoder system to generate stego-constraints using a number of cryptographic and non-cryptographic steps executing in sequence, as shown in Figure 2.7. The demonstrations of the different steps for generating stego-constraints are elaborated as follows:

2.3.1.1 State-matrix formation

In this step, a state matrix is formed using secret design data in the set A. The formation of state matrix is driven through secret stego-key1. The stego-key1 decides the different mode of choosing elements for state-matrix formation. There are total six designer's specified modes of state-matrix formation as shown in Figure 2.6; therefore, the size of stego-key1 is of 3 $\lceil \log_2(6) \rceil$ bits. To form the state matrix, a particular mode of state-matrix formation is chosen based on the stego-key value. Let us assume the chosen stego-key1 value is '001'; hence mode 2 is selected for state-matrix formation. The mode 2 states that every four elements in the set should be chosen and the next four elements should be skipped to form the entire state matrix (based on the mode definition given in Figure 2.6). Therefore, there will be four elements in each row of the state matrix. The state matrix M_S based on chosen value of stego-key1 (i.e. '001') is given as follows:

$$M_S = \begin{bmatrix} 04 & 08 & 09 & 0A \\ 8A & 9A & 15 & 26 \end{bmatrix} \tag{2.1}$$

As shown in the state-matrix M_S, the first four elements of set A are chosen to form the first row. The next four elements of set A are skipped. Then the subsequent four elements are chosen to form the second row. In case during the formation of last row, if a complete quartet is unavailable (remaining elements are less than four) then the row is not formed.

2.3.1.2 Bit manipulation or byte substitution using S-box

Once the state matrix is obtained, non-linear bit manipulation is performed in the matrix elements. To do so, each byte is substituted using forward S-box of AES. The matrix after bit manipulation (M_B) is given as follows:

$$M_B = \begin{bmatrix} F2 & 30 & 01 & 67 \\ 7E & B8 & 59 & F7 \end{bmatrix} \tag{2.2}$$

Security property: The objective of performing bit manipulation is to employ non-linearity in the data using Shannon's property of confusion. This security property incorporates obscurity in the relationship between the input and the final output (i.e. stego-constraints generated).

2.3.1.3 Row diffusion

This step incorporates row diffusion in the matrix using secret stego-key2. The value of stego-key2 controls the amount of diffusion. This is because the key value decides the mode of row diffusion to be applied on each row. A mode decides the number of positions by which circular right shift in each row will be executed. The stego-key2 is a multiple of 2 wherein each pair of two bits decides the mode of row diffusion, for each row of the matrix. Therefore, the size of stego-key2 is equal to 2*(number of rows in the matrix) bits. For each pair of consecutive bits, there are four modes of row diffusion as shown in Figure 2.6. Based on the designer's secret key value, the corresponding mode is applied to each row to perform row diffusion in the matrix.

In this demonstration, the matrix has two rows; therefore, the size of stego-key2 is of 2*2=4 bits. Let us assume the chosen stego-key2 value is '01-00'; hence mode 2 is selected for the first row and mode 1 for the second row (first 2 bits decide the mode for first row and next 2 bits decide the mode for the second row). Based on the definition of the modes given in Figure 2.6, row diffusion is performed in the matrix. The matrix post row diffusion (M_{Rd}) is given as follows:

$$M_{Rd} = \begin{bmatrix} 01 & 67 & F2 & 30 \\ F7 & 7E & B8 & 59 \end{bmatrix} \tag{2.3}$$

Security property: The aim of performing row diffusion in the matrix is to incorporate obscurity in the relationship between input secret design data and the final output (i.e. stego-constraints generated). This security property is also known as Shannon's property of diffusion.

2.3.1.4 Multilayer Trifid-cipher-based encryption

The Trifid-cipher-based encryption is performed on each distinct alphabet of the matrix; hence referred to as 'multilayer Trifid cipher'. This step is driven through the stego-key3. Each distinct alphabet of the matrix is encrypted by an encryption key of 27 characters long. Since 27 characters can have total 27! permutations, the size of encryption key required to encipher each distinct alphabet is $\lceil \log_2(27!) \rceil$ bits. Since a unique key is used for each distinct alphabet, the total size of stego-key3 to encrypt all distinct alphabets is equal to (number of distinct alphabets in the matrix) $*\lceil \log_2(27!) \rceil$.

Let us apply Trifid-cipher-based encryption on the matrix shown in (2.3). There are total three distinct alphabets in the matrix viz. *B*, *E* and *F*. Each distinct alphabet has to be encrypted using an encryption key of 27 characters long. A distinct key is chosen for encrypting *B*, *E* and *F*, respectively. To do the encryption of an alphabet, 27 characters of the chosen key are arranged in three 3×3 matrices. The alphabet to be encrypted belongs to one of the square matrix. The encrypted output for the alphabet is a three digit value *abc* where, *a* indicates the row number, *b* indicates the column number and *c* indicates the square-matrix number. Let us process the Trifid-cipher-based encryption for each alphabet one by one:

1. *Trifid-cipher-based encryption on alphabet 'B'*: First of all, a 27-character-long encryption key is chosen to encrypt alphabet *B*. Let us assume the encryption key is *Q A W S E D R F T G Y H U J I K # O L P Z M X N C B V*.

   The key is arranged in three 3×3 matrices as follows (where *SQ* indicates the square matrix):

$$SQ1 = \begin{bmatrix} Q & A & W \\ S & E & D \\ R & F & T \end{bmatrix} SQ2 = \begin{bmatrix} G & Y & H \\ U & J & I \\ K & \# & O \end{bmatrix} SQ3 = \begin{bmatrix} L & P & Z \\ M & X & N \\ C & \mathbf{B} & V \end{bmatrix}$$

   As shown earlier, the alphabet *B* to be encrypted (highlighted in red) belongs to third row and second column of the third square matrix (*SQ*3). Therefore, the encrypted value *abc* for *B* is '323'.

2. *Trifid-cipher-based encryption on alphabet 'E'*: Let us assume the 27-character-long encryption key to encrypt alphabet *E* is *F T G Y H U J I K O L P Z M X N C B V # Q A W S E D R*.

   The key is arranged in three 3×3 matrices as follows:

$$SQ1 = \begin{bmatrix} F & T & G \\ Y & H & U \\ J & I & K \end{bmatrix} SQ2 = \begin{bmatrix} O & L & P \\ Z & M & X \\ N & C & B \end{bmatrix} SQ3 = \begin{bmatrix} V & \# & Q \\ A & W & S \\ \mathbf{E} & D & R \end{bmatrix}$$

   As shown earlier, the alphabet *E* to be encrypted (highlighted in red) belongs to third row and first column of the third square matrix (*SQ*3). Therefore, the encrypted value for *E* is '313'.

3. *Trifid-cipher-based encryption on alphabet 'F'*: Let us assume the 27-character-long encryption key to encrypt alphabet *F* is *L P Z M X N C B V Q A W S E D R F T G Y H U J I K # O*.

   The key is arranged in three 3×3 matrices as follows:

$$SQ1 = \begin{bmatrix} L & P & Z \\ M & X & N \\ C & B & V \end{bmatrix} SQ2 = \begin{bmatrix} Q & A & W \\ S & E & D \\ R & \mathbf{F} & T \end{bmatrix} SQ3 = \begin{bmatrix} G & Y & H \\ U & J & I \\ K & \# & O \end{bmatrix}$$

As shown earlier, the alphabet *F* belongs to the third row and the second column of the second square matrix (*SQ*2). Therefore, the encrypted value for *F* is '322'.

Security property: The aim of using Trifid-cipher-based encryption is to incorporate following security properties: (i) confusion to obscure the relationship of the stego-keys with generated stego-constraints and (ii) diffusion to obscure the relationship of the input secret design data with the stego-constraints. The Trifid cipher offers these security properties by combining following techniques: fractionation, transposition and substitution.

2.3.1.5 Alphabet substitution

Once all distinct alphabets are encrypted, their encrypted three-digit values are converted into an equivalent value based on a mathematical expression. The mathematical expression to be used is decided by secret stego-key4. Post evaluating mathematical expression, the corresponding alphabet is substituted with the output of the expression. A number of mathematical expressions can be possible, where each distinct mathematical expression defines a mode for alphabet substitution. The mode of alphabet substitution (i.e. the mathematical expression to be used to generate the substituting value) is determined by the stego-key4. Total six kinds of mathematical expressions are defined as shown in Figure 2.6; therefore, total $\lceil \log_2(6) \rceil = 3$ bits are required to determine a mode for each encrypted alphabet. Thus, the total size of stego-key4 is equal to (number of distinct alphabets encrypted using Trifid cipher)*$\lceil \log_2$(number of modes for alphabet substitution)$\rceil$.

Since total three distinct alphabets (*B*, *E* and *F*) have been encrypted in the previous step, therefore total size of stego-key4 is $3 * \lceil \log_2(6) \rceil = 9$ bits. Let us assume the stego-key4 is '001-000-010'. Each group of 3 bits from left to right is used to decide the mode for an alphabet in alphabetic order. Therefore, '001', '000' and '010' decide the mode for *B*, *E* and *F*, respectively, and corresponding mathematical expression (shown in Figure 2.6) to be evaluated are selected. Table 2.2 shows the corresponding mathematical expression for each alphabet to be substituted and the corresponding output value of mathematical expression. Hence, alphabets *B*, *E* and *F* are substituted in the matrix with 8, 9 and 1, respectively. The matrix post alphabet substitution (M_{AS}) is given as follows:

$$M_{AS} = \begin{bmatrix} 01 & 67 & 12 & 30 \\ 17 & 79 & 88 & 59 \end{bmatrix} \tag{2.4}$$

Table 2.2 Details of obtaining equivalent value for alphabet substitution

Alphabets	Encrypted value *abc* (output of Trifid cipher)	Corresponding key bits in stego-key4	Corresponding mathematical expression	Output of mathematical expression
B	323	001	$a+b+c$	8
E	313	000	$a*b*c$	9
F	322	010	$\|a-b-c\|$	1

2.3.1.6 Matrix transposition

The matrix obtained in previous step is transposed. The transposed matrix (M_T) is given as follows:

$$M_T = \begin{bmatrix} 01 & 17 \\ 67 & 79 \\ 12 & 88 \\ 30 & 59 \end{bmatrix} \tag{2.5}$$

2.3.1.7 Mix column diffusion

In this step, each column of the transposed matrix (M_T) is subjected to mix column diffusion by exploiting a circulant MDS (maximum distance separable) matrix. Note: this matrix is also used in AES encryption for column diffusion.

1. Mix column diffusion using MDS matrix for the first column:

$$\begin{bmatrix} B_0^1 \\ B_1^1 \\ B_2^1 \\ B_3^1 \end{bmatrix} = \begin{bmatrix} 02 & 03 & 01 & 01 \\ 01 & 02 & 03 & 01 \\ 01 & 01 & 02 & 03 \\ 03 & 01 & 01 & 02 \end{bmatrix} * \begin{bmatrix} 01 \\ 67 \\ 12 \\ 30 \end{bmatrix} = \begin{bmatrix} 89 \\ C9 \\ 12 \\ 16 \end{bmatrix} \tag{2.6}$$

   where B_j^i indicates the jth byte of ith column after performing mix column operation. Equations for computing each new value of the first column using mix column diffusion are as follows:

$$\begin{aligned} B_0^1 &= (02 * 01) \oplus (03 * 67) \oplus (01 * 12) \oplus (01 * 30) = 89 \\ B_1^1 &= (01 * 01) \oplus (02 * 67) \oplus (03 * 12) \oplus (01 * 30) = \mathrm{C9} \\ B_2^1 &= (01 * 01) \oplus (01 * 67) \oplus (02 * 12) \oplus (03 * 30) = 12 \\ B_3^1 &= (03 * 01) \oplus (01 * 67) \oplus (01 * 12) \oplus (02 * 30) = 16 \end{aligned}$$

2. Mix column diffusion using MDS matrix for the second column:

$$\begin{bmatrix} B_0^2 \\ B_1^2 \\ B_2^2 \\ B_3^2 \end{bmatrix} = \begin{bmatrix} 02 & 03 & 01 & 01 \\ 01 & 02 & 03 & 01 \\ 01 & 01 & 02 & 03 \\ 03 & 01 & 01 & 02 \end{bmatrix} * \begin{bmatrix} 17 \\ 79 \\ 88 \\ 59 \end{bmatrix} = \begin{bmatrix} 74 \\ 3F \\ 8E \\ 7A \end{bmatrix} \tag{2.7}$$

Equations for computing each new value of the second column using mix column diffusion are as follows:

$$\begin{aligned} B_0^2 &= (02 * 17) \oplus (03 * 79) \oplus (01 * 88) \oplus (01 * 59) = 74 \\ B_1^2 &= (01 * 17) \oplus (02 * 79) \oplus (03 * 88) \oplus (01 * 59) = \mathrm{3F} \\ B_2^2 &= (01 * 17) \oplus (01 * 79) \oplus (02 * 88) \oplus (03 * 59) = \mathrm{8E} \\ B_3^2 &= (03 * 17) \oplus (01 * 79) \oplus (01 * 88) \oplus (02 * 59) = \mathrm{7A} \end{aligned}$$

The matrix after performing mix column diffusion (M_{Cd}) is given as follows:

$$M_{Cd} = \begin{bmatrix} 89 & 74 \\ C9 & 3F \\ 12 & 8E \\ 16 & 7A \end{bmatrix} \quad (2.8)$$

Security property: The mix column step enhances the Shannon's property of diffusion by further obscuring the relationship between input secret design data and generated stego-constraints.

2.3.1.8 Byte concatenation

All elements in each column of the matrix M_{Cd} are concatenated to form a sequence of bytes. The byte concatenation step is driven through secret stego-key5. There are a number of possible ways of concatenating all byte of each column. The mode of concatenation is determined by the stego-key5. Six modes of byte concatenation for a column in a matrix are given in Figure 2.6. Since byte concatenation is performed for each column separately based on the selected mode, the total size of stego-key5=(number of columns in the matrix M_{Cd})*⌈log$_2$(number of modes of bytes concatenation)⌉.

In this demonstration, since there are two column in the matrix M_{Cd}, the size of stego-key5 is 2*⌈log$_2$(6)⌉=6 bits. Let us assume that the stego-key5 is '001-000'. The first combination of 3 bits ('001') decides the mode of byte concatenation for the first column, and the second combination of 3 bits ('000') decides the mode of byte concatenation for the second column. For columns 1 and 2, the selected modes (from Figure 2.6) are B0B1B3B2 and B0B1B2B3, respectively. Hence, the final sequence of bytes post concatenation is as follows: $B_0^1B_1^1B_3^1B_2^1B_0^2B_1^2B_2^2B_3^2$ = '89C91612743F8E7A'.

Thus, obtained sequence of bytes is an encrypted byte-stream which is converted into an encrypted bitstream given as follows:

'1000100111001001000101100001001001110100001111111000111001111010'

2.3.1.9 Bitstream truncation

The truncation of encrypted bitstream is performed based on the designer's secret size. For example, following is the truncated bitstream for the chosen stego-constraints size=20:

The truncated bitstream='10001001110010010001'.

The truncated bitstream contains twelve 0s and eight 1s.

2.3.1.10 Bits mapping to the stego-constraints

In order to embed the encrypted bitstream, each bit is mapped to a corresponding stego-constraint. Thus, obtained stego-constraints represent hardware security constraints to be embedded into the cover design data. Mapping rules shown in Figure 2.5 are used to covert bitstream into corresponding stego-constraints. Based on the mapping rules, the mapping of twelve 0s of the truncated bitstream to the stego-constraints is as follows:

$$\langle S0, S2\rangle, \langle S0, S4\rangle, \langle S0, S6\rangle, \langle S0, S8\rangle, \langle S0, S10\rangle, \langle S2, S4\rangle, \langle S2, S6\rangle, \langle S2, S8\rangle,$$
$$\langle S2, S10\rangle, \langle S4, S6\rangle, \langle S4, S8\rangle, \langle S4, S10\rangle$$

where each stego-constraint corresponding to bit 0 has been represented in terms of a constraint (secret) edge to be added additionally into the CIG.

Further, in the mapping of 1s to the stego-constraints, each mapping corresponds to allocation of an operation to a specific FU vendor type. Therefore, the maximum numbers of 1s that can be embedded are equal to the number of operations available in the DSP application. The mapping of eight 1s of the truncated bitstream to the stego-constraints is shown in Table 2.3. Since total available operations in the 4-point DCT core are seven, maximum seven 1s (out of eight) can be mapped.

So far, we have discussed different steps of stego-constraints generation through stego-encoder system. The size of different stego-keys used in the stego-constraints generation process is highlighted in Table 2.4. The total size of stego-key is given as follows:

$$\begin{aligned}\textbf{The total stego} - \textbf{key size} = {} & [3\text{bits}] + [(\text{number of row in state matrix } M_S) * 2] \\ & +[(\text{number of unique alphabets}) * (\log_2(27!))] \\ & +[(\text{number of distinct alphabets encrypted using Trifid cipher}) \\ & *(\log_2(\text{number of modes for alphabet substitution})] \\ & +[(\text{number of columns in the transposed matrix } M_{Cd}) \\ & *(\log_2(\text{number of modes of byte concatenation})] \end{aligned} \tag{2.9}$$

Embedding of Stego-constraints into cover design data (scheduled DFG)

Stego-constraints corresponding to bits 0 and 1 are embedded into scheduled DFG of DSP application during the register allocation and FU vendor allocation phase, respectively. All stego-constraints corresponding to bit '0' are embedded as additional edges into the CIG. Post embedding stego-constraints, the modified mesh network of the CIG contains both default and artificial edges. Thus, modified CIG is shown in Figure 2.9(b). As shown in the figure, the edges constraints ⟨*S*0,*S*2⟩ and ⟨*S*2,*S*4⟩ exist by default in the CIG. In addition, edge constraints ⟨*S*0,*S*6⟩, ⟨*S*2,*S*8⟩, ⟨*S*2,*S*10⟩, ⟨*S*4,*S*6⟩ can be added without any conflict because of different colours of

Table 2.3 Possible allocation of FU vendors to the operations based on mapping of 1s to the stego-constraints

Operation number	**1**	**2**	**3**	**4**	**5**	**6**	**7**
Vendor type of FU	V1	V2	V1	V2	V1	V2	V1

Table 2.4 Size of different stego-keys

Stego-keys	Size (in bits)
Stego-key1	$\lceil\log_2$(total modes of state-matrix M_S formation)$\rceil$
Stego-key2	(Number of rows in matrix M_S)*$\lceil\log_2$(total modes of row diffusion)$\rceil$
Stego-key3	(Number of distinct alphabets)*$\lceil\log_2(27!)\rceil$
Stego-key4	(Number of distinct alphabets encrypted using Trifid cipher)*$\lceil\log_2$(total modes of alphabet substitution)$\rceil$
Stego-key5	(Number of columns in matrix M_{Cd})*$\lceil\log_2$(total modes of byte concatenation)$\rceil$

respective nodes in a node-pair. However, edges ⟨*S*0,*S*4⟩, ⟨*S*0,*S*8⟩, ⟨*S*0,*S*10⟩, ⟨*S*2,*S*6⟩, ⟨*S*4,*S*8⟩ and ⟨*S*4,*S*10⟩ cannot be directly added because of the same colour of both nodes in these node-pairs. In order to add edge constraint ⟨*S*0, *S*4⟩ into the CIG, the colour of one of the nodes is required to be swapped with the colour of another node in the same control step. This is because an edge cannot be added between two nodes of the same colour. Therefore, colour/register of node/storage variable *S*4 (Red) has been swapped with the colour/register of node/storage variable *S*5 (Indigo). Hence, the colour of node *S*4 changes from red to indigo. Now both nodes *S*0 and *S*4 have different colours; therefore, an artificial edge can be added between them. Similarly edge constraint ⟨*S*2,*S*6⟩ is added by swapping the colour of node *S*6 with the node *S*7 in the same control step. However, edge constraints ⟨*S*0,*S*8⟩ and ⟨*S*4,*S*8⟩ cannot be added by swapping of node colour with another node in the same control step. Therefore, an extra colour (register) yellow is used to accommodate storage variable *S*8. Now edge constraints ⟨*S*0,*S*8⟩ and ⟨*S*4, *S*8⟩ can be added into the CIG without any conflict. Further, allocation of storage variable *S*10 also to Yellow register facilitates adding edge constraints ⟨*S*0,*S*10⟩ and ⟨*S*4,*S*10⟩ into the CIG. This discussion highlights that in some cases, extra register may be required to satisfy all edge constraints. However, this is not always true because large size designs have a higher number of registers hence may not require an extra register to accommodate all edge constraints. The register allocation post embedding stego-constraints is shown in Table 2.5. The storages variables subjected to reallocation have been marked shaded in the table. Thus, stego-constraints corresponding to 0s are embedded during register allocation phase of HLS.

Further, stego-constraints corresponding to bit 1 (shown in Table 2.3) are embedded by performing FU vendor reallocation to the operations of the design. The FU vendor reallocation based on the mapping rule of bit 1 is shown in Table 2.3. As shown in the table, operations 1, 3, 5 and 7, being odd operations, should be allocated to the vendor V1, and operations 2, 4, and 6, being even operations, should be allocated to vendor V2. This reallocation is followed for operations 1, 2, 3, 4, 5 and 7. However, the operation 6 is still allocated to vendor V1 (instead of V2). This is because constraint for adder is chosen to be 1

Table 2.5 Register allocation of 4-point DCT post implanting stego-constraints corresponding to bit 0

Control Steps	R	I	G	O	Y
Q0	S0	S1	S2	S3	–
Q1	S5	S4	S2	S3	–
Q2	–	–	S7	S6	S8
Q3	S9	–	S7	–	–
Q4	–	–	–	–	S10

Table 2.6 Final allocation of FU vendors to the operations post embedding 1s as the stego-constraints

Operation number	1	2	3	4	5	6	7
Vendor type of FU	V1	V2	V1	V2	V1	V1	V1

(as mentioned earlier). Since the operation 6 is an addition operation, it is essentially allocated to the only available adder of vendor V1 (i.e. A_1^1). Further it is worth noting that there are total eight 1s to be embedded during FU vendor reallocation; however, only seven are embedded as total available operations are seven. Post embedding stego-constraints corresponding to bit 1, the final allocation of FU vendors to the operations is shown in Table 2.6.

Thus, stego-constraints corresponding to 0s and 1s are embedded into the scheduled DFG during two different phases of HLS. The modified scheduled and allocated DFG of DCT core is shown in Figure 2.10.

2.3.2 Detection of steganography

Detection of embedded steganography information is very crucial in order to validate the authenticity of hardware accelerator designs. Detection of steganography disables the wrong intents of an adversary of claiming authorship fraudulently or counterfeiting designs to earn illegal income. The detection process of embedded stego-constraints which are generated using crypto-based steganography approach is shown in Figure 2.11. The detection process is performed in three major steps as follows:

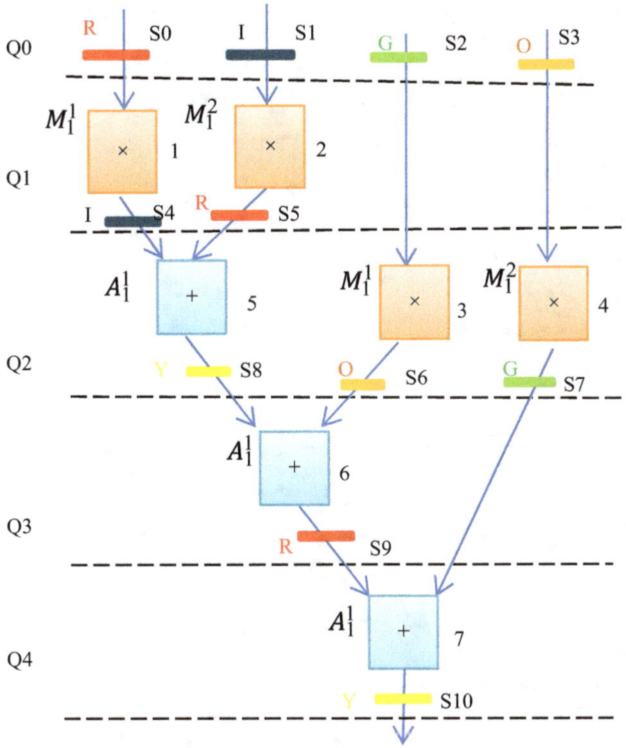

Figure 2.10 Scheduled and hardware allocated 4-point DCT using resource constraints of 1 (+) and 2(*) (Sengupta and Rathor, 2019b)

1. *Stego-constraints regeneration for verification*: In order to verify the presence of stego-constraints (or stego-mark) embedded into the design, they are required to be regenerated. Regeneration of stego-constraints by the IP/IC owner is required so that she/he can prove his rights over the constraints scientifically. This also disables an adversary to claim stego-constraints as his/her own after pirating/copying them. Only the genuine designer/owner is capable of regenerating the stego-constraints through a scientific algorithm and a secret key. An adversary cannot inadvertently regenerate stego-constraints without the knowledge of the stego-constraints generation algorithm and the secret keys employed.
2. *Inspection of stego-constraints into the design*: The design under test is subjected to inspection of embedded stego-constraints. To do so, its RTL structure is analysed. To get information about the stego-constraints corresponding to bit 0, the inputs of Muxes associated with each registers are analysed. This helps in finding the association of storage variables of the design to the register. Further to get information about the stego-constraints corresponding to bit 1, all operations of the design associated with FUs are analysed. Moreover, information about allocation of type of FU vendors to the operations is collected.

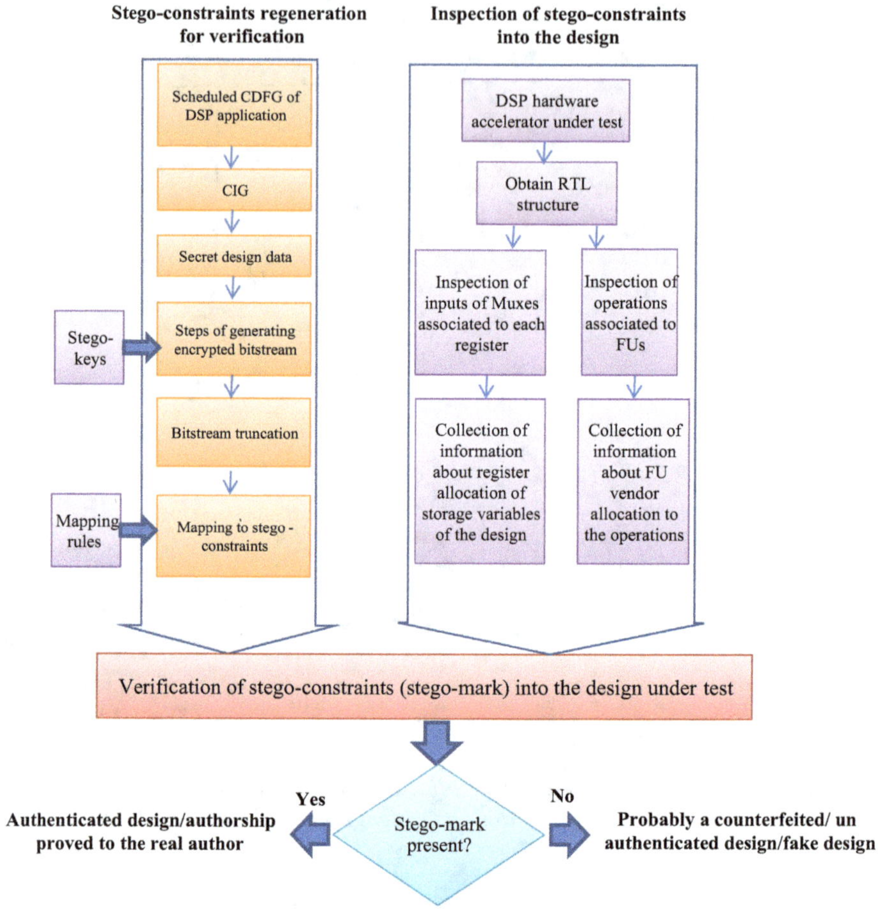

Figure 2.11 Hardware steganography detection in crypto-based steganography approach (Sengupta and Rathor, 2019b)

3. *Verification of the stego-constraints into the design*: The presence of regenerated stego-constraints is verified with the information of register allocation and FU vendor allocation extracted from the design under test. If stego-constraints corresponding to both 0s and 1s are present in to the design, then the design contains the author's stego-mark. The presence of author's stego-mark into the design ascertains the authenticity of the hardware accelerator. Moreover, the presence of author's stego-mark proves the authorship of the author over the design and nullifies the false claim of authorship by an adversary. If the hardware accelerator design does not contain a stego-mark (an authentic mark), then it can probably a counterfeited hence can be separated out from the authentic ones.

2.4 Crypto-stego tool for securing hardware accelerators

The author and his team have developed a ***crypto-stego tool*** to simulate and analyse the functionality of crypto-based steganography approach for DSP hardware accelerators. *This tool provides a friendly graphical interface to users and available for free download publicly at:* http://www.anirban-sengupta.com/Hardware_Security_Tools.php.

A snapshot of the graphical user interface of the tool is shown in Figure 2.12. The left portion of the tool shows the panel for providing required inputs to the tool, and the right portion shows the panel with output buttons to see the intermediate and final outputs of the crypto-based steganography approach. The panel in the middle shows the status of the key-driven steps (i.e. state-matrix formation, row diffusion, Trifid cipher, alphabet substitution and byte concatenation) of the crypto-based steganography approach. Initially, these status bars remain **Red**. Upon applying the stego-key, the respective status bar turns **Blue**. The *crypto-stego* tool accepts the DSP application input in the form CDFG along with module library and resource constraints. The tool shows all the intermediate steps of crypto-based steganography and the finally generated stego-constraints at the output. Further, it also shows scheduling and registers allocation pre and post embedding steganography constraints, onto the output window.

Let us generate all the intermediate and final output of crypto-based steganography approach for 4-point DCT core using the *crypto-stego* tool. We will provide the same inputs and stego-keys used during the demonstration of 4-point DCT core discussed in Section 2.3. Here, we can match the output generated with the tool and with that obtained in the demonstration. First of all, input DFG of 4-point DCT core, resource constraints of 1 adder and 2 multipliers and module library are fed to the tool as shown in Figure 2.13. On clicking on the button 'Reg. Allocation' on output panel, the register allocation table (pre-embedding stego-constraints) becomes available on to the output window. Here, values under the column headings 1, 2, 3 and 4 show the storage variable (*S*) number and the heading of the column show the register number, where Red, Indigo, Green and Orange registers have been denoted by the numbers 1, 2, 3 and 4, respectively. The row headings (0, 1, 2, 3 and 4) show the control step number. This register allocation matches with that demonstrated in Section 2.3. Further, upon clicking on the output button 'secret design data', the secret design data (the same as obtained in demonstration) becomes available onto the output window as shown in Figure 2.13. As again shown in Figure 2.13, the stego-key1='001' (the same as used in demonstration) has been fed. Upon feeding stego-key1, the respective status bar turns blue because stego-key1 is used for state-matrix formation, whereas the remaining status bar are still red as shown in Figure 2.13. The output of state-matrix formation is shown on to the output window upon clicking on the output button 'initial state matrix'. Further, Figure 2.14 shows the output of bit manipulation and row diffusion steps after feeding the stego-key2. Figure 2.15 shows that encryption key (stego-key3) for only alphabets *B, E and F* have been fed as they are the only available alphabets

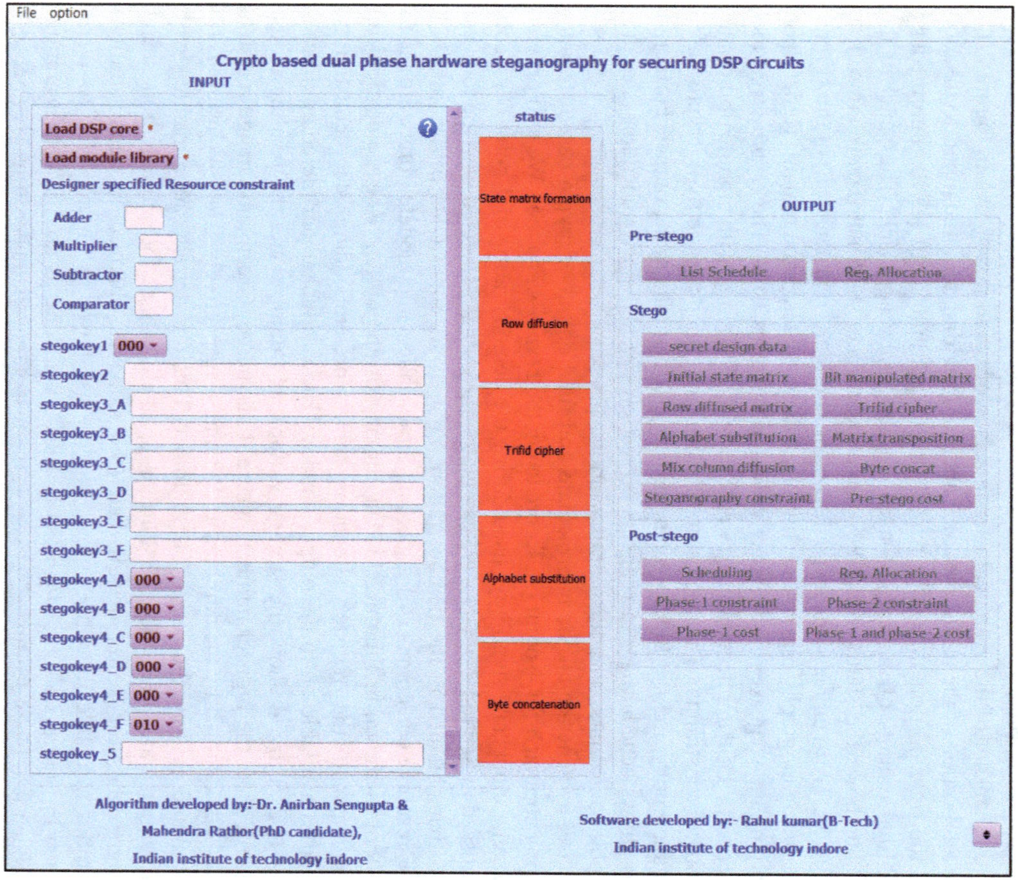

Figure 2.12 A snapshot of the GUI of crypto-stego tool for DSP hardware accelerators

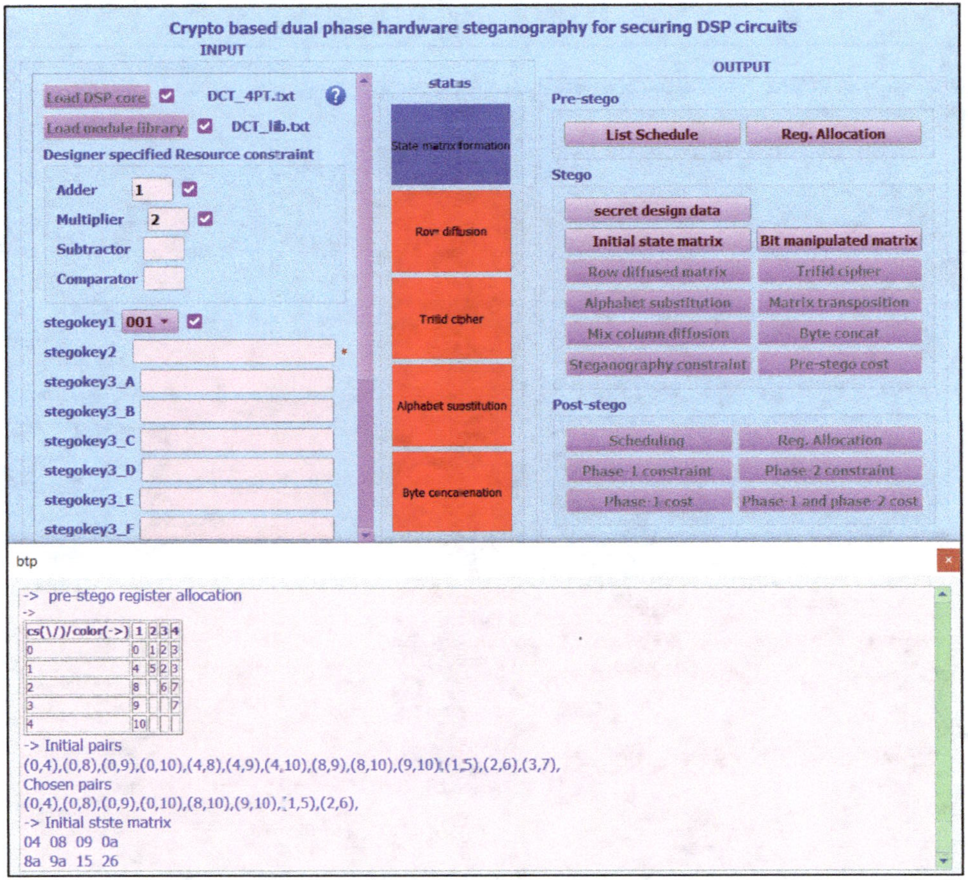

Figure 2.13 *Snapshot of the tool after feeding DFG of 4-point DCT, resource constraints and stego-key1; the output window is shown in the lower portion*

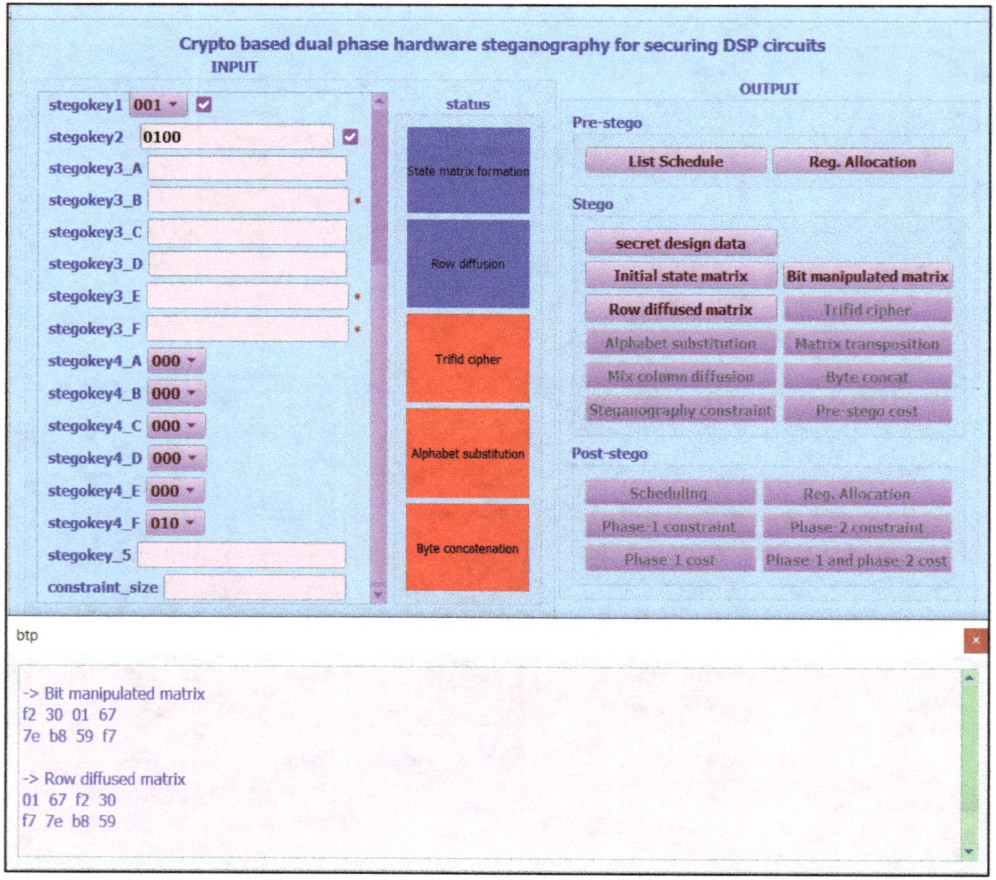

Figure 2.14 Snapshot of the tool post feeding stego-key1 and stego-key2

in the matrix post row diffusion. Moreover, the stego-key4 is fed for alphabet substitution. Therefore, the corresponding status bars (Trifid cipher and alphabet substitution) turn blue. The corresponding outputs are shown on to the output window. Figure 2.15 also shows the transposed matrix. Further, Figure 2.16 shows that the stego-key5 has been fed to the tool for byte concatenation, and the concatenated byte-stream is shown onto the output window. Output of the step before byte concatenation (i.e. mix column diffusion) is also shown in Figure 2.16. Further, Figure 2.17 shows that the constraint size=20 has been fed as input, and the final truncated bitstream is made available on to the output window by clicking on the button 'steganography constraint'. As shown in all the snapshots of the tool, the same inputs and stego-keys that are used in the demonstration on 4-point DCT in Section 2.3 have been fed here. The tool produces the desired outputs that match with the demonstration on 4-point DCT in Section 2.3. Further, Figure 2.17 also shows the register allocation and FU vendor allocation in the scheduling table post-embedding stego-constraints. One difference in the register allocation of the demonstration (in Section 2.3) and that generated using tool is to be noted, which is the reallocation of storage variable $S10$ to a different register. There are two possible ways to reallocate storage variable $S10$ in order to involve adding of edges $\langle S0,S10\rangle$ and $\langle S4,S10\rangle$ in the CIG. It can either be allocated to the Orange register or the Yellow register in the same control step. In the demonstration, the possibility of allocating $S10$ to the Yellow register has been exploited, whereas the register allocation scheme implemented in the tool exploits the possibility of allocating $S10$ to the Orange register (i.e. the register/colour number 4), as shown in the output window of Figure 2.17. Further, the FU vendor reallocation post embedding 1s is shown onto the output window by clicking on the button 'scheduling' under the post-stego section of the output panel. As shown in the 'post-stego operation scheduling' table in the output window, the parameter written at the left to the operator shows the operation number and the parameter written at the right to the operator show the FU vendor type. Thus, the crypto-based steganography approach can be simulated and analysed using the crypto-stego tool developed by the authors. This tool is useful for case studies of various kinds of DSP hardware accelerator applications such as finite impulse response (FIR) filter, infinite impulse response filter, discrete wavelet transform, autoregression filter. In addition, the tool evaluates and shows the design cost pre and post-embedding steganography information into the design.

2.5 Case studies on DSP hardware accelerator applications

Sengupta and Rathor (2019b) analysed crypto-based hardware steganography approach for various DSP hardware accelerators, viz. 8-point DCT, FIR, JPEG IDCT, MPEG, JPEG sample and EWF. The analysis has been performed by assessing security and design cost of the crypto-based hardware steganography approach. The security analysis has been performed in terms of strength of

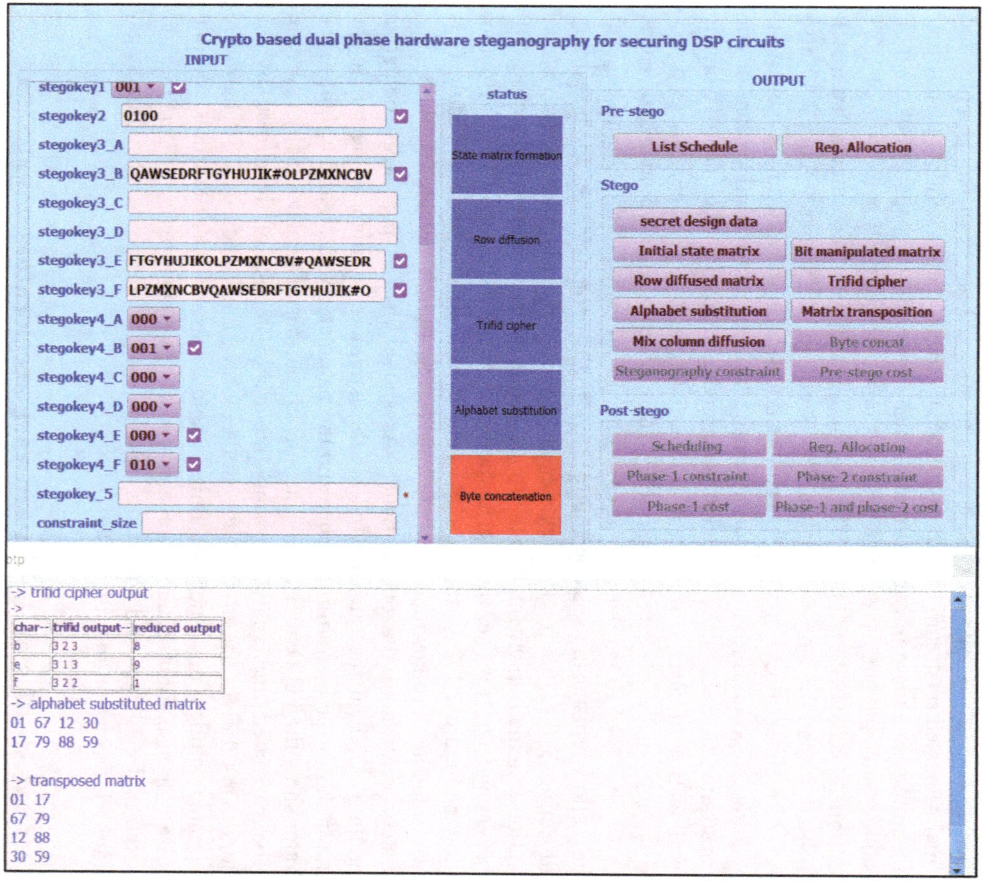

Figure 2.15 Snapshot of the tool post feeding stego-key1, stego-key2, stego-key3 and stego-key4

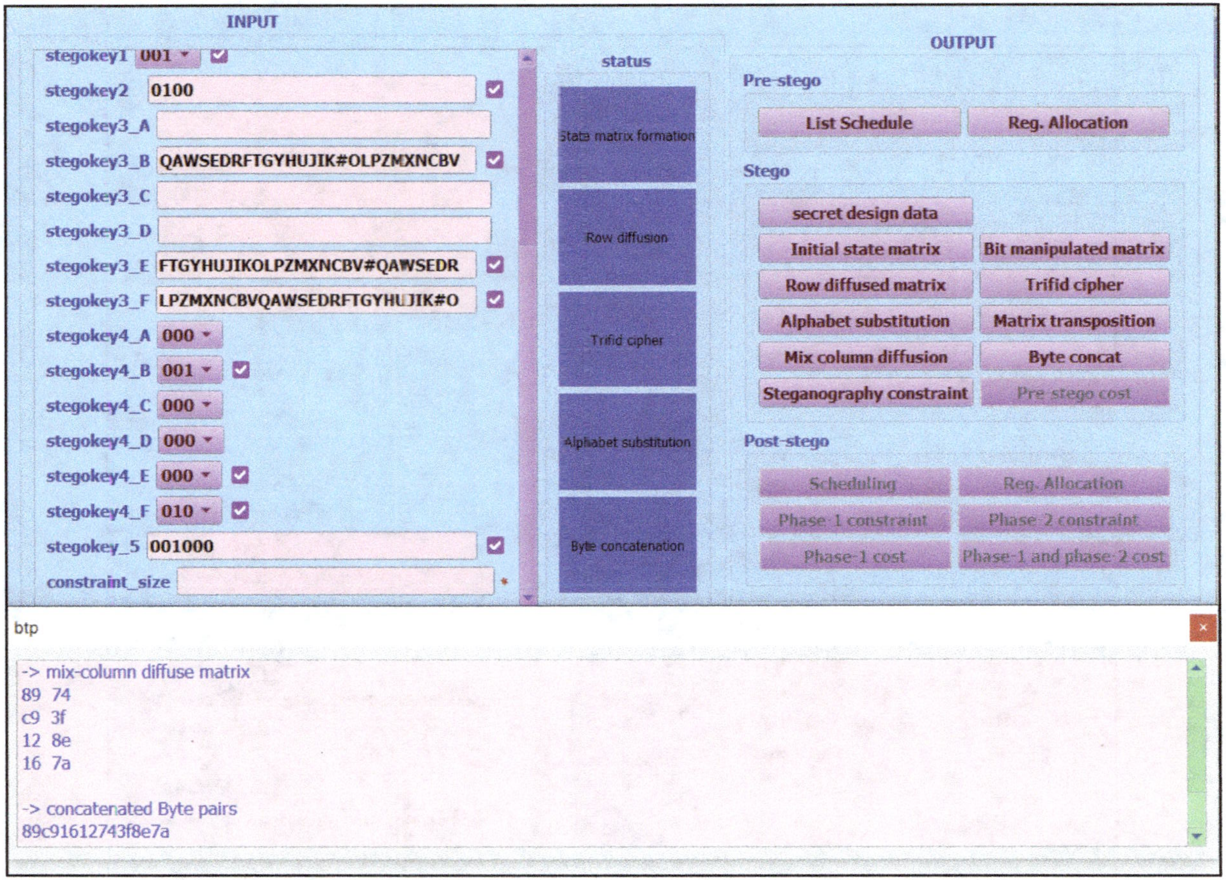

Figure 2.16 Snapshot of the tool post feeding all five stego-keys

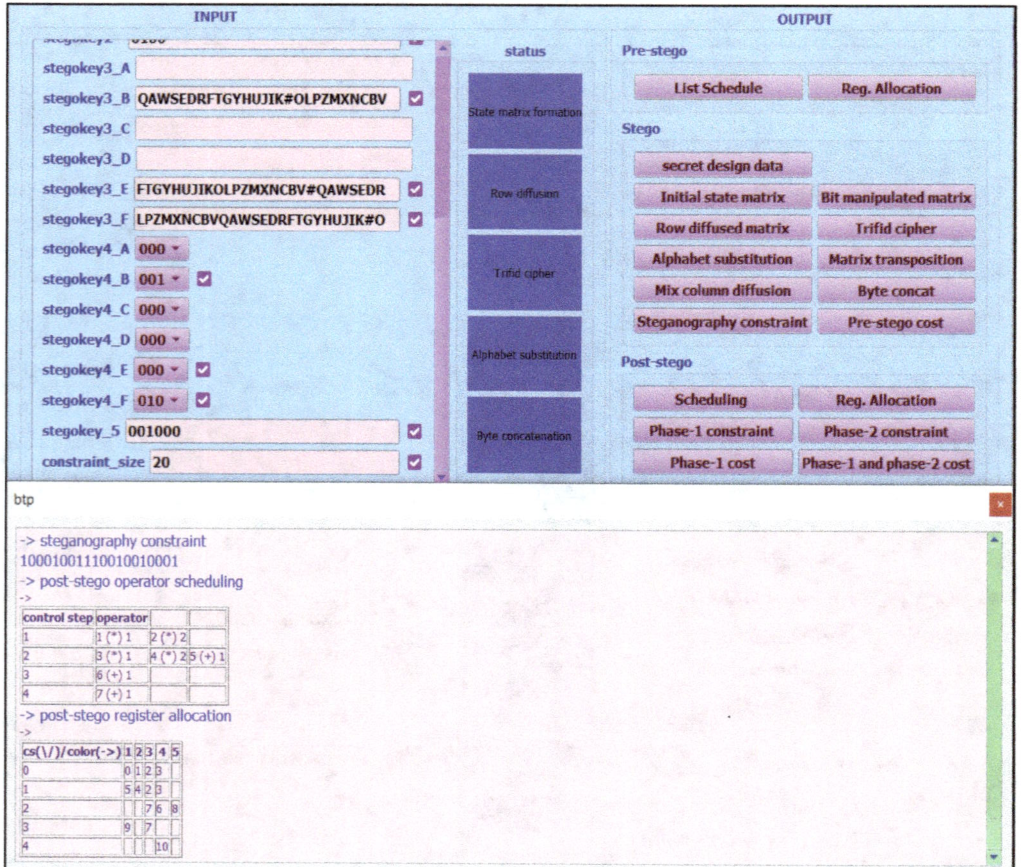

Figure 2.17 Snapshot of the tool post feeding stego-constraints size

authorship proof (digital evidence) and the size of stego-key. Further, the strengths of authorship proof and the key-size have been compared with a related cotemporary approach (Sengupta and Bhadauria, 2016). Additionally, the design cost of the crypto-based hardware steganography approach has been analysed by comparing it with a non-stego-embedded (baseline version) counterpart. The detailed discussion on security and design cost analysis are as follows (Sengupta and Rathor, 2019b).

2.5.1 Security analysis

As discussed earlier, the crypto-based hardware steganography approach aims to secure hardware accelerators against counterfeiting, cloning and false claim of authorship threats. The security against false claim of authorship is ensured using probability of coincidence metric. The strength of authorship proof increases with the decrease in probability of coincidence. Further, the probability of coincidence also indicates the robustness of the embedded stego-mark. A low value of probability of coincidence indicates that higher amount of steganography information (digital evidence) is embedded into the design, thus enhancing the robustness of the stego-mark. This also enhances the resilience against the counterfeiting and cloning threats. This is because the higher robustness of the stego-mark ensures the fail-proof detection of counterfeiting and cloning. Hence, probability of coincidence is an important metric to analyse the security of the crypto-based hardware steganography approach. The mathematical formulation of probability of coincidence is given as follows (Sengupta and Rathor, 2019b):

$$Pc = \left(1 - \frac{1}{h}\right)^{k1} * \left(1 - \frac{1}{\pi_{j=1}^{m} N(U_j)}\right)^{k2} \tag{2.10}$$

where Pc indicates the probability of coincidence, h indicates the number of colours/registers in the CIG before steganography and $k1$ indicates the number of stego-constraints embedded during the register allocation phase (i.e. number of 0s embedded). Further, $k2$ indicates the number of stego-constraints embedded during the FU vendor allocation phase (i.e. effective number of 1s embedded), $N(U_j)$ indicates the number of resources of FU type U_j, and m indicates the total types of FU resources required in the design of a hardware accelerator. The lower the Pc metric, the lower the probability of coincidently detecting the same stego-mark in an unsecured (non-stego-embedded) version, which also signifies the false-positive rate. Therefore, a designer always aims to achieve lower value of Pc metric. The Pc value reduces with the increase in the number of stego-constraints $k1+k2$ (i.e. number of 0s and 1s embedded into register and FU vendor allocation phase, respectively). For various DSP applications, the stego-constraints embedded in the crypto-based steganography approach during both register allocation and FU vendor allocation phase are shown in Figure 2.18. Further, Figure 2.19 shows the probability of the coincidence value of the crypto-based steganography approach and compares with a contemporary security approach (Sengupta and Bhadauria, 2016)

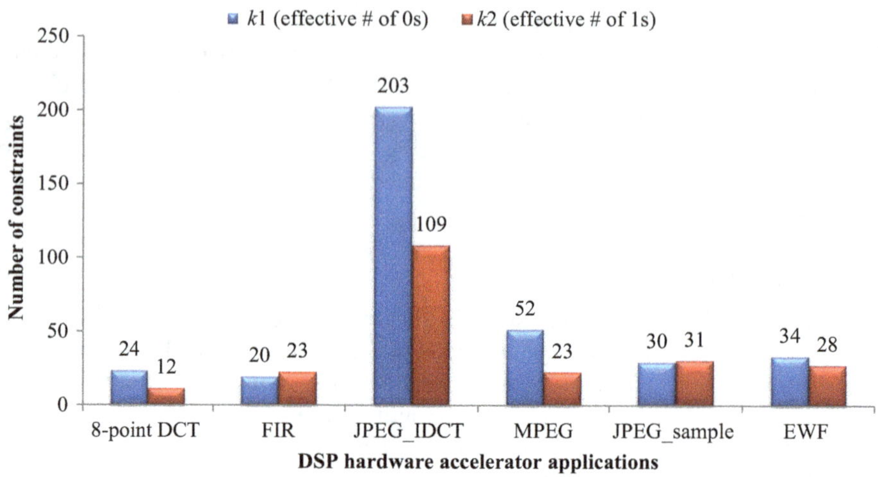

Figure 2.18 Total stego-constraints (k1+k2) embedded into DSP applications (Sengupta and Rathor, 2019b)

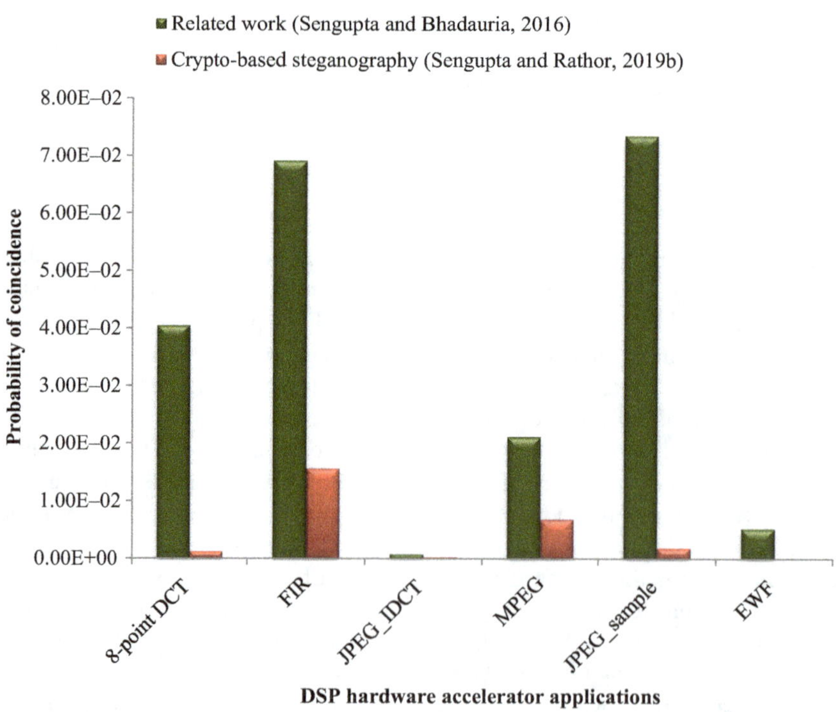

Figure 2.19 Comparison of crypto-steganography approach with watermarking in terms of proof of authorship (Pc) (Sengupta and Rathor, 2019b)

for the same number of constraints implanted into the register allocation phase. As evident from the figure, the crypto-based steganography approach achieves a very low value of *Pc* because of embedding of stego-constraints into two different phases of HLS process. More explicitly, the crypto-based steganography approach more deeply (and uniformly) embeds the steganography information (digital evidence) than the contemporary approach because of embedding into the FU vendor allocation phase also. Moreover, the contemporary approach performed embedding of secret constraints into a single phase only, i.e. register allocation phase. Hence, the crypto-based steganography approach embeds higher digital evidence into the design of hardware accelerators and achieves a stronger proof of authorship in contrast to the contemporary approach.

Additionally, the crypto-based steganography approach employs a number of security mechanisms to generate stego-constraints. Further, a very large size stego-key has been involved in the process of stego-constraints generation. This involvement of stego-key renders the back engineering of stego-constraints highly intricate for an attacker; hence she/he fails to regenerate or extract the stego-constraints to prove ownership. This provides very high security to the generated stego-constraints. Therefore, only a designer or vendor who is aware of the scientific algorithm and stego-keys involved in the stego-constraints generation process can regenerate the stego-constraints during detection. Hence, the piracy of the stego-constraints by an attacker does not help him/her as she/he cannot prove as his/her meaningful right over them. Table 2.7 shows the size of individual sub-keys (i.e. stego-key1 to stego-key5) and the total size of stego-key for various DSP applications. The size of individual sub-keys indicates the contribution in security by different major intermediate steps of stego-constraints generation process. Further, it is evident from the table that the crypto-based steganography approach (Sengupta and Rathor, 2019b) requires a very large size stego-key, whereas the contemporary approach (Sengupta and Bhadauria, 2016) does not involve any crypto-key to generate secret constraints. Hence, the crypto-based steganography approach offers very high security in terms of larger key size, complex involvement of various security properties in the stego-constraints generation algorithm and as

Table 2.7 Stego-key size in bits for different DSP applications

DSP applications	Key size (stego-strength) in bits					
	Stego-key1	**Stego-key2**	**Stego-key3**	**Stego-key4**	**Stego-key5**	**Total key size**
8-Point DCT	3	10	564	18	15	610
FIR	3	16	564	18	24	625
JPEG_IDCT	3	80	564	18	120	785
MPEG	3	14	564	18	21	620
JPEG_sample	3	32	564	18	48	665
EWF	3	22	564	18	33	640

well as higher strength of authorship proof (digital evidence) in contrast to the contemporary approach (Sengupta and Bhadauria, 2016).

2.5.2 Design cost analysis

The employment of a security mechanism to secure a design against various hardware threats should be realistic. In other words, employing security mechanism should not incur excessive design overhead. Otherwise, the security mechanism will be lesser effective even if it offers higher security. Therefore, the design cost of the crypto-based steganography approach needs to be analysed. The following equation is used to evaluate the design cost:

$$C_d(U_i) = \rho_1 \frac{L_d}{L_m} + \rho_2 \frac{A_d}{A_m} \tag{2.11}$$

where $C_d(U_i)$ is the design cost on FU constraints U_i, L_d and L_m are the design latency at specified FU constraints and maximum design latency, respectively, A_d and A_m are the design area at specified FU constraints and maximum area, respectively, and ρ_1, ρ_2 are the weights which are kept at 0.5 to fix equal priority for both.

The design costs of the crypto-based steganography approach (Sengupta and Rathor, 2019b) post-phase 1 (i.e. register allocation phase) and post-phase 2 (i.e. FU vendor allocation phase) have been shown in Figure 2.20. Further, the design cost of baseline (i.e. cost before embedding steganography) is also

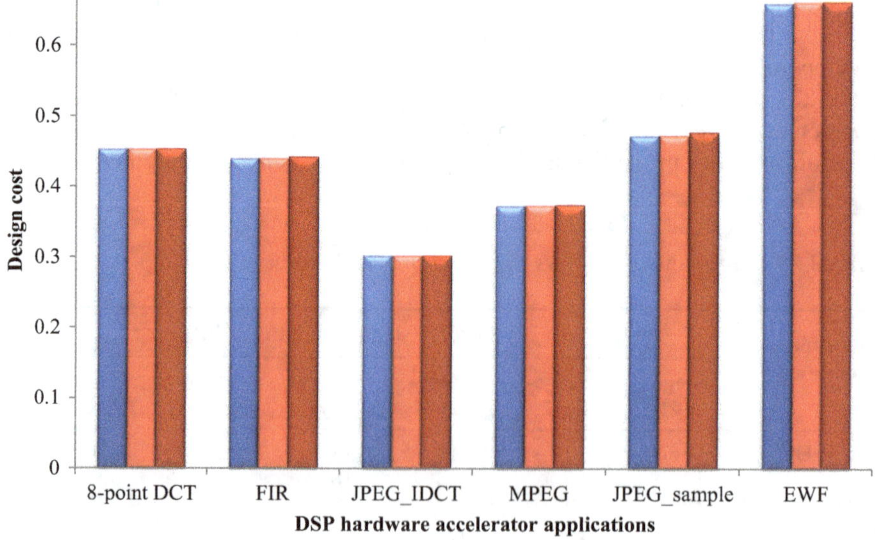

Figure 2.20 Design cost comparison of crypto-steganography approach with respect to baseline (Sengupta and Rathor, 2019b)

shown in Figure 2.20. It is obvious from the figure that the design cost post-phase 1 and phase 2 either remains the same as the baseline design cost or increases by a very marginal value. This signifies that the almost zero design overhead is incurred because of embedding crypto-based dual phase steganography, which is a desirable feature of any security algorithm employed. The reason behind incurring a negligible design overhead for some applications is the requirement of extra registers to embed all secret edge constraints into the CIG. As the size of DSP application increases, the chances of register overhead significantly reduces because a larger DSP application already comprises a larger number of registers, and there is a very high probability that the available registers would accommodate all the secret edge constraints without incurring any register overhead. In conclusion, it can be inferred that the crypto-based steganography approach (Sengupta and Rathor, 2019b) works more efficiently for larger DSP applications (i.e. incurs zero overhead and simultaneously offering higher security).

2.6 Conclusion

Hardware accelerators play a crucial role in modern age electronics systems. This chapter has focused on the DSP hardware accelerators which are typically employed to speed up the processing of DSP applications such as image processing, audio and video processing applications. Because of wide utility of DSP hardware accelerators in electronics products, their security perspective has also been highlighted in this chapter. To secure DSP hardware accelerators against counterfeiting, cloning and false claim of authorship threats, crypto-based hardware steganography approach has been discussed in this chapter. The crypto-based hardware steganography approach exploits various cryptographic and non-cryptographic properties to generate stego-constraints through stego-encoder process. Further, the involvement of stego-keys in the constraints generations process enhances the security level. In addition, stego-information has been embedded into two distinct phases of HLS, resulting in deeper, more uniform and more distributed embedding of security evidence. Comparative perspective of crypto-based steganography contemporary security approaches has also been discussed in this chapter.

At the end of this chapter, a reader understands the following concepts:

1. Utility of DSP hardware accelerators
2. Advantages of hardware steganography over hardware watermarking
3. Stego-encoder or stego-constraints generation process in crypto-based steganography
4. Various security properties/mechanisms such as bit manipulation, row diffusion, Trifid-cipher-based encryption, alphabet substitution, mix column diffusion, byte concatenation and bit mapping.
5. Demonstration of crypto-based steganography on 4-point DCT hardware accelerator
6. Detection process of crypto-based steganography

7. Highlights of a *crypto-stego tool* developed by the authors of crypto-based steganography approach
8. Case studies of crypto-based steganography on various DSP hardware accelerators

2.7 Questions and exercise

1. What is the difference between steganography and watermarking?
2. What is entropy threshold?
3. Explain the process to computer entropy threshold.
4. What is a hardware accelerator?
5. What is the significance of hardware accelerator in the context of DSP and image processing applications?
6. Give some examples of hardware accelerators.
7. How do you control the amount of secret stego-information added in a DSP design?
8. What is a stego-encoder?
9. What inputs does a stego-encoder accept in a crypto-steganography process?
10. Why is HLS crucial for designing a hardware accelerator?
11. How is a single-phase watermarking different than a multiphase watermarking?
12. Mention the cryptographic and non-cryptographic security properties incorporated in the stego-constraints generation process.
13. Explain the mapping rules used in stego-embedding process of a DSP hardware accelerator.
14. What is the cover design data in hardware steganography?
15. Calculate the stego-key strengths used in crypto-driven hardware steganography.
16. What is the role of Trifid cipher in hardware steganography?
17. How is state-matrix formation important in secret stego-constraint generation process?
18. How do you determine the number of layers of Trifid cipher application?
19. What is the security property used in S-Box?
20. What is the security property used in row/column diffusion?
21. How do you decide the key-size in Trifid-cipher-based encryption?
22. What is the role of byte concatenation in crypto-driven hardware steganography?
23. How does a designer choose stego-constraint strength before implanting into a hardware accelerator?
24. Explain the FU vendor allocation process.
25. Calculate the total stego-key size of a 8-point DCT core.

References

E. Castillo, U. Meyer-Baese, A. Garcia, L. Parilla and A. Lloris (2007), 'IPP@HDL: efficient intellectual property protection scheme for IP cores,' *IEEE Trans. Very Large Scale Integr. (VLSI) Syst.*, vol. 15(5), pp. 578–590.

F. Koushanfar, I. Hong, and M. Potkonjak (2005), 'Behavioral synthesis techniques for intellectual property protection,' *ACM Trans. Des. Autom. Electron. Syst.*, vol. 10(3), pp. 523–545.

B. Le Gal and L. Bossuet (2012), 'Automatic low-cost IP watermarking technique based on output mark insertions,' *Des. Autom. Embedded Syst.*, vol. 16(2), pp. 71–92.

H. R. Mahdiany, A. Hormati and S. M. Fakhraie (2001), 'A hardware accelerator for DSP system design,' *Proc. ICM,* pp. 141–144.

M. C. McFarland, A. C. Parker and R. Camposano (1988), 'Tutorial on high-level synthesis,' *DAC '88 Proceedings of the 25th ACM/IEEE Design Automation*, vol. 27 (1), pp. 330–336.

V. K. Mishra and A. Sengupta (2014), 'MO-PSE: Adaptive multi-objective particle swarm optimization based design space exploration in architectural synthesis for application specific processor design,' *Adv. Eng. Software*, vol. 67, pp. 111–124.

C. Pilato, S. Garg, K. Wu, R. Karri and F. Regazzoni (2018), 'Securing hardware accelerators: a new challenge for high-level synthesis,' *IEEE Embedded Syst. Lett.*, vol. 10(3), pp. 77–80.

S. M. Plaza and I. L. Markov (2015), 'Solving the third-shift problem in IC piracy with test-aware logic locking,' *IEEE Trans. Comput.-Aided Des. Integr. Circuits Syst.*, vol. 34(6), pp. 961–971.

D. Roy and A. Sengupta (2019), 'Multilevel Watermark for Protecting DSP Kernel in CE Systems [hardware matters],' *IEEE Consum. Electron. Mag.*, vol. 8(2), pp. 100–102.

R. Schneiderman (2010), 'DSPs evolving in consumer electronics applications,' *IEEE Signal Process. Mag.*, vol. 27(3), pp. 6–10.

A. Sengupta (2016), 'Intellectual property cores: protection designs for CE products,' *IEEE Consum. Electron. Mag.*, vol. 5(1), pp. 83–88.

A. Sengupta (2017), 'Hardware security of CE devices [hardware matters],' *IEEE Consum. Electron. Mag.*, vol. 6(1), pp. 130–133.

A. Sengupta and S. Bhadauria (2016), 'Exploring low cost optimal watermark for reusable IP cores during high level synthesis,' *IEEE Access*, vol. 4, pp. 2198–2215.

A. Sengupta, E. R. Kumar and N. P. Chandra (2019), 'Embedding digital signature using encrypted-hashing for protection of DSP cores in CE,' *IEEE Trans. Consum. Electron.*, vol. 3, pp. 398–407.

A. Sengupta and S. P. Mohanty (2019a), 'Advanced encryption standard (AES) and its hardware watermarking for ownership protection', *IP Core Protection and Hardware-Assisted Security for Consumer Electronics*, e-ISBN: 9781785618000, pp. 317–335.

A. Sengupta and S. P. Mohanty (2019b), 'IP core and integrated circuit protection using robust watermarking', *IP Core Protection and Hardware-Assisted Security for Consumer Electronics*, e-ISBN: 9781785618000, pp. 123–170.

A. Sengupta and M. Rathor (2019a), 'IP core steganography for protecting DSP kernels used in CE systems,' *IEEE Trans. Consum. Electron.*, vol. 65(4), pp. 506–515.

A. Sengupta and M. Rathor (2019b), 'Crypto-based dual-phase hardware steganography for securing IP cores,' *Lett. IEEE Comput. Soc.*, vol. 2(4), pp. 32–35.

A. Sengupta, D. Roy and S. P. Mohanty (2019), 'Low-overhead robust RTL signature for DSP core protection: new paradigm for smart CE design,' *Proc. 37th IEEE International Conference on Consumer Electronics (ICCE),* pp. 1–6.

A. Sengupta and D. Roy (2017), 'Antipiracy-aware IP chipset design for CE devices: a robust watermarking approach [hardware matters],' *IEEE Consum. Electron. Mag.*, vol. 6(2), pp. 118–124.

A. Sengupta and D. Roy (2018), 'Multi-phase watermark for IP core protection,' *Proc. 36th IEEE International Conference on Consumer Electronics (ICCE)*, pp. 1–3.

A. Sengupta, D. Roy and S. P. Mohanty (2018), 'Triple-phase watermarking for reusable IP core protection during architecture synthesis,' *IEEE Trans. Comput.-Aided Des. Integr. Circuits Syst.*, vol. 37(4), pp. 742–755.

A. Sengupta, R. Sedaghat and Z. Zeng (2010), 'A high level synthesis design flow with a novel approach for efficient design space exploration in case of multi-parametric optimization objective,' *Microelectron. Reliab.*, vol. 50(3), pp. 424–437.

D. Ziener and J. Teich (2008), 'Power signature watermarking of IP cores for FPGAs,' *J. Signal Process. Syst.*, vol. 51(1), pp. 123–136.

Chapter 3
Double line of defence to secure JPEG codec hardware for medical imaging systems

Anirban Sengupta[1]

The chapter describes a double line of defence mechanism for securing a JPEG codec hardware accelerator used in medical imaging systems. The chapter starts with the background/motivation of JPEG codec for imaging modalities, followed by discussing the dual line of security based on structural obfuscation and crypto-steganography for the image compression hardware, and highlighting the results on case studies in terms of security and overhead.

The chapter is organized as follows: Section 3.1 introduces the chapter; Section 3.2 discusses on the motivation of using JPEG compression in medical imaging systems; Section 3.3 presents the salient features of the chapter; Sections 3.4 and 3.5 explain the process of double line of defence for a JPEG codec hardware accelerator; Section 3.6 presents the analysis on case studies; Section 3.7 concludes the chapter; Section 3.8 presents some exercise for readers.

3.1 Introduction

The modern age healthcare systems heavily rely upon electronics and internet technology to enable accurate, rapid diagnosis and advanced treatments, where electronics hardware are critical in healthcare systems for processing of medical data, e.g. compression, decompression and filtering. Further, internet technology plays a pivotal role in transmitting medical data for teleradiology and telepathology (Koff and Shulman, 2006). Thereby, the role of electronics and internet technology in medical systems has led more accurate, easy/rapid diagnosis and treatment of critical diseases.

The discussion of this chapter mainly concerns with the imaging modalities or medical imaging systems such as computed tomography (CT) scanner and magnetic resonance imaging (MRI) scanner, where images of patient's internal organs are captured for disease diagnosis. However, the size of medical data (images) generated from MRI or CT scan is very large therefore requires large storage

[1]Computer Science and Engineering, Indian Institute of Technology Indore, Indore, India

capacity to store and process them locally (Koff and Shulman, 2006). Moreover, when a large size data of medical images is transmitted over the internet for remote diagnosis, it needs a larger bandwidth. In short, excessively large size medical data cannot be efficiently stored and transmitted. This can be more understood through an illustration of CT abdomen images. A whole data set of CT abdomen images comprises 200–400 images, where each slice of images contains 512×512 pixels. For 16 bits size of each pixel, the whole data set of CT abdomen images requires around 150-MB data storage (Gokturk, 2001; Gokturk *et al.*, 2001). This demands a compression of medical images for low capacity storage and low bandwidth transmission.

As compression of images can be of two types, lossy and lossless, lossy compression under the acceptable limit of a compression ratio can be performed for medical images. However, the acceptable limit of the compression ratio varies for various imaging modalities and body organs (Chen and Ti, 2004; Koff and Shulman, 2006; Chen, 2007; Agarwal *et al.*, 2019). Further in some medical imaging techniques, both lossy and lossless image compressions are simultaneously employed. Such techniques are referred to as hybrid compression techniques where different regions of a same image are subjected to lossy or lossless compression depending upon whether those regions are diagnostically important or not. The lossless compression is applied on diagnostically important regions which are also referred to as region of interest (ROI) and the other regions are subjected to the lossy compression. This is because a high image quality may be needed for the ROIs for profound analysis. Thus, hybrid compression of medical images is performed by applying lossless compression (for high quality) in ROIs and lossy compression in other regions (Gokturk, 2001; Gokturk *et al.*, 2001).

In the case of both hybrid and lossy compressions in an acceptable limit, the Joint Photographic Experts Group (JPEG) compression can be applied. However, stringent performance and power constraints entail using of a dedicated processor for compression and decompression. Therefore, a dedicated JPEG compression–decompression (codec) processor is employed to facilitate compression and decompression of images in medical imaging systems (Sengupta *et al.*, 2018; Pilato *et al.*, 2018; Mahdiany *et al.*, 2001). Thereby, multimedia hardware accelerators such as JPEG codec processors have claimed their role in the medical imaging systems. Further, the security standpoint of JPEG codec hardware accelerator is concurrently important (Sengupta, 2016, 2017; Sengupta *et al.*, 2018; Sengupta and Mohanty, 2019a). The motivation behind employing the security for JPEG codec processor is discussed in detail in Section 3.2.

This chapter discusses a design process of secured JPEG codec processor to be used for compression in medical imaging systems. The security mechanism to be discussed in this chapter is performed by employing double line of defence (Sengupta and Rathor, 2020) against prevalent hardware threats such as counterfeiting, cloning and Trojan insertion. The double line of defence-based security technique proposed by Sengupta and Rathor (2020) offers both preventive and detective control over the hardware threats so that a highly robust protection can be ensured. The first line of defence has been deployed using a structural transformation-based obfuscation technique, and the second line of defence has been deployed by performing crypto-based dual-phase steganography. The need of enhancing the protection using a double line of defence has surged because should the preventive measures against the hardware threats are made ineffective or nullified by a highly advanced attacker, then there should remain a feasibility for detective control of fake designs. The second line of defence using crypto-based steganography discussed in this chapter serves this purpose of detective control over fake (counterfeited/cloned) designs. Apart from the double line of defence-based security mechanism (Sengupta and Rathor, 2020), there are some other approaches that provide either preventive control or detective control to secure the digital signal processing (DSP) hardware accelerator designs against aforementioned threats. A brief discussion on those contemporary approaches is as follows: Koushanfar *et al.* (2005), Le Gal and Bossuet (2012), Sengupta and Bhadauria (2016) and Sengupta and Roy (2017) proposed high-level synthesis (HLS)-based water-marking techniques to provide detective control on counterfeiting and cloning threats for DSP hardware accelerators. Further Sengupta and Rathor (2019a, 2019b) proposed hardware steganography techniques for enabling the detection of counterfeiting and cloning threats. However, these approaches did not consider preventive counter-measures against Trojan insertion, counterfeiting and cloning threats. Further, Sengupta *et al.* (2017a) and Sengupta and Rathor (2019c) protect DSP hardware accelerators against Trojan insertion by applying structural obfuscation. Sengupta *et al.* (2017a) employed compiler-driven transformation-based structural obfuscation, whereas Sengupta and Rathor (2019c) employed hologram-inspired obfuscation to prevent against Trojan insertion. These approaches did not discuss the design process of securing a whole JPEG compression processor. However, the structural obfuscation proposed by Sengupta *et al.* (2017a) and Sengupta and Rathor (2019c) has been applied on 8-point discrete cosine transform (DCT) core which is a part of JPEG compression hardware. Further, Sengupta *et al.* (2018) employed structural obfusca-tion on a complete JPEG codec processor to thwart Trojan insertion. However, this approach only took preventive measure against Trojan insertion in consideration. There was the absence of detective measures against counterfeiting and cloning. However, the chances of deobfuscation of an obfuscated design by an adversary cannot be neglected. If a potential attacker deobfuscates the design or deduces original structure or functionality then she/he can infringe or tamper them, thus defeating the goal of structural obfuscation. Therefore, detective control also plays an important role in case the first line of defence is compromised. Sengupta and Rathor (2020) proposed the second line of defence using crypto-based steganography following the first line of defence using structural obfuscation. This double line of defence technique of securing JPEG codec hardware accelerators has been discussed in detail in this chapter.

3.2 Why secure JPEG codec processors used in medical imaging systems?

The integrity and correctness of medical data in compressed medical images (generated from JPEG codec hardware) is highly desirable in order to avoid wrong diagnosis of diseases. However, compressed images generated from a fake or non-authenticated JPEG codec hardware (counterfeited, cloned or infected with malicious logic insertion such as hardware Trojans) may not be fully trusted. This is because the genuine diagnostically important pixel post-compression information of medical images may be altered or corrupted using non-authenticated JPEG compression processors. Thus, generated corrupted medical data can mislead a healthcare professional during the diagnosis process, hence leading to false diagnosis of diseases and wrong treatment of patients.

In order to keep intact the correctness of the generated compressed medical data, the underlying JPEG compression hardware needs to be authentic and secured. The security and authenticity of a JPEG compression hardware accelerator design needs to be ensured against malicious logic insertion (resulting from reverse engineering (RE) attack) (Zhang and Tehranipoor, 2011; Sengupta *et al.*, 2017b, 2017c), counterfeiting and cloning threats (Sengupta *et al.*, 2019; Sengupta and Rathor, 2019a, 2019b; Sengupta, 2020) within the design supply chain. This is because the hardware accelerator design process has to undergo various design phases which are accomplished in different offshore design houses and foundry in order to satisfy the design-to-market and economic constraints. Because of participation of offshore entities (which may not be trustworthy) in the design process, the JPEG codec hardware accelerator design may become prone to aforementioned hardware threats. A secure JPEG codec hardware accelerator used in medical imaging modalities ensures that the computed compressed/decompressed medical data (digital pixels) generated remains in its genuine or authentic form (not corrupted), hence resulting into correct diagnosis.

In order to achieve a stronger resilience against hardware threats of counterfeiting, cloning and Trojan insertion, both preventive and detective measures can be taken. Considering both preventive and detective measures against hardware threats, Sengupta and Rathor (2020) proposed a double line of defence mechanism to secure JPEG codec hardware accelerator. The first line of defence ensures the prevention against RE (and Trojan insertion), counterfeiting and cloning, whereas the second line ensures the detection of counterfeiting and cloning threats.

3.3 Salient features of the chapter

In this chapter, the discussion on the double line of defence using structural obfuscation and crypto-steganography for JPEG codec hardware accelerators orbits around the following salient features (Sengupta and Rathor, 2020):

1. The key discussion of this chapter is on ensuring the correctness of the compressed data (pixels) of medical images computed from JPEG codec hardware accelerator used in medical imaging modalities.
2. Discussion on securing an underlying JPEG compression processor by deploying a double line of defence to ensure the correctness of the computed compressed data of medical images.

3. Discussion on the combined process of structural obfuscation and crypto-based steganography to offer a double line of defence to secure the JPEG compression processor.
4. Detailed discussion on crypto-based steganography, the second line of defence, using the demonstration on 8-point DCT core employed underneath the JPEG compressor hardware accelerator. The second line of defence in the form of detective measure enables the detection of non-authenticated JPEG compression hardware accelerators and hence ensures that only authentic designs are integrated in the medical imaging systems.

3.4 Securing JPEG compression hardware using a double line of defence

The overall process of securing JPEG compression hardware using double line of defence is discussed under the following subsections.

3.4.1 A high-level perspective of the process (Sengupta and Rathor, 2020)

As discussed earlier, the JPEG codec hardware accelerator has wide utility in healthcare systems for generating (computing) compressed medical data (digital pixel values) in order to store them in limited memory space and transmit them in restricted bandwidth. However, because of potential hardware threats to the JPEG codec processor, the computed compressed medical data may not remain accurate and hence mislead healthcare professionals. Therefore, the security and authenticity of underlying compression hardware is vital. In order to handle this security issue, Sengupta and Rathor (2020) proposed a double line of defence to secure JPEG codec processor against hardware threats. In the double line of defence mechanism, crypto-based steganography has been deployed on the top of the structural-obfuscation-based security against Trojan insertion, counterfeiting and cloning threats. Figure 3.1 depicts a generic diagram to show the possible threats to the compression hardware for medical images and countermeasures to secure it. As shown in the figure, an unsecure JPEG codec processor may not be authentic and hence can generate/compute a corrupted or altered compressed pixel value of medical images due to the presence of malicious logic within the compression hardware. On the contrary, as shown in the lower part of the figure, a JPEG codec processor secured using a double line of defence ensures the generation of a genuine compressed pixel value of a medical image of a patient's internal organ.

A high-level view of the double line of defence mechanism is shown in Figure 3.2. As shown in the figure, the double line of defence mechanism has been integrated with the HLS (Sengupta *et al.*, 2010) process. This turns the HLS process into a security-aware HLS process. A high-level description such as C/C++ code or transfer function of a JPEG codec processor is the primary input to the security aware HLS process. Along with the high-level description of the JPEG codec processor, resource constraints, module library and stego-key are fed as inputs. Post

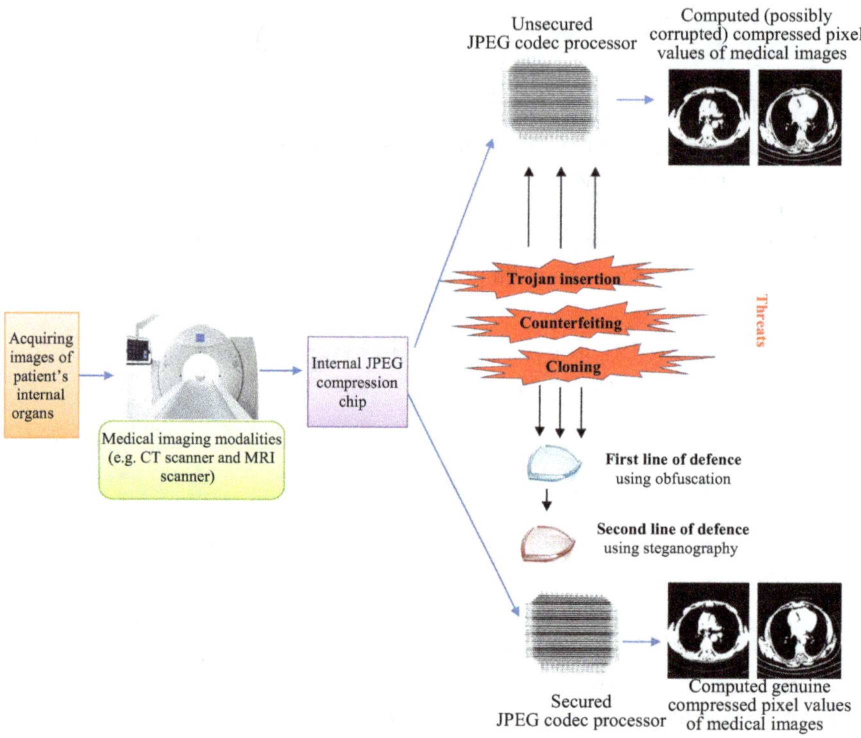

Figure 3.1 A generic diagram showing the hardware threats and its countermeasure for compressed medical images (Sengupta and Rathor, 2020)

performing a double line of defence during HLS process, a secure JPEG codec processor design is generated at the output. The major steps of securing the JPEG codec processor are as follows: (i) conversion of the high-level description of the JPEG codec processor into an intermediate representation in the form of control data flow graph (CDFG); (ii) transformation of the CDFG using tree height transformation (THT)-based structural obfuscation that acts as first line of defence; (iii) scheduling of structurally obfuscated CDFG followed by resource allocation using designer's specified resource constraints and module library and (iv) employing crypto-based steganography on the scheduled and resource-allocated CDFG that acts as a second line of defence. More highlights on the first and second line of defence are as follows.

3.4.1.1 First line of defence – structural-obfuscation-based preventive control (Sengupta and Rathor, 2020)

Performing structural obfuscation in design architectures ensures thwarting and preventive mechanism against RE, thus hindering backdoor insertion of Trojan.

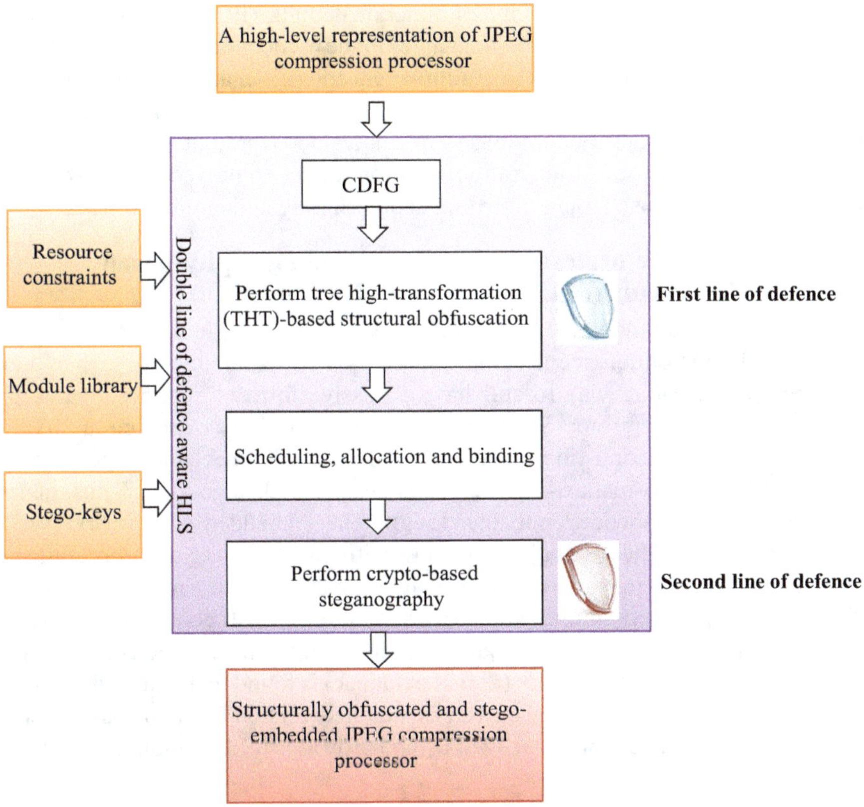

Figure 3.2 High-level view of a double line of defence-based security mechanism for securing JPEG codec hardware accelerators (Sengupta and Rathor, 2020)

Further, it impedes counterfeiting and cloning. This is because to insert a Trojan in the form of a hidden malicious logic or to counterfeit/clone a design, RE is launched by an attacker. By performing RE, the attacker tries to realize the correct functionality or structure of the design. If the attacker successfully reverse engineers the design, she/he becomes able to hide malicious logic into the design in an appropriate location (such that rate triggering occurs to evade detection), to produce Trojan-infected circuits or can generate counterfeited or cloned designs. The structural obfuscation thwarts the attacker's malicious intents of inserting Trojans, counterfeiting and cloning by obscuring the structure and functionality of the circuit. Thereby, the obfuscated circuit becomes harder to reverse engineer by an adversary, thus thwarting the Trojan insertion, counterfeiting and cloning.

Sengupta and Rathor (2020) applied THT-based structural obfuscation on CDFG design representation of JPEG codec hardware accelerator in order to deploy a first line of defence against hardware threats. The THT-based structural obfuscation causes substantial transformation in the structure of the design such as

(i) changes in the interconnectivity of functional units (FUs) resources such as adders, multipliers and subtractors; (ii) changes in the number of interconnect binding resources such as multiplexers and demultiplexers and (iii) changes in the number of storage resources such as registers and latches. The aforementioned changes in the design architecture render it unobvious to be understood (through RE) for an adversary. Thus, the THT-based structural obfuscation, being employed as a first line of defence, impedes against Trojan insertion, counterfeiting and cloning threats.

3.4.1.2 Second line of defence – crypto-steganography-based detective control (Sengupta and Rathor, 2020)

A designer or owner cannot keep a full reliance only on a single line of defence. In case, the single line of defence is compromised by a potential adversary, then there should be an alternative way to still have a passive form of security against the hardware threats. Therefore, Sengupta and Rathor (2020) deployed a crypto-steganography as a second line of defence. This defence mechanism generates a designer's robust stego-mark using secret design data and stego-keys. Thus, generated stego-mark is embedded into the design. The embedded stego-mark (secret digital evidence) becomes the basis of counterfeiting and cloning detection. Since a counterfeited design is just an imitation of an original design, it cannot contain vendor's genuine secret stego-mark, while an original design will contain the authentic secret stego-mark. This is how the counterfeited designs can be detected. Further, if vendor's genuine secret stego-mark (digital evidence) is found in the same design of a different brand, then it can be considered as a cloned version of the original design. This is how cloning detection is realized using hardware steganography.

3.4.2 Hardware threats and protection scenario (Sengupta and Rathor, 2020)

Hardware threats and protection scenarios of double line of defence mechanism for securing JPEG codec hardware accelerators are highlighted in Figure 3.3. As shown in the figure, the security of JPEG codec hardware accelerators has been handled against following hardware threats: (i) Trojan insertion, (ii) counterfeiting and (iii) cloning. Protection scenarios against these threats have also been highlighted in the figure. Preventive control (thwarting)-based security against Trojan insertion, counterfeiting and cloning has been ensured using structural obfuscation technique. Further, detective control (detection)-based security against counterfeiting and cloning has been ensured using crypto-hardware steganography technique. Basic properties of structural obfuscation and crypto-hardware steganography are also highlighted in Figure 3.3.

3.4.3 Structural obfuscation and crypto-based steganography for securing JPEG compression processor design

The details of structural obfuscation and crypto-steganography-based double line of defence for securing JPEG compression processor are discussed in this section. The

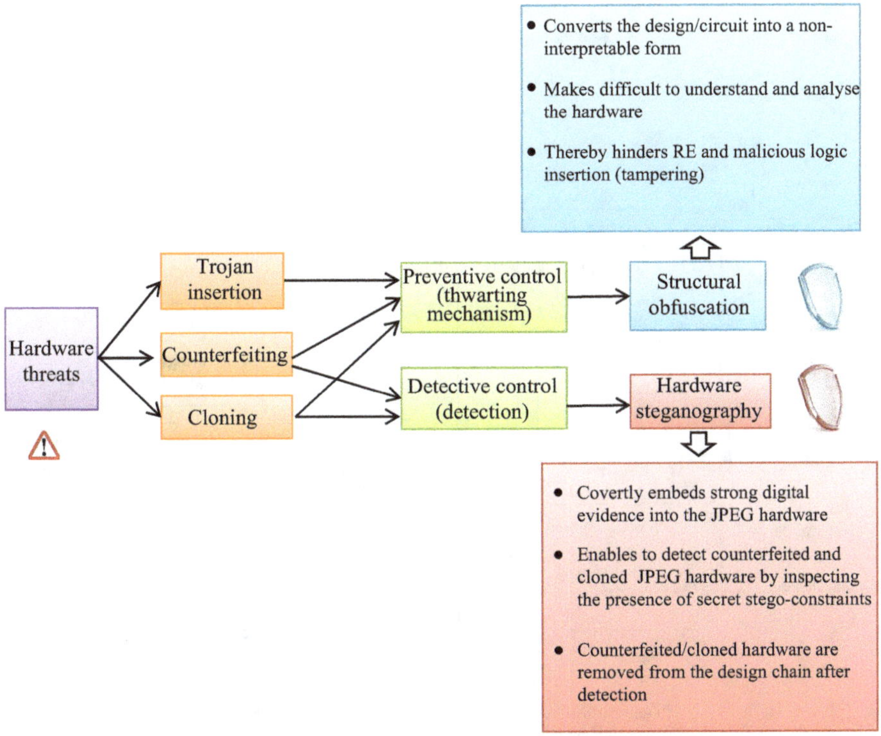

Figure 3.3 Hardware threats and protection scenarios of double line of defence mechanism for securing JPEG codec hardware accelerators (Sengupta and Rathor, 2020)

complete flow of applying double line of defence and generating a stego-embedded obfuscated JPEG compression processor is shown in Figure 3.4. As shown in the figure, the algorithmic description of JPEG compression application is first converted to an equivalent CDFG representation. Thus, obtained CDFG is subjected to the first line of defence using structural obfuscation. The structural obfuscation is performed using THT-based structural transformation. The THT is applied by breaking the sequential execution flow in the CDFG and performing some sub-computations concurrently. Thus, the CDFG is structurally obfuscated. The obfuscated CDFG is subjected to scheduling and resource allocation of HLS process based on designer's specified resource constraints and module library. Thus, obtained scheduled and allocated CDFG represents an obfuscated JPEG codec design in an intermediate form (which is convertible to its data path design).

The scheduled and allocated CDFG design of JPEG codec processor is used for performing crypto-based steganography as a second line of defence. The crypto-steganography approach produces secret stego-constraints using a stego-encoder system. The generated stego-constraints are implanted as secret stego-mark into the design. During the detection of counterfeiting and cloning, the implanted secret

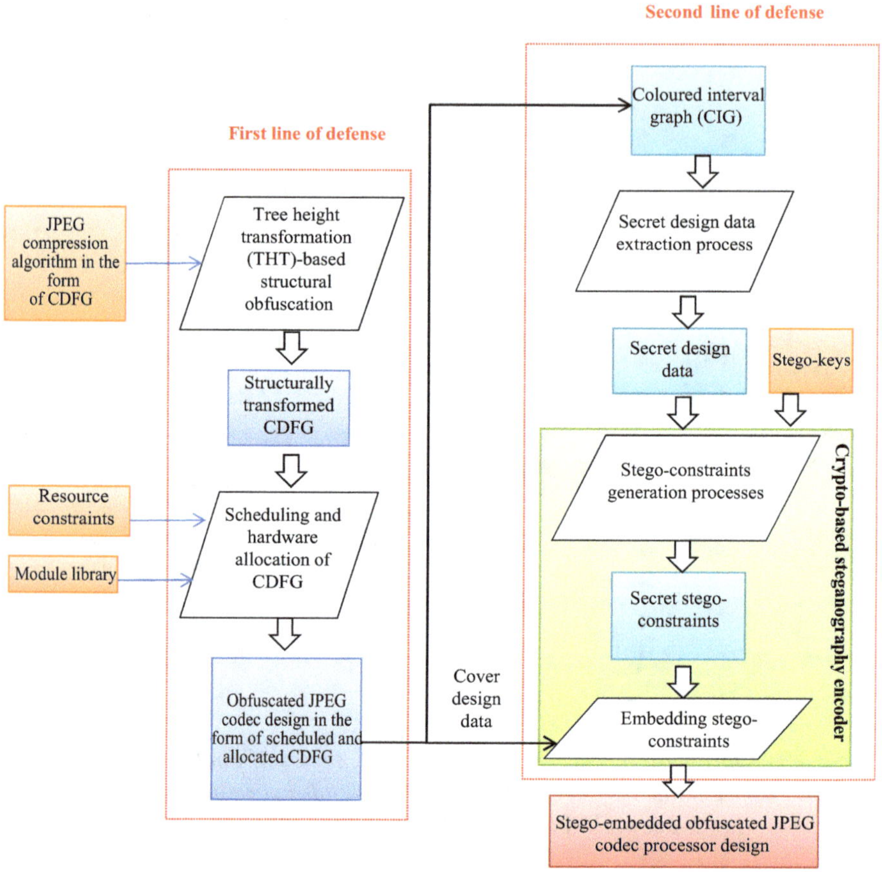

Figure 3.4 Flow of the process of securing a JPEG codec processor using structural obfuscation (first line of defence) and crypto-based steganography encoder (second line defence) (Sengupta and Rathor, 2020)

stego-mark is detected using a stego-decoder system. Highlights of the stego-encoder and stego-decoder system are as follows.

3.4.3.1 Stego-encoder system (Sengupta and Rathor, 2020)

The stego-encoder system of crypto-based steganography requires the following inputs to generate a stego-embedded design: (i) secret design data, (ii) stego-key and (iii) cover design data. The stego-key is a user input and the secret design data is generated using a process discussed as follows: first the scheduled and allocated CDFG is converted into a coloured interval graph (CIG) representation as shown in Figure 3.4. In the CIG, nodes indicate storage variables (S), and the colour of a node indicates its assignment to a register. The total number of distinct colours used in the CIG is equal to the minimum number of registers required to store all storage variables. Further, edges between nodes represent the overlapping of lifetime of storage variables. Thus, obtained CIG is leveraged to extract the secret design data. The secret design data is a collection or set of indices (i, j) of such node pairs (S_i, S_j) of CIG which are of the same colours. Thus, obtained secret design data is fed as primary inputs to the stego-encoder system.

The stego-constraints generation process of stego-encoder system executes multifarious steps, which are as follows: (i) state matrix formation, (ii) bit manipulation, (iii) row diffusion, (iv) Trifid-cipher-based encryption, (v) alphabet substitution, (vi) matrix transposition, (vii) mix-column diffusion, (viii) byte concatenation, (ix) bit-stream truncation and (v) bit-mapping. Of the aforementioned ten steps, five steps are driven through stego-key1 to stego-key5 as shown in Figure 3.5. The size of each stego-key, different modes of applying each stego-key and definition of each mode have already been discussed in Chapter 2. Further, different steps of stego-constraints generation process accomplish certain security properties which have been discussed in Chapter 2. Furthermore, the basic functions of these steps have also been discussed in Chapter 2. However, this chapter discusses the demonstration of these steps for stego-constraints generation for

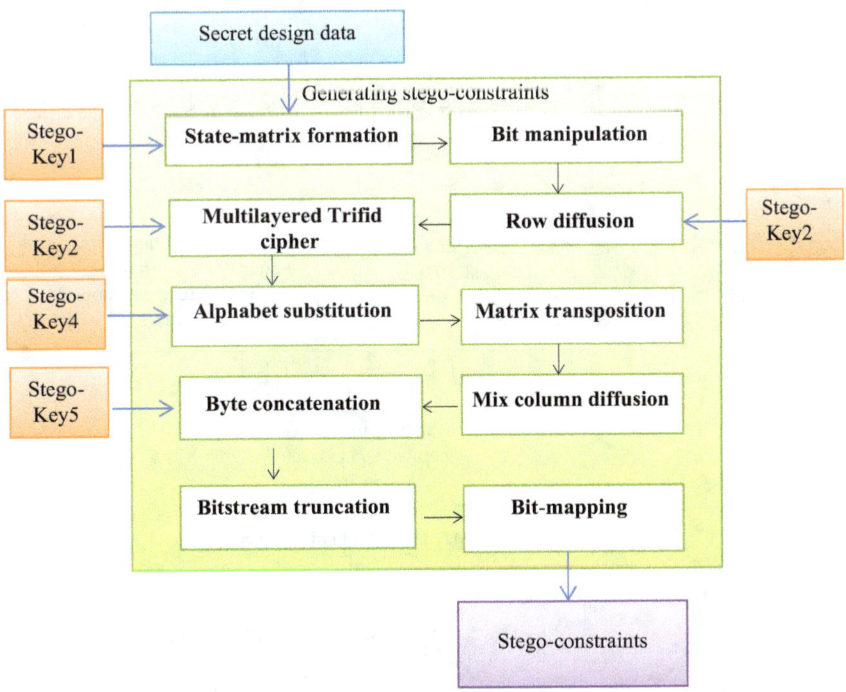

Figure 3.5 Steps of stego-constraints generation process of crypto-based steganography encoder system (Sengupta and Rathor, 2020)

embedding steganography in an 8-point DCT core used in JPEG as well as JPEG compression processor design.

Further in the stego-encoder system, the generated stego-constraints are embedded into the cover design data. The scheduled and allocated CDFG of obfuscated JPEG compression hardware is used as cover design data to embed the stego-constraints, as shown in Figure 3.4. The stego-constraints are embedded during two distinct phases of HLS by performing register reallocation and resource reallocation. Post embedding stego-constraints, a stego-embedded obfuscated JPEG compression hardware accelerator design at register transfer level (RTL) is generated at the output.

3.4.3.2 Stego-decoder system (Sengupta and Rathor, 2020)

The stego-decoder system of crypto-based steganography approach enables the detection of counterfeiting and cloning during forensic detection. Figure 3.6 depicts the stego-decoder system of crypto-based steganography approach. Inputs to the stego-decoder system are as follows: (i) secret design data which is the same as that was used in stego-encoder system, (ii) stego-key which is again the same as that was used in stego-encoder and (iii) stego-embedded JPEG codec processor RTL design. The major three processes of decoding steganography information are as follows: (i) secret stego-constraints generation process which generates stego-constraints using the same algorithm as stego-encoder used, (ii) hidden

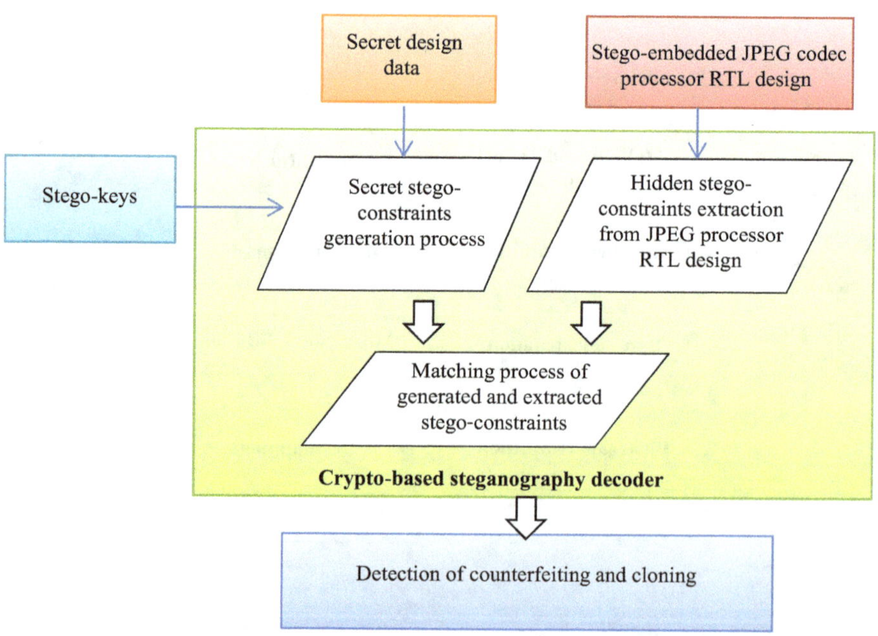

Figure 3.6 Detection of counterfeiting and cloning using a steganography decoder (Sengupta and Rathor, 2020)

stego-constraints extraction from stego-embedded JPEG processor by inspecting the RTL data path of the design and (iii) matching process of generated and extracted stego-constraints which confirms the presence of steganography information embedded into the design. If the presence of stego-information is found in the JPEG compression processor design of the same (original) brand, then it is ensured that the design is authentic (not counterfeited). However, if the presence of secret stego-information of the genuine vendor is found in the JPEG compression processor design of different brand, then the design is considered to be a cloned version.

The flow chart of the overall process of securing JPEG compression hardware using a double line of defence is shown in Figure 3.7. Further, before discussing the demonstration of the double line of defence approach for JPEG compression processor, let us discuss the details of crypto-steganography-based second line of

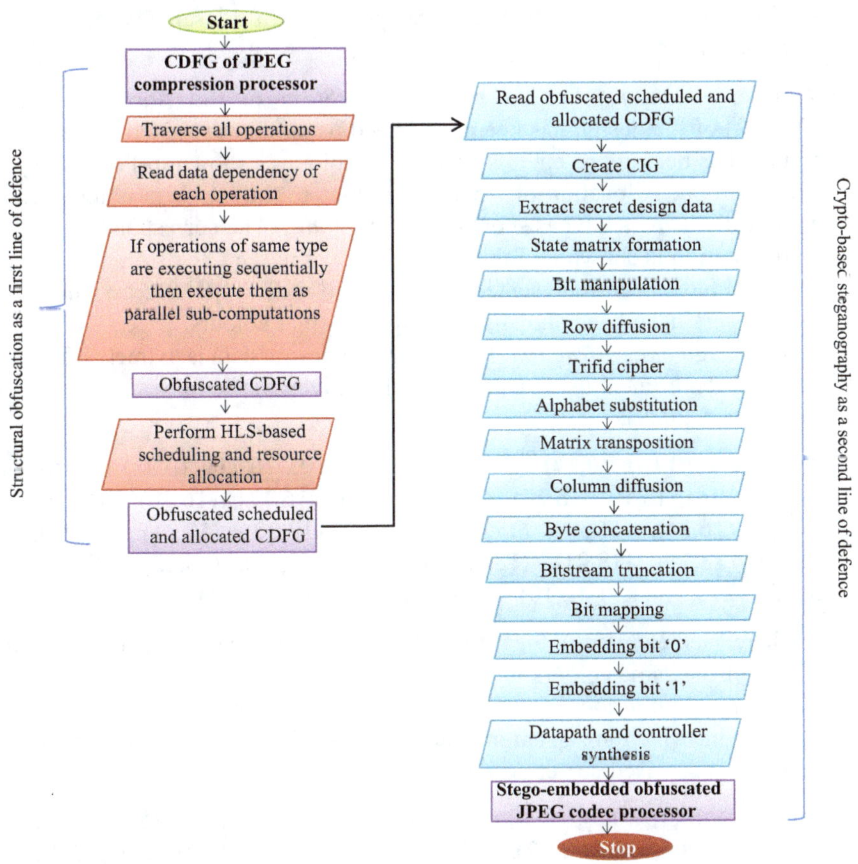

Figure 3.7 Flow chart of the double line of defence approach for securing JPEG compression processor used in medical imaging systems (Sengupta and Rathor, 2020)

defence for securing an 8-point DCT core which is used underneath the JPEG compression process. The role of an 8-point DCT core in JPEG compression processor is to transform the images from spatial domain to frequency domain. This discussion gives a deeper insight about the complex process of crypto-steganography-based double line of defence.

Following are the steps of employing crypto-steganography-based double line of defence in 8-point DCT cores (Sengupta and Rathor, 2020):

1. *Scheduling and resource allocation:* A CDFG representation of 8-point DCT is first scheduled and resource allocated based on resource constraints of four multipliers and one adder. Figure 3.8 shows the scheduled and resource allocated DFG of an 8-point DCT core. As shown in the figure, total 15 operations of the 8-point DCT application have been scheduled in nine control steps (Q0–Q8). Further resource allocation has been performed using two-vendor allocation scheme (i.e. two vendors of the same resource type have been used for allocating resources to two or more operations of the same type in a same control step).

1. *Register allocation*: Further register allocation to the storage variables (*S*0–*S*22) of the design has been performed using eight registers, where each register has been represented by a distinct colour as shown in Figure 3.8.
2. *CIG creation*: Allocation of storage variables to the registers is represented graphically using a CIG as shown in Figure 3.9(a). The corresponding tabular representation is shown in Table 3.1. As shown in the CIG, storage variables (*S*0–*S*22) have been represented as nodes and their respective assignment to the registers have been shown using eight distinct colours.
3. *Secret design data extraction*: Further, secret design data is extracted from the CIG and represented using a set *A* which is given in the following (Sengupta and Rathor, 2020):

   A = {(0,8), (0,16), (0,17), (0,18), (0,19), (0,20), (0,21), (0,22), (8,16), (8,17), (8,18), (8,19), (8,20), (8,21), (8,22), (16,17), (16,18), (16,19), (16,20), (16,21), (16,22), (17,18), (17,19), (17,20), (17,21), (17,22), (18,19), (18,20), (18,21), (18,22), (19,20), (19,21), (19,22), (20,21), (20,22), (21,22), (1,9), (2,10), (3,11), (4,12), (5,13), (6,14), (7,15)}

   where each element in the set is representing the indices (i, j) of node pairs (S_i, S_j) of the same colours in the CIG.
4. *State matrix formation*: In order to form a state matrix using set *A*, first of all those digits which are **greater than 15** are subjected to **modulo 15 operations.** Thus, revised secret design data is as follows:

   A = {(0,8), (0,1), (0,2), (0,3), (0,4), (0,5), (0,6), (0,7), (8,1), (8,2), (8,3), (8,4), (8,5), (8,6), (8,7), (1,2), (1,3), (1,4), (1,5), (1,6), (1,7), (2,3), (2,4), (2,5), (2,6), (2,7), (3,4), (3,5), (3,6), (3,7), (4,5), (4,6), (4,7), (5,6), (5,7), (6,7), (1,9), (2,10), (3,11), (4,12), (5,13), (6,14), (7,15)}

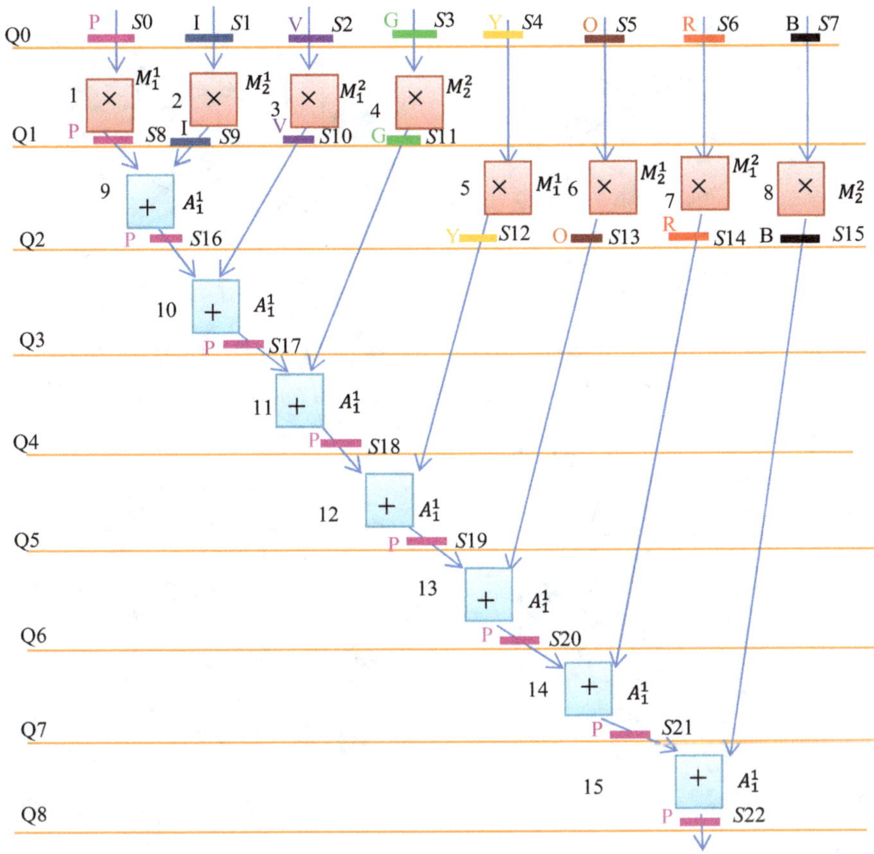

Figure 3.8 Scheduled and hardware-allocated 8-point DCT using 1A and 4M before steganography (Sengupta and Rathor, 2020) Q is the control step; M_1^1 is the first instance of multiplier of vendor V1; M_2^1 is the second instance of multiplier of vendor V1; M_1^2 is the first instance of multiplier of vendor V2; M_2^2 is the second instance of multiplier of vendor V2; A_1^1 is the first instance of adder of vendor V1; S0–S22 are the 23 storage variables; P, I, V, G, Y, O, R, B are the eight distinct colours representing eight distinct registers

Subsequently, each digit is converted to equivalent hexadecimal notation. Thus, further revised secret design data is as follows (Sengupta and Rathor, 2020):

A = {(0,8), (0,1), (0,2), (0,3), (0,4), (0,5), (0,6), (0,7), (8,1), (8,2), (8,3), (8,4), (8,5), (8,6), (8,7), (1,2), (1,3), (1,4), (1,5), (1,6), (1,7), (2,3), (2,4), (2,5), (2,6), (2,7), (3,4), (3,5), (3,6), (3,7), (4,5), (4,6), (4,7), (5,6), (5,7), (6,7), (1,9), (2,A), (3,B), (4,C), (5,D), (6,E), (7,F)}

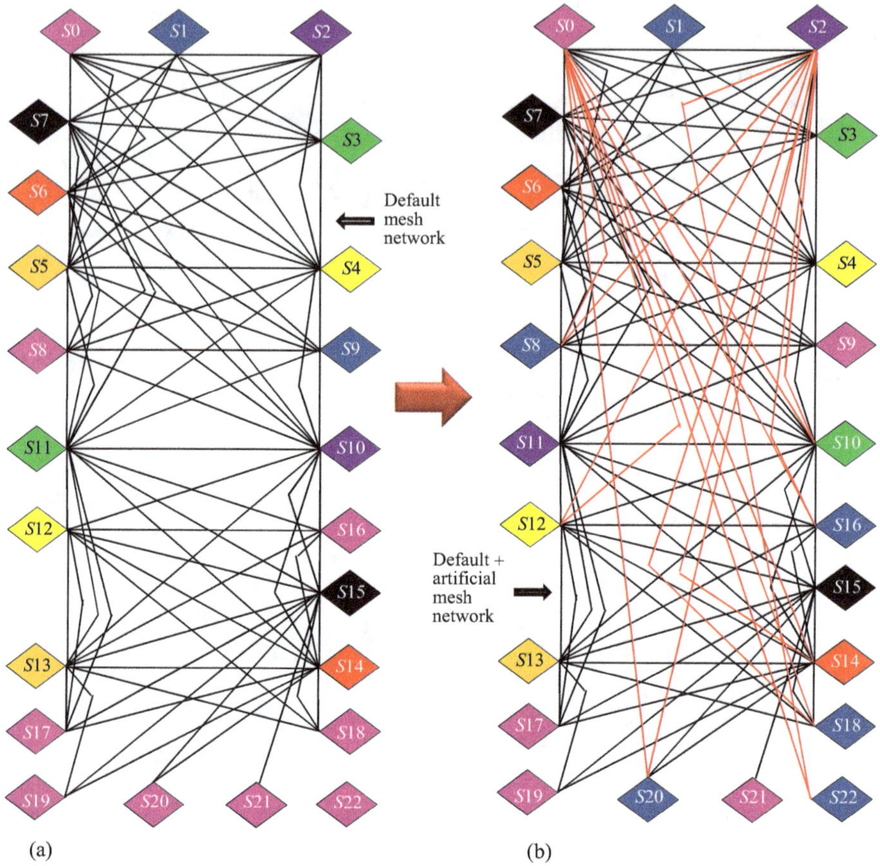

Figure 3.9 *CIG (a) pre embedding stego-constraints (b) post embedding stego-constraints (Sengupta and Rathor, 2020)*

This set A is used to form a state matrix based on designer's chosen value of stego-key1. For stego-key1='001', mode 2 of state matrix formation is applied (different modes of state matrix formation based on stego-key1 and their definitions have been given in Chapter 2). According to this mode, consecutive four elements of set A are chosen and the next four elements are discarded to form the rows of the state matrix. The state matrix M_S is given in the following (Sengupta and Rathor, 2020):

$$M_S = \begin{bmatrix} 08 & 01 & 02 & 03 \\ 81 & 82 & 83 & 84 \\ 13 & 14 & 15 & 16 \\ 26 & 27 & 34 & 35 \\ 47 & 56 & 57 & 67 \end{bmatrix} \tag{3.1}$$

Table 3.1 Register/colour allocations of storage variables (S0–S22) in an 8-point DCT before embedding steganography

Control steps	Pink	Indigo	Violet	Green	Yellow	Orange	Red	Black
Q0	S0	S1	S2	S3	S4	S5	S6	S7
Q1	S8	S9	S10	S11	S4	S5	S6	S7
Q2	S16	–	S10	S11	S12	S13	S14	S15
Q3	S17	–	–	S11	S12	S13	S14	S15
Q4	S18	–	–	–	S12	S13	S14	S15
Q5	S19	–	–	–	–	S13	S14	S15
Q6	S20	–	–	–	–	–	S14	S15
Q7	S21	–	–	–	–	–	–	S15
Q8	S22	–	–	–	–	–	–	–

5. *Bit manipulation*: To apply non-linear bit manipulation, each element or byte in the matrix M_S is substituted on the basis of forward S-box. The matrix M_B post byte-substitution or bit manipulation is as follows (Sengupta and Rathor, 2020):

$$M_B = \begin{bmatrix} 30 & 7C & 77 & 7B \\ 0C & 13 & EC & 5F \\ 7D & FA & 59 & 47 \\ F7 & CC & 18 & 96 \\ A0 & B1 & 5B & 85 \end{bmatrix} \tag{3.2}$$

6. *Row-diffusion*: Row diffusion is performed on the basis of stego-key2. For designer's chosen stego-key2='01 00 10 00 11', mode 2, mode 1, mode 3, mode 1 and mode 4 of row diffusion are performed for row 1 to row 5, respectively (different modes of row diffusion based on stego-key2 and their definitions have been given in Chapter 2). According to the chosen modes, circular right shift by two positions, one position, three positions, one position and four positions are performed in row 1 to row 5, respectively. The matrix M_{Rd} post row diffusion is given in the following (Sengupta and Rathor, 2020):

$$M_{Rd} = \begin{bmatrix} 77 & 7B & 30 & 7C \\ 5F & 0C & 13 & EC \\ FA & 59 & 47 & 7D \\ 96 & F7 & CC & 18 \\ A0 & B1 & 5B & 85 \end{bmatrix} \tag{3.3}$$

7. *Encryption using multilayered Trifid cipher*: Each distinct alphabet in matric M_{Rd} is subjected to Trifid-cipher-based encryption based on stego-key3. The encryption key contains 27 unique characters (26 alphabets+1 special character) which are arranged in three square matrices of size 3×3. The encrypted value of alphabet is a three-digit value given by *abc*, where *a*, *b*, *c* indicate row, column and square matrix number, respectively.
   (i) Encryption of alphabet *A*:

   Suppose encryption key for alphabet *A* is *V# Q A W S E D R F T G Y H U J I K O L P Z M X N C B*'. The arrangement of encryption key in three square matrices is as follows (Sengupta and Rathor, 2020):

$$SQ1 = \begin{bmatrix} V & \# & Q \\ A & W & S \\ E & D & R \end{bmatrix} SQ2 = \begin{bmatrix} F & T & G \\ Y & H & U \\ J & I & K \end{bmatrix} SQ3 = \begin{bmatrix} O & L & P \\ Z & M & X \\ N & C & B \end{bmatrix}$$

   Hence, the encrypted value of alphabet *A* is 211.

   (ii) Encryption of alphabet *B*:

   Suppose encryption key for alphabet *B* is *Q A W S E D R F T G Y H U J I K # O L P Z M X N C B V*. The arrangement of encryption key in three square matrices is as follows (Sengupta and Rathor, 2020):

$$SQ1 = \begin{bmatrix} Q & A & W \\ S & E & D \\ R & F & T \end{bmatrix} SQ2 = \begin{bmatrix} G & Y & H \\ U & J & I \\ K & \# & O \end{bmatrix} SQ3 = \begin{bmatrix} L & P & Z \\ M & X & N \\ C & B & V \end{bmatrix}$$

   Hence, the encrypted value of alphabet *B* is 323.

   (iii) Encryption of alphabet *C*:

   Suppose encryption key for alphabet *C* is *O L P Z M X N C B V # Q A W S E D R F T G Y H U J I K*. The arrangement of encryption key in three square matrices is as follows (Sengupta and Rathor, 2020):

$$SQ1 = \begin{bmatrix} O & L & P \\ Z & M & X \\ N & C & B \end{bmatrix} SQ2 = \begin{bmatrix} V & \# & Q \\ A & W & S \\ E & D & R \end{bmatrix} SQ3 = \begin{bmatrix} F & T & G \\ Y & H & U \\ J & I & K \end{bmatrix}$$

   Hence, the encrypted value of alphabet *C* is 321.

   (iv) Encryption of alphabet *D*:

   Suppose encryption key for alphabet *D* is *G Y H U J I K # O L P Z M X N C B V Q A W S E D R F T*. The arrangement of encryption key in three square matrices is as follows (Sengupta and Rathor, 2020):

$$SQ1 = \begin{bmatrix} G & Y & H \\ U & J & I \\ K & \# & O \end{bmatrix} SQ2 = \begin{bmatrix} L & P & Z \\ M & X & N \\ C & B & V \end{bmatrix} SQ3 = \begin{bmatrix} Q & A & Q \\ S & E & D \\ R & F & T \end{bmatrix}$$

   Hence, the encrypted value of alphabet D is 233.

(v) Encryption of alphabet E:

Suppose encryption key for alphabet E is $F\ T\ G\ Y\ H\ U\ J\ I\ K\ O\ L\ P\ Z\ M\ X\ N\ C\ B\ V\ \#\ Q\ A\ W\ S\ E\ D\ R$'. The arrangement of encryption key in three square matrices is as follows (Sengupta and Rathor, 2020):

$$SQ1 = \begin{bmatrix} F & T & G \\ Y & H & U \\ J & I & K \end{bmatrix} SQ2 = \begin{bmatrix} O & L & P \\ Z & M & X \\ N & C & B \end{bmatrix} SQ3 = \begin{bmatrix} V & \# & Q \\ A & W & S \\ E & D & R \end{bmatrix}$$

Hence, the encrypted value of alphabet E is 313.

(vi) Encryption of alphabet F:

Suppose encryption key for alphabet F is L P Z M X N C B V Q A W S E D R F T G Y H U J I K # O'. The arrangement of encryption key in three square matrices is as follows (Sengupta and Rathor, 2020):

$$SQ1 = \begin{bmatrix} L & P & Z \\ M & X & N \\ C & B & V \end{bmatrix} SQ2 = \begin{bmatrix} Q & A & W \\ S & E & D \\ R & F & T \end{bmatrix} SQ3 = \begin{bmatrix} G & Y & H \\ U & J & I \\ K & \# & O \end{bmatrix}$$

Hence, the encrypted value of alphabet F is 322.

8. *Alphabet substitution*: The alphabets in matrix M_{Rd} are substituted using an equivalent value that is obtained by applying a mathematical expression on the encrypted value of alphabets (obtained from previous step). The mathematical expression to be chosen for computing an equivalent value of alphabet depends on stego-key4. For designer's chosen stego-key4='001 001 000 010 101 010', mode 2, mode 2, mode 1, mode 3, mode 6 and mode 3 of alphabet substitution are applied for alphabets A, B, C, D, E and F, respectively (different modes of alphabet substitution based on stego-key4 and their definitions have been given in Chapter 2). According to the chosen mode, following mathematical expressions are chosen for alphabets A, B, C, D, E and F, respectively: $a + b + c$, $a + b + c$, $a * b * c$, $|a - b - c|$, $(c + a) * b$, $|a - b - c|$. These mathematical expressions are applied on corresponding encrypted value of each alphabet. Table 3.2 highlights the encrypted value of each alphabet, corresponding mode of alphabet substitution and equivalent value to be used for alphabet substitution (i.e. output of mathematical expression). The matrix post alphabet substitution is given in the following (Sengupta and Rathor, 2020):

$$M_{AS} = \begin{bmatrix} 77 & 78 & 30 & 76 \\ 51 & 06 & 13 & 66 \\ 14 & 59 & 47 & 74 \\ 96 & 17 & 66 & 18 \\ 40 & 81 & 58 & 85 \end{bmatrix} \tag{3.4}$$

Table 3.2 Details of obtaining equivalent value for alphabet substitution

Alphabets	Encrypted value	Mode of alphabet substitution	Selected mathematic expression	Equivalent value to be used to substitute corresponding alphabet
A	211	2	$a + b + c$	4
B	323	2	$a + b + c$	8
C	321	1	$a * b * c$	6
D	233	3	$\|a - b - c\|$	4
E	313	6	$(a + c) * b$	6
F	322	3	$\|a - b - c\|$	1

9. *Matrix transposition*: The matrix post transposition is given in the following (Sengupta and Rathor, 2020):

$$M_T = \begin{bmatrix} 77 & 51 & 14 & 96 & 40 \\ 78 & 06 & 59 & 17 & 81 \\ 30 & 13 & 47 & 66 & 58 \\ 76 & 66 & 74 & 18 & 85 \end{bmatrix} \tag{3.5}$$

10. *Mix-column diffusion*: Each column of the transposed matrix is subjected to a transformation using a maximum distance separable (MDS) matrix in order to achieve mix-column diffusion. The transformation of each column using MDS matrix is as follows (Sengupta and Rathor, 2020):

    For first column:

$$\begin{bmatrix} B_0^1 \\ B_1^1 \\ B_2^1 \\ B_3^1 \end{bmatrix} = \begin{bmatrix} 02 & 03 & 01 & 01 \\ 01 & 02 & 03 & 01 \\ 01 & 01 & 02 & 03 \\ 03 & 01 & 01 & 02 \end{bmatrix} * \begin{bmatrix} 77 \\ 78 \\ 30 \\ 76 \end{bmatrix} = \begin{bmatrix} 20 \\ A1 \\ F5 \\ 3D \end{bmatrix} \tag{3.6}$$

$$B_0^1 = (02 * 77) \oplus (03 * 78) \oplus (01 * 30) \oplus (01 * 76) = 20$$

$$B_1^1 = (01 * 77) \oplus (02 * 78) \oplus (03 * 30) \oplus (01 * 76) = A1$$

$$B_2^1 = (01 * 77) \oplus (01 * 78) \oplus (02 * 30) \oplus (03 * 76) = F5$$

$$B_3^1 = (03 * 77) \oplus (01 * 78) \oplus (01 * 30) \oplus (02 * 76) = 3D$$

Note: Computations are performed using Rijndael's Galois (finite) field arithmetic (GF(2^8)). In Galois (finite) field arithmetic, multiplying a number by 01 yields the same number; multiplying a number by 02 means left shifting the number by 1 bit; multiplying a number by 03 indicates left shift the number by 1 bit, followed by adding by the original number.

For the second column:

$$\begin{bmatrix} B_0^2 \\ B_1^2 \\ B_2^2 \\ B_3^2 \end{bmatrix} = \begin{bmatrix} 02 & 03 & 01 & 01 \\ 01 & 02 & 03 & 01 \\ 01 & 01 & 02 & 03 \\ 03 & 01 & 01 & 02 \end{bmatrix} * \begin{bmatrix} 51 \\ 06 \\ 13 \\ 66 \end{bmatrix} = \begin{bmatrix} DD \\ 0E \\ DB \\ 2A \end{bmatrix} \tag{3.7}$$

$$B_0^2 = (02 * 51) \oplus (03 * 06) \oplus (01 * 13) \oplus (01 * 66) = DD$$

$$B_1^2 = (01 * 51) \oplus (02 * 06) \oplus (03 * 13) \oplus (01 * 66) = 0E$$

$$B_2^2 = (01 * 51) \oplus (01 * 06) \oplus (02 * 13) \oplus (03 * 66) = DB$$

$$B_3^2 = (03 * 51) \oplus (01 * 06) \oplus (01 * 13) \oplus (02 * 66) = 2A$$

For the third column:

$$\begin{bmatrix} B_0^3 \\ B_1^3 \\ B_2^3 \\ B_3^3 \end{bmatrix} = \begin{bmatrix} 02 & 03 & 01 & 01 \\ 01 & 02 & 03 & 01 \\ 01 & 01 & 02 & 03 \\ 03 & 01 & 01 & 02 \end{bmatrix} * \begin{bmatrix} 14 \\ 59 \\ 47 \\ 74 \end{bmatrix} = \begin{bmatrix} F0 \\ 1B \\ 5F \\ CA \end{bmatrix} \tag{3.8}$$

$$B_0^3 = (02 * 14) \oplus (03 * 59) \oplus (01 * 47) \oplus (01 * 74) = F0$$

$$B_1^3 = (01 * 14) \oplus (02 * 59) \oplus (03 * 47) \oplus (01 * 74) = 1B$$

$$B_2^3 = (01 * 14) \oplus (01 * 59) \oplus (02 * 47) \oplus (03 * 74) = 5F$$

$$B_3^3 = (03 * 14) \oplus (01 * 59) \oplus (01 * 47) \oplus (02 * 74) = CA$$

For the fourth column:

$$\begin{bmatrix} B_0^4 \\ B_1^4 \\ B_2^4 \\ B_3^4 \end{bmatrix} = \begin{bmatrix} 02 & 03 & 01 & 01 \\ 01 & 02 & 03 & 01 \\ 01 & 01 & 02 & 03 \\ 03 & 01 & 01 & 02 \end{bmatrix} * \begin{bmatrix} 96 \\ 17 \\ 66 \\ 18 \end{bmatrix} = \begin{bmatrix} 70 \\ 0A \\ 65 \\ E0 \end{bmatrix} \tag{3.9}$$

$$B_0^4 = (02 * 96) \oplus (03 * 17) \oplus (01 * 66) \oplus (01 * 18) = 70$$

$$B_1^4 = (01 * 96) \oplus (02 * 17) \oplus (03 * 66) \oplus (01 * 18) = 0A$$

$$B_2^4 = (01 * 96) \oplus (01 * 17) \oplus (02 * 66) \oplus (03 * 18) = 65$$

$$B_3^4 = (03 * 96) \oplus (01 * 17) \oplus (01 * 66) \oplus (02 * 18) = E0$$

For the fifth column:

$$\begin{bmatrix} B_0^5 \\ B_1^5 \\ B_2^5 \\ B_3^5 \end{bmatrix} = \begin{bmatrix} 02 & 03 & 01 & 01 \\ 01 & 02 & 03 & 01 \\ 01 & 01 & 02 & 03 \\ 03 & 01 & 01 & 02 \end{bmatrix} * \begin{bmatrix} 40 \\ 81 \\ 58 \\ 85 \end{bmatrix} = \begin{bmatrix} C5 \\ 34 \\ E5 \\ 08 \end{bmatrix} \tag{3.10}$$

$$B_0^5 = (02 * 40) \oplus (03 * 81) \oplus (01 * 58) \oplus (01 * 85) = C5$$

$$B_1^5 = (01 * 40) \oplus (02 * 81) \oplus (03 * 58) \oplus (01 * 85) = 34$$

$$B_2^5 = (01 * 40) \oplus (01 * 81) \oplus (02 * 58) \oplus (03 * 85) = E5$$

$$B_3^5 = (03 * 40) \oplus (01 * 81) \oplus (01 * 58) \oplus (02 * 85) = 08$$

The matrix M_{Cd} post mix-column diffusion is as follows:

$$M_{Cd} = \begin{bmatrix} 20 & DD & F0 & 70 & C5 \\ A1 & 0E & 1B & 0A & 34 \\ F5 & DB & 5F & 65 & E5 \\ 3D & 2A & CA & E0 & 08 \end{bmatrix} \tag{3.11}$$

11. *Byte concatenation*: Bytes of each column of the matrix M_{Cd} are concatenated on the basis of stego-key5. For designer's chosen stego-key5='001 000 010 101 000', mode 2, mode 1, mode 3, mode 6 and mode 1 of byte concatenation are applied for column 1 to column 5, respectively (different modes of byte concatenation based on stego-key5 and their definitions have been given in Chapter 2). For the chosen modes, the concatenated byte stream is as follows (Sengupta and Rathor, 2020):

    $B_0^1B_1^1B_3^1B_2^1B_0^2B_1^2B_2^2B_3^2B_0^3B_2^3B_1^3B_3^3B_0^4B_3^4B_2^4B_1^4B_0^5B_1^5B_2^5B_3^5$

    20A13DF5DD0EDB2AF05F1BCA70E0650AC534E508

12. *Conversion into bitstream*: Thus, the obtained byte stream is converted into a bitstream by replacing each digit with its equivalent binary notation. The bitstream is as follows:

    '0010000010100001001111011111010111011101000011101101101100
    1010101111000001011111000110111100101001110000111000000 1100
    1010000101011000101 0 011010011100101000 01000'

13. *Bitstream truncation*: The bitstream is truncated based on designer's chosen size of stego-constraints. For stego-constraints size=48, the truncated bitstream is as follows (Sengupta and Rathor, 2020):

    '001000001010000100111101111101011101110100001110'

    The truncated bitstream contains twenty-four 0s and twenty-four 1s.

14. *Bit mapping*: Bit '0' and bit '1' of the truncated bitstream are mapped to stego-constraints (hardware security constraints to be embedded into design) based on the following mapping rules (Sengupta and Rathor, 2020):
    Bit '0': Embeds an edge between node pair (even, even) of CIG (causing register reallocation during HLS)
    Bit '1': Odd operations are assigned to resources of vendor type 1 (V1) and even operations are assigned to resources of vendor type 2 (V2) (resource reallocation during HLS)
15. *Embedding bit '0'*: Embedding of stego-constraints represented by bit '0' is performed by adding artificial edges into the CIG. To do so, first of all each '0' bit in the bitstream is mapped to stego-constraints based on the aforementioned mapping rule. Table 3.3 shows the stego-constraints in the form of artificial edges to be embedded corresponding to each '0' bit in the bitstream (Sengupta and Rathor, 2020). The CIG post embedding stego-constraints is shown in Figure 3.9(b). The added artificial edges corresponding to bit '0' have been shown using red lines in the figure. As shown in the figure, and effective insertion of constraint edges ⟨*S*0,*S*2⟩, ⟨*S*0,*S*4⟩, ⟨*S*0,*S*6⟩, ⟨*S*2,*S*4⟩, ⟨*S*2, *S*6⟩, ⟨*S*4,*S*6⟩, ⟨*S*4,*S*8⟩ and ⟨*S*4,*S*10⟩ are not necessary as theses constraint edges already exist in the CIG. Further, constraint edges ⟨*S*0,*S*8⟩, ⟨*S*0,*S*10⟩, ⟨*S*0,*S*12⟩, ⟨*S*0,*S*14⟩, ⟨*S*0,*S*16⟩, ⟨*S*0,*S*18⟩, ⟨*S*0,*S*20⟩, ⟨*S*0,*S*22⟩, ⟨*S*2,*S*8⟩, ⟨*S*2, *S*10⟩, ⟨*S*2,*S*12⟩, ⟨*S*2,*S*14⟩, ⟨*S*2,*S*16⟩, ⟨*S*2,*S*18⟩, ⟨*S*2,*S*20⟩, ⟨*S*2,*S*22⟩ do not exist by default and hence are artificially added. However, embedding of some edges in the CIG requires colour (register) reallocation of some nodes (storage variables). This is due to the fact that two nodes connected through an edge should essentially have distinct colours. Since both nodes in the node pairs of constraint edges ⟨*S*0,*S*8⟩, ⟨*S*0,*S*16⟩, ⟨*S*0,*S*18⟩, ⟨*S*0,*S*20⟩, ⟨*S*0,*S*22⟩ and ⟨*S*2,*S*10⟩ are initially assigned to same colours, the colour (register) swapping of one of the nodes in each pair has to be performed with another colour being used within the same control step. As shown in the CIG (Figure 3.9(b)) post embedding constraint edges, colour of node *S*8 (pink) has been swapped with *S*9 (indigo) to embed the constraint edge ⟨*S*0,*S*8⟩. Since colour of *S*8 is now indigo, the edge ⟨*S*0,*S*8⟩ can be added. Similarly, colour of *S*16, *S*18, *S*20 and *S*22 has been changed from pink to indigo to embed constraint edges ⟨*S*0, *S*16⟩, ⟨*S*0,*S*18⟩, ⟨*S*0,*S*20⟩ and ⟨*S*0,*S*22⟩. In addition, colour of *S*10 has been swapped with *S*11 to enable embedding of ⟨*S*2,*S*10⟩. The information of register reallocation post embedding stego-constraints is also shown in Table 3.4. Thus, stego-constraints represented by bit '0' in the encrypted bitstream are embedded by performing register reallocation during HLS (Sengupta and Rathor, 2020).

1. *Embedding bit '1'*: Embedding of stego-constraints represented by bit '1' in the bitstream is performed by reallocating resources in the scheduled and allocated CDFG. To do so, first of all each '1' bit in the bitstream is mapped to stego-constraints on the basis of aforementioned mapping rule. Table 3.5 shows the stego-constraints in the form of possible resource reallocation corresponding to

Table 3.3 Stego-constraints in the form of artificial edges corresponding to each '0' bit in the bitstream

Position in bitstream	Bit value	Mapped stego-constraints	Remarks
1	0	⟨S0,S2⟩	Effective insertion of edge is not necessary
2	0	⟨S0,S4⟩	Effective insertion of edge is not necessary
4	0	⟨S0,S6⟩	Effective insertion of edge is not necessary
5	0	⟨S0,S8⟩	Effective edge inserted
6	0	⟨S0,S10⟩	Effective edge inserted
7	0	⟨S0,S12⟩	Effective edge inserted
8	0	⟨S0,S14⟩	Effective edge inserted
10	0	⟨S0,S16⟩	Effective edge inserted
12	0	⟨S0,S18⟩	Effective edge inserted
13	0	⟨S0,S20⟩	Effective edge inserted
14	0	⟨S0,S22⟩	Effective edge inserted
15	0	⟨S2,S4⟩	Effective insertion of edge is not necessary
17	0	⟨S2,S6⟩	Effective insertion of edge is not necessary
18	0	⟨S2,S8⟩	Effective edge inserted
23	0	⟨S2,S10⟩	Effective edge inserted
29	0	⟨S2,S12⟩	Effective edge inserted
31	0	⟨S2,S14⟩	Effective edge inserted
35	0	⟨S2,S16⟩	Effective edge inserted
39	0	⟨S2,S18⟩	Effective edge inserted
41	0	⟨S2,S20⟩	Effective edge inserted
42	0	⟨S2,S22⟩	Effective edge inserted
43	0	⟨S4,S6⟩	Effective insertion of edge is not necessary
44	0	⟨S4,S8⟩	Effective insertion of edge is not necessary
48	0	⟨S4,S10⟩	Effective insertion of edge is not necessary

each '1' bit in the bitstream (Sengupta and Rathor, 2020). As shown in the table, effective resource reallocations for operations 1, 4, 5, 8, 9, 11, 13 and 15 are not necessary as they satisfy the constraints by default. This is because among the operations 1, 4, 5, 8, 9, 11, 13 and 15, odd operations are already allocated to vendor V1 and even operations are already allocated to vendor V2. Further, effective resource reallocations for operations 2, 3, 6 and 7 have been performed according to the mapped stego-constraints. Earlier operations 2, 3, 6 and 7 were assigned to vendors V1, V2, V1 and V2, respectively (odd operations to even vendor V2 and even operations to odd vendor V1). However, post embedding bit '1' of the bitstream, odd operations 3 and 7 are now allocated to odd vendor V1 and even operations 2 and 6 are now allocated to even vendor

Table 3.4 Register/colour allocations of storage variables (S0–S22) in an 8-point DCT post embedding steganography

Control steps	Pink	Indigo	Violet	Green	Yellow	Orange	Red	Black
Q0	S0	S1	S2	S3	S4	S5	S6	S7
Q1	S9	S8	S11	S10	S4	S5	S6	S7
Q2	–	S16	S11	S10	S12	S13	S14	S15
Q3	S17	–	S11	–	S12	S13	S14	S15
Q4	–	S18	–	–	S12	S13	S14	S15
Q5	S19	–	–	–	–	S13	S14	S15
Q6	–	S20	–	–	–	–	S14	S15
Q7	S21	–	–	–	–	–	–	S15
Q8	–	S22	–	–	–	–	–	–

V2. However, effective resource reallocations for operations 10, 12 and 14 are not feasible. This is because, these operations are additional operations and the chosen adder constraint is 1. Therefore, only one adder of vendor V1 (i.e. A_1^1) can be availed for resource allocation. Hence, operations 10, 12 and 14 cannot be assigned to vendor V2 according to mapped constraints. Moreover, mapping of stego-constraints corresponding to only fifteen 1s in the bitstream is possible. This is because only fifteen operations are present in the 8-point DCT application. Therefore, remaining 1s in the bitstream are left unmapped to stego-constraints and hence cannot be embedded.

The scheduled and allocated CDFG post embedding all '0' and '1' bits of the bitstream is shown in Figure 3.10. Thus, stego-embedded 8-point DCT design is achieved by performing crypto-steganography-based second line of defence (Sengupta and Rathor, 2020).

3.5 Process of securing JPEG compression processor using double line of defence

3.5.1 Designing a secure JPEG codec processor using first line of defence

Before discussing the process of employing the first line of defence, let us understand the background on JPEG compression process in brief. The JPEG compression process performs the compression by first converting the input images from spatial representation to frequency representation. Therefore, the underlying JPEG compression process is centred on 8-point DCT core which is responsible for segregating entire image into portions of distinct frequencies. Further, the actual

Table 3.5 Stego-constraints corresponding to each '1' bit in the bitstream

Position in bitstream	Bit value	Mapped stego-constraints	Remarks
3	1	Operation 1→V1	Effective resource reallocation is not necessary
9	1	Operation 2→V2	Effective resource reallocation is performed
11	1	Operation 3→V1	Effective resource reallocation is performed
16	1	Operation 4→V2	Effective resource reallocation is not necessary
19	1	Operation 5→V1	Effective resource reallocation is not necessary
20	1	Operation 6→V2	Effective resource reallocation is performed
21	1	Operation 7→V1	Effective resource reallocation is performed
22	1	Operation 8→V2	Effective resource reallocation is not necessary
24	1	Operation 9→V1	Effective resource reallocation is not necessary
25	1	Operation 10→V2	Effective resource reallocation is not feasible
26	1	Operation 11→V1	Effective resource reallocation is not necessary
27	1	Operation 12→V2	Effective resource reallocation is not feasible
28	1	Operation 13→V1	Effective resource reallocation is not necessary
30	1	Operation 14→V2	Effective resource reallocation is not feasible
32	1	Operation 15→V1	Effective resource reallocation is not necessary
33	1	Not applicable	Not applicable
34	1	Not applicable	Not applicable
36	1	Not applicable	Not applicable
37	1	Not applicable	Not applicable
38	1	Not applicable	Not applicable
40	1	Not applicable	Not applicable
45	1	Not applicable	Not applicable
46	1	Not applicable	Not applicable
47	1	Not applicable	Not applicable

compression phase is performed by quantization process which discards less important frequency components and keeps only the most important frequency components. Therefore, the compressed image/data contains lesser information (but most important for decompression/reconstruction process) than original hence requires less memory space to store and less bandwidth to transmit. Figure 3.11 shows the block diagram of JPEG compression processor hardware. The step-by-step process of JPEG compression is summarized as follows:

1. An input image is transformed into an $N\times N$ matrix of pixels, where N indicates the size of square matrix. A value of a pixel indicates nothing but its intensity at respective position in the image. The intensity values of pixels range between 0 and 255, where 0 and 255 indicate pure dark (black) and pure bright (white) pixel, respectively (for a greyscale image).
2. Further, the $N\times N$ pixels of input image are partitioned into non-overlapping 8×8 matrices (or blocks). This is because an 8-point DCT operates on an 8×8 matrix at a time.
3. Further, the subtraction of 128 from each pixel value of each 8×8 block is performed to bring the pixel values within the range of -128 to $+127$. This is because the 8-point DCT requires the values within this range to operate upon.

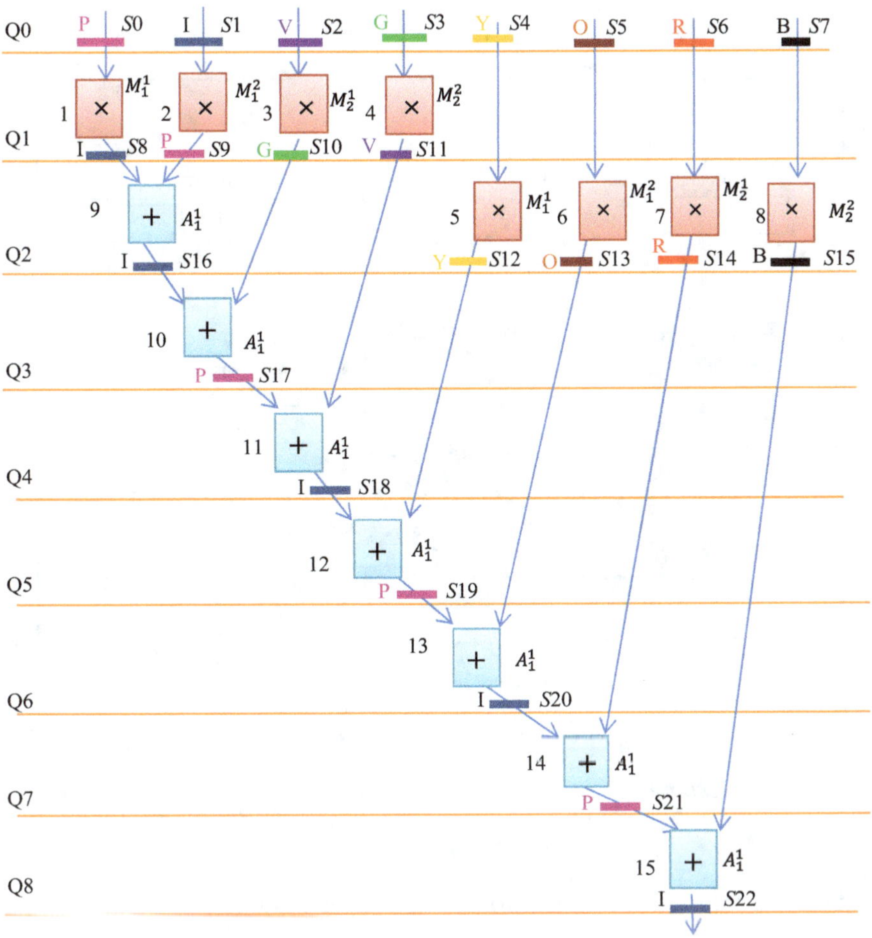

Figure 3.10 Scheduled and hardware-allocated 8-point DCT using 1A and 4M post crypto-based steganography (Sengupta and Rathor, 2020)

4. Subsequently, each 8×8 block of pixels is subjected to DCT transformation as shown in Figure 3.11. The 8-point DCT transformation is performed using a two-dimensional (2D)-DCT coefficient matrix (Sengupta and Rathor, 2020) represented by *B*. The underlying equation is as follows:

$$W = (B * P) * B' \tag{3.12}$$

where *W* indicates the DCT-transformed 8×8 block of image pixels, *P* indicates an 8×8 block of image pixels and *B'* represents the transpose of 2D-DCT coefficient matrix *B*. The equation of computing the first pixel value *W*11 of

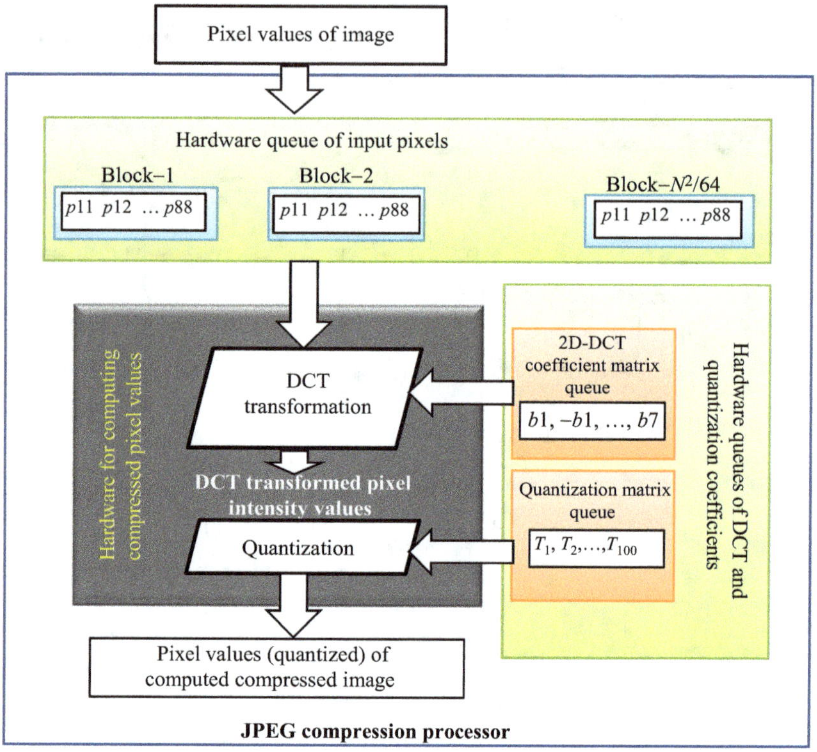

Figure 3.11 Design of JPEG compression hardware accelerator (Sengupta and Rathor, 2020)

the DCT-transformed 8×8-matrix W is given as follows:

$$W11 = b4 * g11 + b4 * g12 + b4 * g13 + b4 * g14 + b4 * g15 + b4 * g16 + b4 * g17 + b4 * g18 \tag{3.13}$$

where $g11$–$g18$ indicate the elements in the first row of the matrix $[B^*P]$, and $b4$ indicates the elements in the first column (repeating throughout the column) of matrix B'. The equation of computing the element $g11$ of the matrix $[B^*P]$ is as follows:

$$g11 = b4 * p11 + b4 * p21 + b4 * p31 + b4 * p41 + b4 * p51 + b4 * p61 + b4 * p71 + b4 * p81 \tag{3.14}$$

where $p11$–$p81$ indicate the pixel values in the first column of the input matrix P, and $b4$ indicates the elements in the first row (repeating throughout the row) of matrix B. Likewise, calculations of remaining elements ($g12$–$g88$) of matrix $[B * P]$ are performed.

Thereby, the DCT-transformed matrix *W* is computed by calculating all the elements using (3.12).

5. The next step in the JPEG compression process is the quantization which is performed using quantization matrix *T*. By choosing suitable quantization matrices, different levels of compression and quality can be achieved. The quality levels range between 1 and 100. To achieve the highest compression at the cost of the poorest quality, the quality level 1 is chosen, while to achieve the best quality at the cost of the lowest compression, the quality level 100 is chosen. Depending on the required level of compression and quality of image, suitable quantization matrix can be exploited.

   In order to perform quantization, each pixel value in the DCT-transformed matrix *W* is divided by the respective value (*t*) in the quantization matrix *T*. Post division, each value is rounded off to the closest integer value. Sengupta and Rathor (2020) exploited the quantization matrix T_{90} (i.e. quality level 90) to perform compression (within the acceptable compression ratio) of the medical images obtained from CT scan.
6. Further, the computed compressed data of an image in the matrix (2D form) is converted into 1D array by performing zigzag scanning.
7. Eventually, the compressed data in the 1D array is subjected to run-length encoding for storage into memory.

Thus, an input medical image is compressed using JPEG compression process and stored into memory. It is evident from the previous discussion that the DCT transformation and quantization is highly computational intensive processes of the JPEG compression processor. Therefore, it is performance-wise efficient to perform them using a hardware accelerator. The hardware realization of JPEG compression process using HLS process has been proposed by Sengupta *et al.* (2018). To generate a JPEG codec hardware accelerator using HLS, following are the inputs: algorithmic description of JPEG compression processor, resource constraints and module library. Further, algorithmic description of computational intensive (DCT transformation and quantization processes) portion of the JPEG compression processor is converted into a DFG representation. The entire DFG form is referred to as macro-IP which generates computed compressed pixel values of an image. The macro-IP uses micro-IP underneath to perform a part of DCT transformation (Sengupta *et al.*, 2018). The DFG representation of JPEG compression processor (as a macro-IP) is shown in Figure 3.12. The micro-IP used under the macro-IP has been highlighted by zooming in the figure. Further as shown in the figure, operation 135 produces the first pixel value (*W*11) computed post DCT transformation. To generate the first pixel of compressed JPEG image (*W*11′), operation 136 in the DFG performs quantization on the DCT-transformed value (output of operation 135) using respective element *t* of the quantization matrix T_{90}. Thus, the macro-IP representing the DFG of JPEG compression process computes compressed pixel values by performing DCT transformation and quantization. Now, let us discuss the process of employing the structural-obfuscation-based first line of defence in a JPEG compression processor.

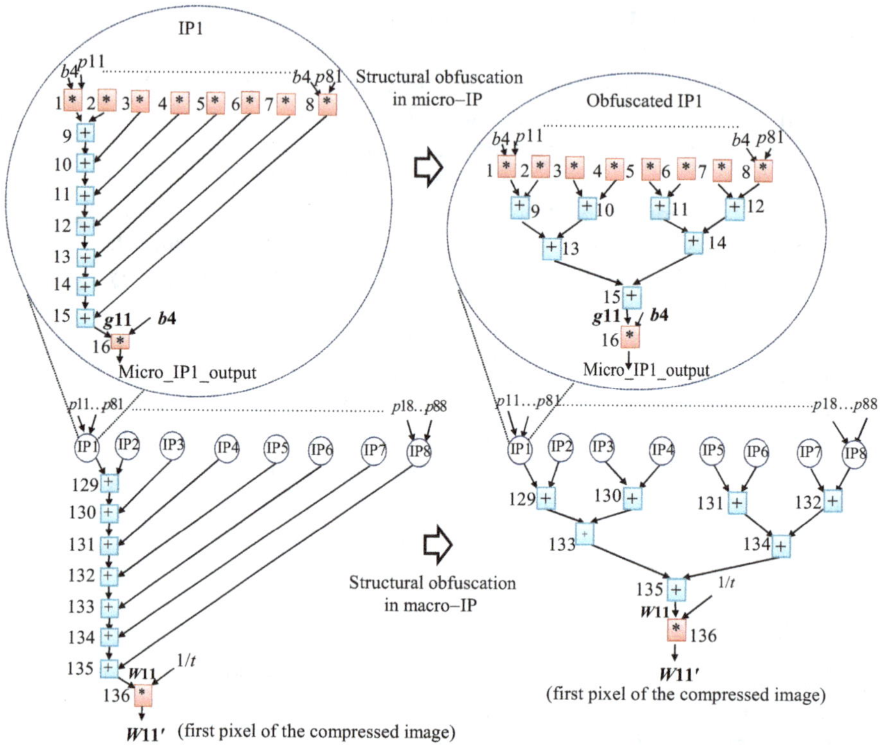

Figure 3.12 THT-based structural transformation of DFG form of JPEG compression algorithm (Sengupta and Rathor, 2020)

Sengupta and Rathor (2020) performed structural obfuscation of a JPEG compression processor by transforming its architecture at an early level of design process, i.e. behavioural or high level. The structural obfuscation during behavioural level is employed by performing rigorous high-level transformations in the DFG. Sengupta and Rathor (2020) performed THT-based structural transformation to achieve structural obfuscation. This transformation is applied on macro-IP as well as on each micro-IP used underneath. To perform THT-based transformation, sequential execution flow in the DFG is broken into concurrently executable sub-computations without affecting the functionality. Performing forced concurrent execution rather than sequential execution leads to THT of the DFG of a JPEG compression processor. The structural transformation of the DFG representing a JPEG compression processor is shown in Figure 3.12. Further, scheduling and resource allocation of the structurally transformed DFG results in a structurally obfuscated JPEG compression processor design in the form of scheduled and allocated DFG. Post-performing data path synthesis, an RTL representation of the design is obtained where structural obfuscation manifests in terms of the following modifications: changes in the interconnectivity of FU resources such as adders, multipliers and subtractors; changes in the number of interconnect binding resources such as multiplexers and demultiplexers; changes in the number of storage resources such as registers and latches. These changes render the design architecture vastly unobvious to be understood (through RE) for an adversary. Thus, the THT-based structural obfuscation, being employed as a first line of defence, impedes against RE (thus Trojan insertion) and piracy threats.

3.5.2 Designing a secure JPEG codec processor using double line of defence

Once the JPEG compression processor is secured using structural-obfuscation-based first line of defence, it is subjected to crypto-steganography-based second line of defence to further enhance the security level. To embed crypto-based steganography in the structurally obfuscated JPEG compression processor, the following two primary inputs are required: (i) structurally obfuscated JPEG compression processor in the form of scheduled and allocated DFG and (ii) stego-keys. To obtain scheduled and allocated DFG, all 136 operations of the structurally transformed DFG of JPEG compression processor (shown in Figure 3.12) have been scheduled in 30 control steps and allocated to FUs on the basis of resource constraints of say: three multipliers and three adders. Resource allocation has been performed using two-vendor allocation scheme, where two instances from vendor V1 and one instance from vendor V2 of the same FU resource type have been chosen for the allocation. Further, there are total 209 storage variables (S0–S208) in the design, which have been assigned to 73 registers (R1–R73). Now, let us discuss the process of performing crypto-steganography on a structurally obfuscated JPEG compression processor using approach in Sengupta and Rathor (2020).

In the process of performing crypto-steganography, first a CIG is created from the structurally obfuscated scheduled and allocated DFG. Thus, obtained CIG contains 209 nodes (storage variables) and 73 colours (registers). This CIG is used to extract secret design data which is fed as inputs along with the stego-keys to the stego-encoder system of crypto-steganography process. In the stego-encoder system, following steps are performed to generate stego-constraints using secret design data and stego-keys: state matrix formation, bit manipulation, row diffusion, Trifid-cipher-based encryption, alphabet substitution, matrix transposition, mix mix-column diffusion, byte concatenation, bitstream truncation and bit-mapping. The value and size of stego-key1 to stego-key5 used in the constraints generation process are highlighted in Table 3.6. The total size of the stego-key is the sum of the size of individual sub-keys (i.e. 3+76+564+18+114=775 bits). The stego-key1 to stego-key5 are used to drive the following steps, respectively: state matrix formation, row diffusion, Trifid-cipher-based encryption, alphabet substitution and byte concatenation. Post byte-concatenation step of the stego-constraints generation process, the generated byte stream is converted to a bitstream. Further, the bitstream truncation is performed for the designer's chosen size of stego-constraints=400. This truncated bitstream contains 197 times '0' bits and 203 times '1'

Table 3.6 Values and size of all five stego-keys used in crypto-steganography of JPEG compression processor

Stego-keys		Key value	Key size
Stego-key1		'001'	3 bits
Stego-key2		'11 10 00 01 00 10 10 10 11 10 00 00 10 01 11 11 11 11 10 00 00 10 10 11 01 11 11 01 11 01 00 11 11 11 00 11 01 11'	2 * 38 = 76 bits
Stego-key3	Alphabet 'A'	V#QAWSEDRFTGYHUJIKOLPZMXNCB	6 * ($\log_2(27!)$) bits = 564 bits
	Alphabet 'B'	QAWSEDRFTGYHUJIK#OLPZMXNCBV	
	Alphabet 'C'	OLPZMXNCBV#QAWSEDRFTGYHUJIK	
	Alphabet 'D'	GYHUJIK#OLPZMXNCBVQAWSEDRFT	
	Alphabet 'E'	FTGYHUJIKOLPZMXNCBV#QAWSEDR	
	Alphabet 'F'	LPZMXNCBVQAWSEDRFTGYHUJIK#O	
Stego-key4		'010 001 100 101 011 001'	6 * 3 = 18 bits
Stego-key5		'000 001 010 011 100 101 001 011 010 100 100 000 100 100 011 010 001 000 100 101 011 010 001 000 101 011 001 000 100 101 011 010 001 011 101 011 011 100'	38 * 3 = 114 bits

bits. Further by using mapping rules, all '0' bits are converted into constraint edges to be added into the CIG and all '1' bits are converted into resource reallocation constraints to be applied on scheduled and allocated DFG. A portion of register allocation of storage variables (*S*0–*S*208) pre and post embedding constraint edges (stego-constraints represented by bit '0') is shown in Tables 3.7 and 3.8, respectively. As shown in Table 3.8, storage variables *S*196, *S*202 and *S*208 have been **reallocated to register R2 from R1** and storage variable *S*138 has been **reallocated to R4 from R3** to accommodate all constraint edges. The storage variables which were subjected to register reallocation have been marked shaded in Table 3.8. Further to perform embedding of bit '1', odd operations are allocated to odd vendor type, whereas even operations are allocated to even vendor type according to the mapping rule of bit '1'. Post embedding stego-constraints represented by bit '1', the allocations of operations to the multiplier and adder resources have been shown in Tables 3.9 and 3.10, respectively. Since there are maximum 136 operations available in the JPEG compression processor design, maximum 136 number of '1' bits (out of 203) can possibly be embedded. However, embedding of only 111 number of '1' bits is effectively possible. This is because sometimes even/odd vendor types are not available to be allocated to even/odd operation number (due to the lack of resources) on the basis of chosen resource constraints. For example, operation 6 being an even operation should be allocated to even vendor type V2. However, it is eventually allocated to the odd vendor V1 as shown in Table 3.9. This is because in the control step Q2, there are two even multiplication operations (4 and 6) scheduled and only one instance of even vendor type V2 is available due to resource constraints (i.e. two instances of multiplier/adder from odd vendor type V1 and one instance of multiplier/adder from even vendor type V2). Since operation 4 has been allocated to the only available instance of multiplier from vendor type V2 (i.e. M_1^2), operation 6 is allocated to the remaining other instance of vendor type V1 (i.e. M_2^1), as shown in Table 3.9. Therefore, bit '1' is not effectively embedded for some operations; thus, the number of effectively embedded '1' bits is lesser than the total possible embedding of bits '1' (Sengupta and Rathor, 2020).

Hence, crypto-based steganography is embedded into the structurally obfuscated JPEG compression processor to secure it via a double line of defence. Similarly, a JPEG decompression processor can be secured using a double line of defence.

Compression and decompression of CT scan images using a secure JPEG codec processor: Authors have employed the stego-embedded obfuscated JPEG compression processor to obtain the computed compressed pixel value (data) of medical images of CT scan. The compression of CT scan medical images has been performed on the basis of quantization matrix T_{90} to ensure the acceptable level of image quality. Further, the medical images have been reconstructed from the compressed data using stego-embedded obfuscated JPEG decompression processor by performing de-quantization and inverse DCT transformation. Figure 3.13 shows a tested original CT scan image (available at *CT medical Images, Kegal*, www.kaggle.com/kmader/siim-medical-images/home, 2019), its quantized version and

Table 3.7 Register allocation to storage variables in JPEG compression processor before steganography

CS	**Q0**	**Q1**	**Q2**	**Q3**	**Q4**	**...**	**Q22**	**Q23**	**Q24**	**Q25**	**Q26**	**Q27**	**Q28**	**Q29**	**Q30**
R1	*S*0	*S*73	*S*137	*S*137	*S*141	...	*S*187	*S*196	*S*196	*S*202	*S*202	*S*202	*S*202	*S*207	*S*208
R2	*S*1	*S*74	NA	NA	NA	...	NA	NA	NA	NA	NA	NA	NA	NA	NA
R3	*S*2	*S*75	*S*75	*S*138	NA	...	NA	NA	NA	NA	NA	NA	NA	NA	NA
R4	*S*3	*S*3	*S*76	NA	NA	...	NA	NA	NA	NA	NA	NA	NA	NA	NA
⋮	⋮	⋮	⋮	⋮	⋮	⋮	⋮	⋮	⋮	⋮	⋮	⋮	⋮	⋮	⋮
R36	*S*35	*S*35	*S*35	*S*35	*S*35	...	NA	NA	NA	NA	NA	NA	NA	NA	NA
R37	*S*36	*S*36	*S*36	*S*36	*S*36	...	NA	NA	NA	NA	NA	NA	NA	NA	NA
⋮	⋮	⋮	⋮	⋮	⋮	⋮	⋮	⋮	⋮	⋮	⋮	⋮	⋮	⋮	⋮
R72	*S*71	*S*71	*S*71	*S*71	*S*71	...	*S*71	*S*71	*S*71	NA	NA	NA	NA	NA	NA
R73	*S*72	*S*72	*S*72	*S*72	*S*72	...	*S*72	*S*72	*S*72	*S*72	*S*72	*S*72	*S*72	*S*72	NA

Note: NA indicates no allocation.

Table 3.8 Register allocation to storage variables in JPEG compression processor after steganography

CS	Q0	Q1	Q2	Q3	Q4	...	Q22	Q23	Q24	Q25	Q26	Q27	Q28	Q29	Q30
R1	*S*0	*S*73	*S*137	*S*137	*S*141	...	*S*187	NA	NA	NA	NA	NA	NA	*S*207	NA
R2	*S*1	*S*74	NA	NA	NA	...	NA	*S*196	*S*196	*S*202	*S*202	*S*202	*S*202	NA	*S*208
R3	*S*2	*S*75	*S*75	NA	NA	...	NA	NA	NA	NA	NA	NA	NA	NA	NA
R4	*S*3	*S*3	*S*76	*S*138	NA	...	NA	NA	NA	NA	NA	NA	NA	NA	NA
⋮	⋮	⋮	⋮	⋮	⋮	⋮	⋮	⋮	⋮	⋮	⋮	⋮	⋮	⋮	⋮
R36	*S*35	*S*35	*S*35	*S*35	*S*35	...	NA	NA	NA	NA	NA	NA	NA	NA	NA
R37	*S*36	*S*36	*S*36	*S*36	*S*36	...	NA	NA	NA	NA	NA	NA	NA	NA	NA
⋮	⋮	⋮	⋮	⋮	⋮	⋮	⋮	⋮	⋮	⋮	⋮	⋮	⋮	⋮	⋮
R72	*S*71	*S*71	*S*71	*S*71	*S*71	...	*S*71	*S*71	*S*71	*S*71	NA	NA	NA	NA	NA
R73	*S*72	*S*72	*S*72	*S*72	*S*72	...	*S*72	*S*72	*S*72	*S*72	*S*72	*S*72	*S*72	*S*72	NA

Note: NA indicates no allocation.

Table 3.9 Scheduling and allocation of multiplication operations of JPEG compression processor post performing crypto-steganography

Control steps	Operations (O) assigned to M_1^1	Operations (O) assigned to M_2^1	Operations (O) assigned to M_1^2
Q1	O1	O3	O2
Q2	O5	O6	O4
Q3	O7	O17	O8
Q4	O19	O20	O18
Q5	O21	O23	O22
Q6	O33	O34	O24
Q7	O35	O37	O36
Q8	O39	O40	O38
Q9	O49	O51	O50
Q10	O53	O54	O52
Q11	O55	O65	O56
Q12	O67	O68	O66
Q13	O69	O71	O70
Q14	O81	O82	O72
Q15	O83	O85	O84
Q16	O87	O88	O86
Q17	O97	O99	O98
Q18	O101	O102	O100
Q19	O103	O113	O104
Q20	O115	O116	O114
Q21	O117	O119	O118
Q22	O32	O120	O16
Q23	O64	O80	O48
Q24	O112	NA	O96
Q25	NA	NA	NA
Q26	NA	NA	O128
Q27	NA	NA	NA
Q28	NA	NA	NA
Q29	NA	NA	NA
Q30	NA	O136	NA

Note: NA indicates no assignment.

eventually reconstructed/decompressed image. The secure JPEG codec processor ensures that the computed genuine information of the medical data (pixel value) does not get altered or corrupted due to the compression and decompression process. For various CT scan test-images compressed using quantization level of 90, the variations in peak signal-to-noise ratio (PSNR) and mean square error (MSE) (Sengupta and Mohanty, 2019b) have been shown in Figures 3.14 and 3.15, respectively (Sengupta and Rathor, 2020). PSNR is used as a quality measurement between the original and a compressed image. The higher is the PSNR, the better the quality of the compressed or reconstructed image. The MSE represents the cumulative squared error between the compressed and the original images, whereas

Table 3.10 *Scheduling and allocation of addition operations of JPEG compression processor post performing crypto-steganography*

Control steps	**Operations (O) assigned to A_1^1**	**Operations (O) assigned to A_2^1**	**Operations (O) assigned to A_1^2**
Q1	NA	NA	NA
Q2	O9	NA	NA
Q3	O11	NA	O10
Q4	O13	NA	O12
Q5	O25	O26	O14
Q6	O15	O27	O29
Q7	O41	NA	O28
Q8	O42	NA	O30
Q9	O43	O45	O44
Q10	O31	O57	O46
Q11	O59	O47	O58
Q12	O61	NA	O60
Q13	O73	O74	O62
Q14	O63	O75	O77
Q15	O89	NA	O76
Q16	O90	NA	O78
Q17	O91	O93	O92
Q18	O79	O105	O94
Q19	O95	O107	O106
Q20	NA	O109	O108
Q21	O121	O122	O110
Q22	O111	O123	O125
Q23	O129	NA	O124
Q24	O130	NA	O126
Q25	O127	O131	O133
Q26	NA	NA	NA
Q27	NA	NA	O132
Q28	NA	NA	O134
Q29	O135	NA	NA
Q30	NA	NA	NA

Note: NA indicates no assignment.

PSNR represents a measure of the peak error. The lower the value of MSE, the lower the error. As evident, the results obtained for both these metrics in Sengupta and Rathor (2020) were of acceptable value.

3.6 Analysis on case studies

This section analyses the security achieved by a double line of defence mechanism (structural obfuscation as a first line of defence and crypto-steganography as a second line of defence) for JPEG compression hardware accelerators and the

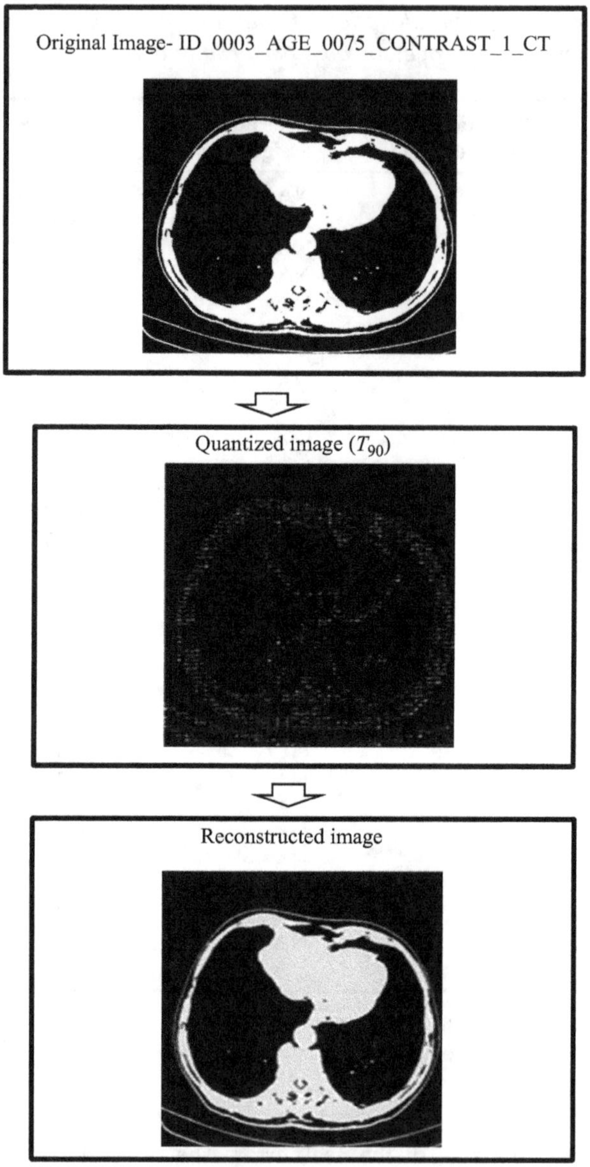

Figure 3.13 Compression and decompression of a CT scan image using a secure JPEG codec processor (Sengupta and Rathor, 2020)

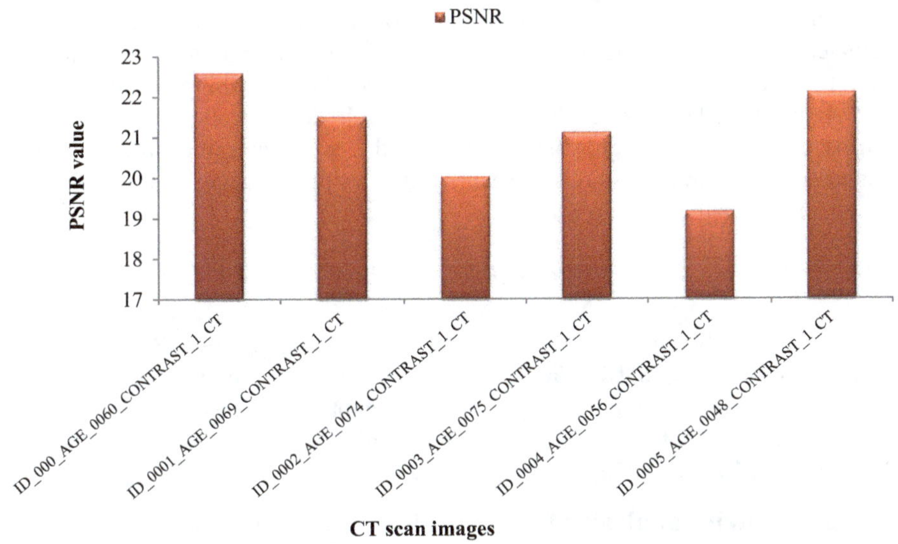

Figure 3.14 PSNR values of compressed CT scan images (Sengupta and Rathor, 2020)

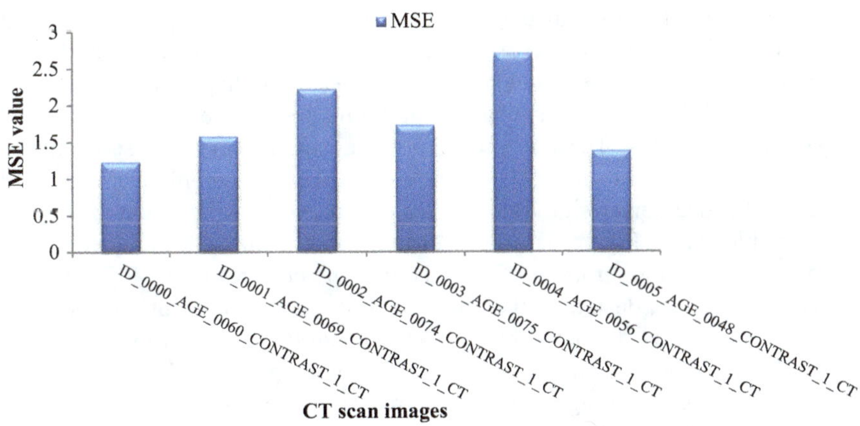

Figure 3.15 MSE values of compressed CT scan images (Sengupta and Rathor, 2020)

impact of employing security over the design cost or overhead. The security achieved and its impact over design cost have been analysed for various design solutions (or resource constraints) chosen for designing a JPEG compression processor. This gives an insight as to how chosen design solutions affect the security and design overhead. The security achieved through structural obfuscation has been evaluated using the strength of obfuscation metric. Further, security achieved through crypto-steganography has been evaluated using probability of coincidence metric and stego-key size. Additionally, for each design solution, variation in security metric and design cost are observed for varying the size of stego-constraints. Detailed discussions on security and design cost analysis are as follows (Sengupta and Rathor, 2020).

3.6.1 Analysis in terms of security (Sengupta and Rathor, 2020)

Since in the double line of defence mechanism (Sengupta and Rathor, 2020), security has been employed using structural-obfuscation-based preventive control and hardware-steganography-based detective control. Therefore, the security achieved through both the lines of defence has been discussed one by one (Sengupta and Rathor, 2020).

3.6.1.1 Security analysis of structural-obfuscation-based preventive control (the first line of defence)

The structural obfuscation ensures the preventive control over the hardware threats such as back door Trojan insertion, illegal counterfeiting and cloning by making the process of RE arduous for a potential attacker. The THT-based structural obfuscation employed by Sengupta and Rathor (2020) as the first line of defence causes the following modifications in the RTL structure of the JPEG processor design: changes in the interconnectivity of FU resources such as adders, multipliers and subtractors; changes in the number of interconnect binding resources such as multiplexers and demultiplexers; changes in the number of storage resources such as registers and latches. These changes confound the attacker in discovering the original structure by performing RE of the design. The RTL components of JPEG compression processor design pre- and post-structural obfuscation have been shown graphically in Figure 3.16. The changes in the size and number of RTL components are obvious from the figure. Higher the changes in the RTL structure of the JPEG processor design, farther the attacker is to deduce the original structure through RE and hence the probability of malfunction the design by the attacker becomes lesser. The structurally obfuscated architecture (using THT) of a JPEG compression processor design can be referred from Sengupta *et al.* (2018) and Sengupta and Mohanty (2019b) for more understanding.

Further, THT-based structural obfuscation also incurs massive change in the number of gates affected post obfuscation. Post performing THT-based structural obfuscation on JPEG compression processor, total 10,064 gates are affected (Sengupta and Rathor, 2020). The number of affected gates (due to the change in interconnectivity of gates and change in overall gate count) is also a measure of strength of structural obfuscation. Further, it is noteworthy that the gates are affected due to the change in the number and size of RTL components; hence, the change in number of gates does not follow any particular pattern. Therefore, an attempt to analyse any pattern in the change of gate count does not help an attacker

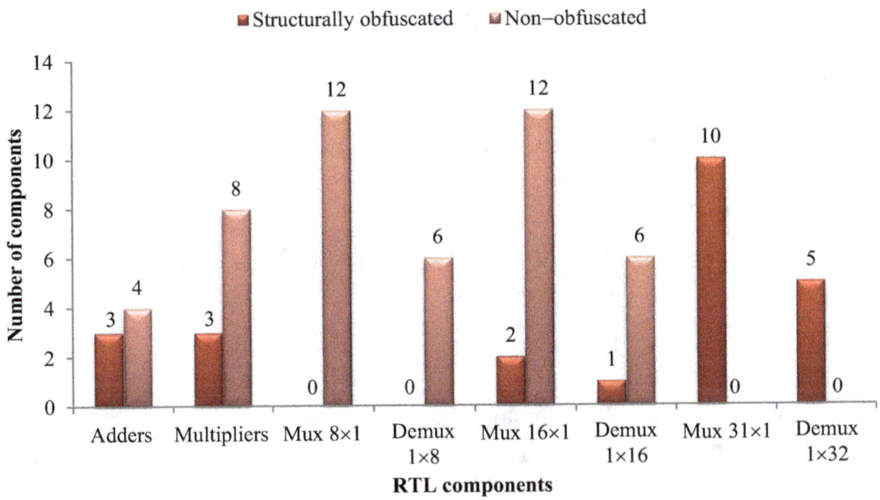

Figure 3.16 Change in components of JPEG compression processor post structural obfuscation (Sengupta and Rathor, 2020)

in deducing the original design. Thus, THT-based obfuscation is capable of providing a robust preventive control against the malicious intents of Trojan insertion, counterfeiting and cloning.

3.6.1.2 Security analysis of crypto-steganography-based detective control (the second line of defence)

The crypto-based steganography hides digital evidence into the design by implanting stego-constraints. Since the stego-constraints are implanted in the early design phase, they are distributed throughout the design post synthesis. This leads to the distribution of digital evidence into the design without giving any inkling to an attacker. Therefore, the attacker's effort to clone the JPEG compression processor design becomes unsuccessful as she/he inadvertently copies the owner's stego-information embedded into the original design. Hence, the cloned versions (with different brands) of the original designs inadvertently contain the owner's stego-information. Since the amount and location of embedded digital evidence are only known to the designer/owner, cloning can be detected by finding the owner's stego-information or digital evidence in the cloned designs. Moreover, the attacker's effort to counterfeit the design also fails. This is because the attacker is unaware of embedded stego-information into the design and hence while producing counterfeited versions, the attacker cannot embed genuine owner's stego-information during the imitation of the original design. Therefore, owner's stego-information is absent in the counterfeited designs. Hence, the absence of the owner's stego-information or digital evidence in his/her own brand of designs indicates that the designs are counterfeited. Thus, counterfeiting and cloning can be detected during forensic detection by analysing the covertly embedded stego-information (digital evidence of authenticity). This ensures that only genuine and authentic JPEG compression processors are integrated into medical imaging systems, hence avoiding wrong diagnosis of diseases.

The robustness of the secretly embedded digital evidence is evaluated using the probability of coincidence (*Pc*) metric which is given in the following (Sengupta and Rathor, 2020):

$$Pc = Pc(\text{post embedding`0' bits}) * Pc(\text{post embedding`1'bits})$$

$$Pc = \left(1 - \frac{1}{h}\right)^{k1} * \left(1 - \frac{1}{\pi_{j=1}^{m} N(U_j)}\right)^{k2} \tag{3.15}$$

where h indicates the number of colours/registers in the CIG of JPEG compression processor design before steganography and $k1$ indicates the number of stego-constraints embedded during the register allocation phase (i.e. number of 0s embedded). Further, $k2$ indicates the number of stego-constraints embedded during the resource allocation phase (i.e. effective number of 1s embedded), $N(U_j)$ indicates the number of resources of FU type U_j and m indicates the total types of FU resources present in the JPEG compression processor design. Here, *Pc* indicates the probability of coincidence post embedding both '0' and '1' bits. The amount of digital evidence (stego-constraints corresponding to 0s and 1s) to be embedded can be augmented by increasing the value of $k1$ and $k2$. As evident from (3.15), the probability of coincidence reduces with the augmentation in embedded digital evidence. Therefore, the lower *Pc* indicates that the higher digital evidence is embedded into the design. For varying design solutions of JPEG compression processor, the numbers of digital evidence embedded in the form of '0' and '1' bits are shown in Figures 3.17–3.19. The numbers of '0' bits embedded for design solutions (3*A*, 3*M*), (5*A*, 5*M*) and (9*A*, 9*M*) have been shown in Figures 3.17(a), 3.18(a) and 3.19(a), respectively. And, the number of effectively embedded '1' bits for design solutions (3*A*, 3*M*), (5*A*, 5*M*) and (9*A*, 9*M*) have been shown in Figures 3.17(b), 3.18(b) and 3.19(b), respectively. Further for each design solution, the number of '0' and '1' bits embedded has been shown for the increasing size of stego-constraints from 100 to 400.

Further, Figures 3.20–3.22 show the probability of coincidence value post embedding '0' and '1' bits for varying design solutions of a JPEG compression processor. For each design solution, the *Pc* value has been reported for the increasing size of stego-constraints from 100 to 400. The *Pc* post embedding '0' bits for design solutions (3*A*, 3*M*), (5*A*, 5*M*) and (9*A*, 9*M*) has been shown in Figures 3.20(a), 3.21(a) and 3.22(a), respectively. And, the *Pc* post embedding both '0' and '1' bits for design solutions (3*A*, 3*M*), (5*A*, 5*M*) and (9*A*, 9*M*) has been shown in Figures 3.20(b), 3.21(b) and 3.22(b), respectively. In Figures 3.20–3.22, the vertical axis shows the *Pc* value in the decreasing direction. As shown in these figures, more reduction in *Pc* is achieved as the stego-constraints size increases from 100 to 400. Further, it can be observed that more reduction in *Pc* value is

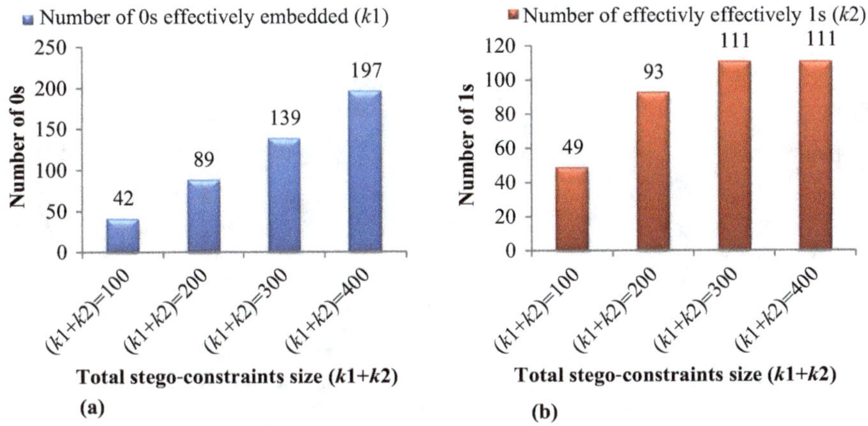

Figure 3.17 Stego-constraints for design solution 3A, 3M: (a) variation in the number of effectively embedded 0s for varying size of stego-constraints and (b) variation in the number of effectively embedded 1s for varying size of stego-constraints (Sengupta and Rathor, 2020)

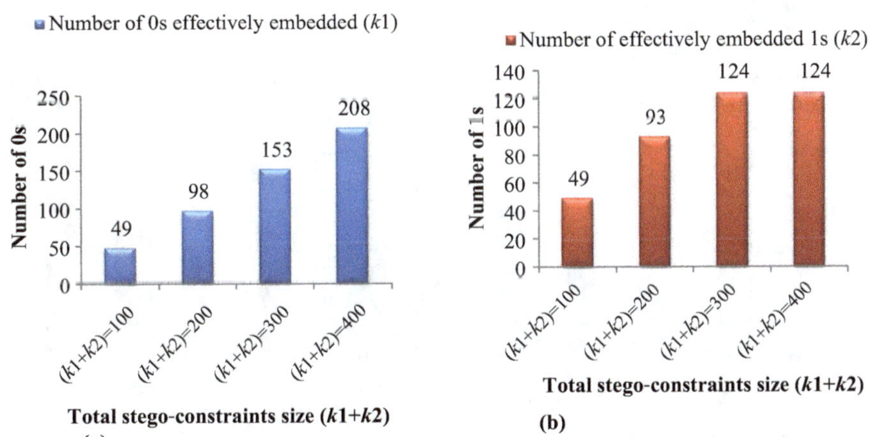

Figure 3.18 Stego-constraints for design solution 5A, 5M: (a) variation in the number of effectively embedded 0s for varying size of stego-constraints and (b) variation in the number of effectively embedded 1s for varying size of stego-constraints (Sengupta and Rathor, 2020)

incurred post embedding both '0' and '1' bits than embedding only '0' bits. Hence, it can be inferred that the larger constraints size should be chosen for achieving desirable lower probability of coincidence and higher robustness of steganography.

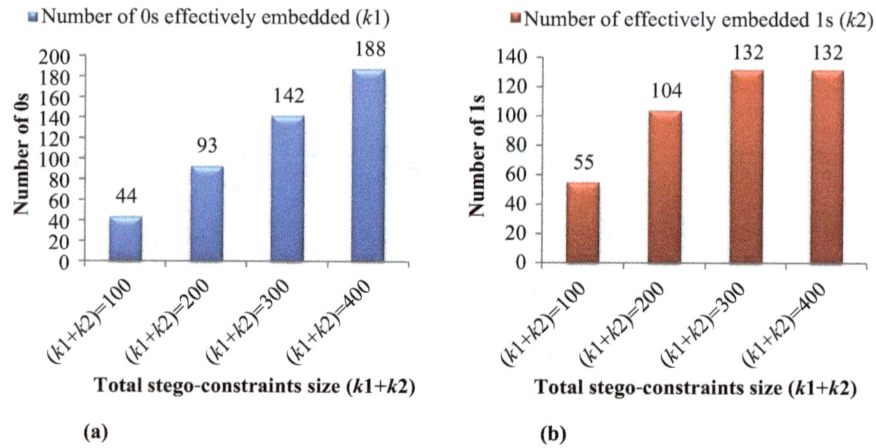

Figure 3.19 Stego-constraints for design solution 9A, 9M: (a) variation in the number of effectively embedded 0s for varying size of stego-constraints and (b) variation in the number of effectively embedded 1s for varying size of stego-constraints (Sengupta and Rathor, 2020)

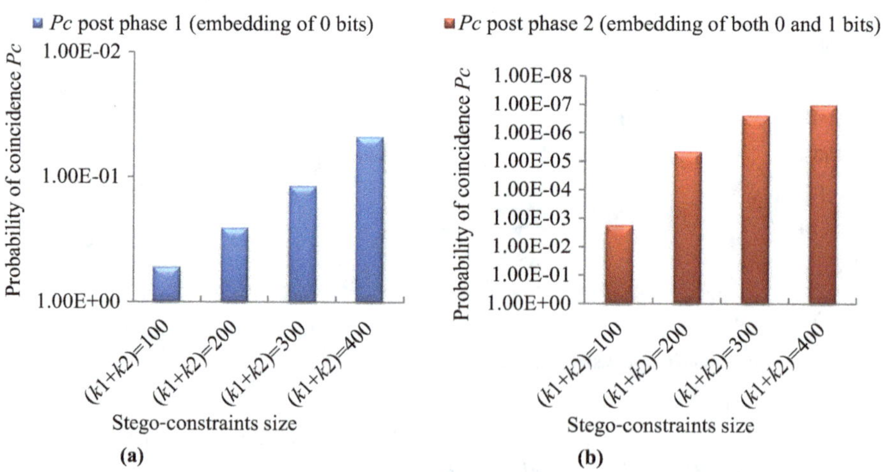

Figure 3.20 Probability of coincidence for design solution 3A, 3M: (a) variation in Pc post embedding '0' bits for varying size of stego-constraints and (b) variation in Pc post embedding '0' and '1' bits for varying size of stego-constraints (Sengupta and Rathor, 2020)

In addition, the *Pc* value is also affected by the chosen design solution (resource constraints) of JPEG compression processor design as shown in Figures 3.20–3.22. This is due to the fact that the chosen design solution determines the secret design data to be used for generating stego-constraints. Therefore, for the same size of stego-constraints, different numbers of ‘0’ and ‘1’ bits may be present in the stego-constraints for different design solutions as shown in Figures 3.17–3.19. Hence, the robustness of steganography is also dependent on the choice of the suitable design solution. It is evident from Figures 3.20–3.22 that the higher robustness of steganography (i.e. lower *Pc*) is obtained for the design solution (3*A*, 3*M*) in contrast to other design solutions. More information on the analysis of *Pc* for other design solutions such as (3*A*, 5*M*), (7*A*, 9*M*) and (11*A*, 11*M*) can be found in Sengupta and Rathor (2020).

Additionally, crypto-steganography-based detective control enhances the security level by making regeneration or extraction of stego-constraints highly convoluted by an adversary. This is because the stego-encoder system exploits various security mechanisms and a very large size stego-key to generate stego-constraints. For an attacker, it is almost infeasible to backtrack the stego-constraints generation process and find the stego-key value. Therefore, it is highly unlikely that the attacker will extract/regenerate the stego-constraints and use them in his/her counterfeited designs to evade counterfeiting detection. Thus, the crypto-based steganography ensures the high secrecy of generated stego-constraints and hence avoids the chances of misuse of owner’s stego-constraints by an adversary.

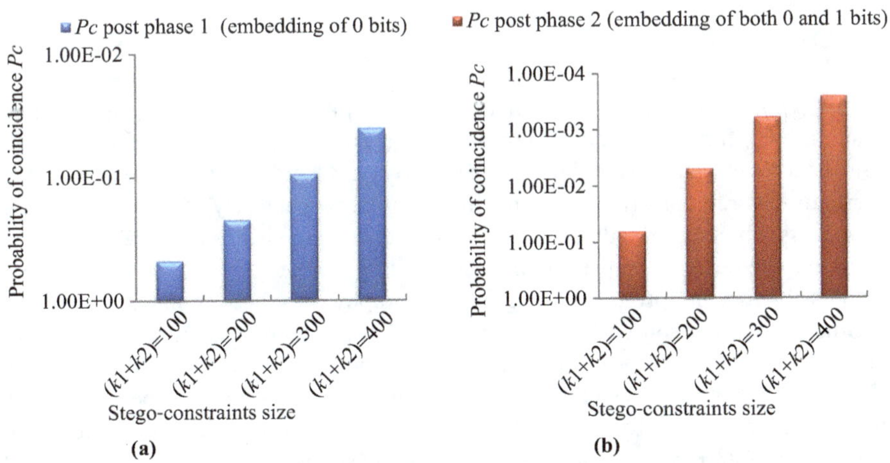

Figure 3.21 Probability of coincidence for design solution 5A, 5M: (a) variation in Pc post embedding ‘0’ bits for varying size of stego-constraints and (b) variation in Pc post embedding ‘0’ and ‘1’ bits for varying size of stego-constraints (Sengupta and Rathor, 2020)

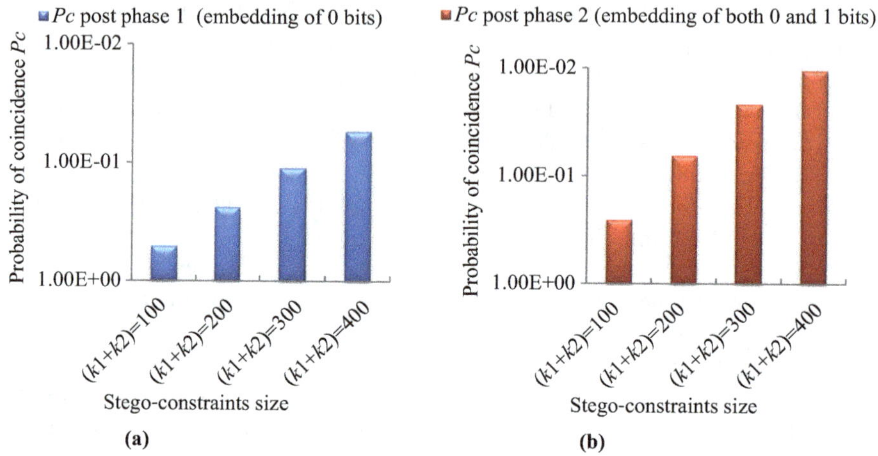

Figure 3.22 Probability of coincidence for design solution 9A, 9M: (a) variation in Pc post embedding '0' bits for varying size of stego-constraints and (b) variation in Pc post embedding '0' and '1' bits for varying size of stego-constraints (Sengupta and Rathor, 2020)

3.6.2 Analysis based on design cost/overhead (Sengupta and Rathor, 2020)

This subsection discusses the impact of employing security on the design cost. Following function is used to evaluate the design cost (Sengupta and Rathor, 2020):

$$C_d(U_i) = \rho_1 \frac{L_d}{L_m} + \rho_2 \frac{A_d}{A_m} \tag{3.16}$$

where $C_d(U_i)$ is the design cost of a JPEG compression processor for resource constraints U_i, further L_d and L_m are the design latency at specified resource constraints and maximum design latency, respectively, A_d and A_m are the design area at specified resource constraints and maximum area, respectively, and ρ_1 and ρ_2 are the weights which are fixed at 0.5. The design cost analysis has been performed by comparing the design cost pre and post embedding steganography. (Note: the JPEG compression design pre-embedding steganography has been considered as the baseline for comparison with the design cost of JPEG compression processor post-embedding steganography.) The design cost has been compared for various design solutions; and for each design solution, variation in design cost has been analysed for varying sizes of stego-constraints from 100 to 400. For a design solution (3*A*, 3*M*), Figure 3.23 compares the design cost of baseline with the design cost post phase-1 steganography (post embedding '0' bits) and Figure 3.24 compares the design cost of baseline with the design cost post phase-2 steganography (post embedding both '0' and '1' bits). Similarly, Figures 3.25 and 3.26 show the design cost comparison for design solution (5*A*, 5*M*). Figures 3.27 and 3.28 show the

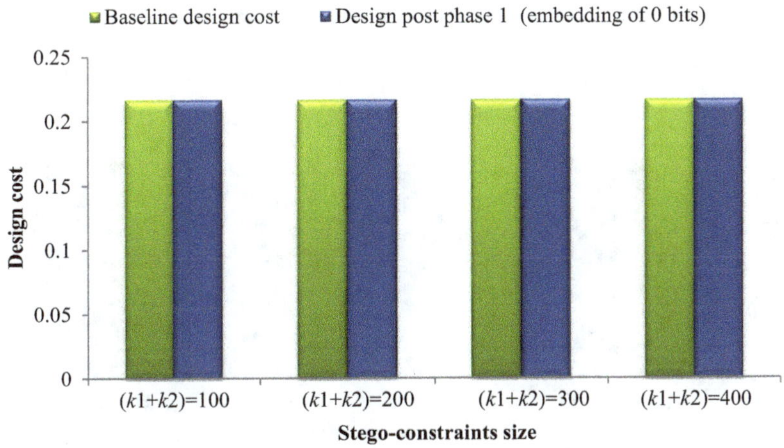

Figure 3.23 Comparison of design cost between baseline and post embedding '0' bits for varying size of total stego-constraints for 3A, 3M (Sengupta and Rathor, 2020)

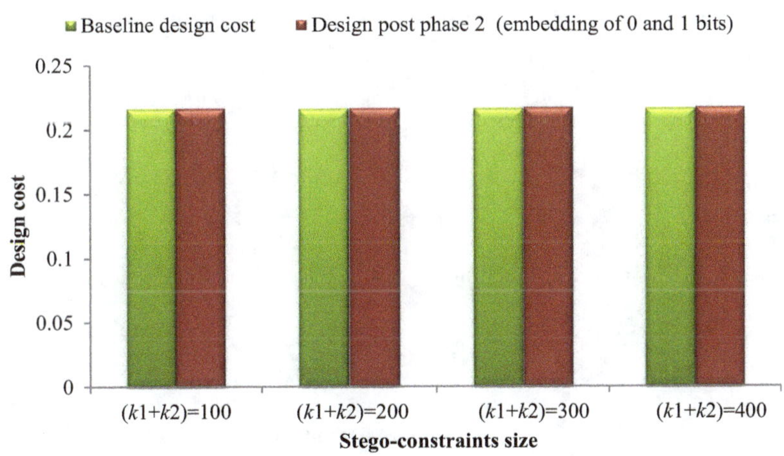

Figure 3.24 Comparison of design cost between baseline and post embedding '0' and '1' bits for varying size of total stego-constraints for 3A, 3M (Sengupta and Rathor, 2020)

design cost comparison for design solution (9*A*, 9*M*). Let us discuss the comparison of design cost of baseline and design cost post embedding only '0' bits for varying constraints size. It can be observed from Figures 3.23, 3.25 and 3.27, the design cost post embedding '0' bits remains the same as baseline cost for all design solutions (3*A*, 3*M*), (5*A*, 5*M*) and (9*A*, 9*M*). This is because embedding of stego-

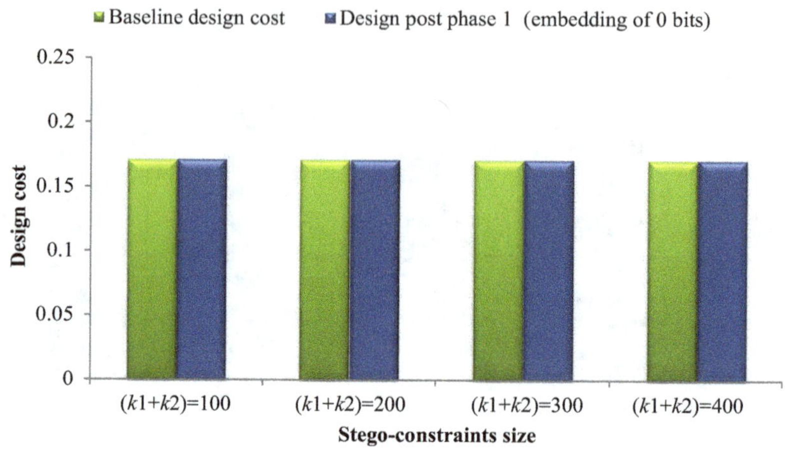

Figure 3.25 Comparison of design cost between baseline and post embedding '0' bits for varying size of total stego-constraints for 5A, 5M (Sengupta and Rathor, 2020)

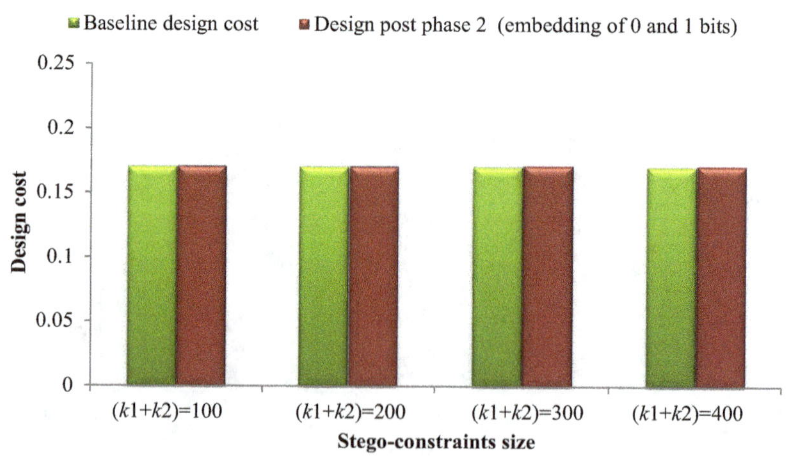

Figure 3.26 Comparison of design cost between baseline and post embedding '0' and '1' bits for varying size of total stego-constraints for 5A, 5M (Sengupta and Rathor, 2020)

constraints corresponding to '0' bits in the CIG do not result into extra requirement of colours (registers); therefore, no design overhead incurs. Now let us discuss the comparison of design cost of baseline and design cost post embedding both '0' and '1' bits for varying constraints size. It can be observed from Figures 3.24, 3.26 and 3.28, the design cost post embedding 'both '0' and '1' bits may increase negligibly

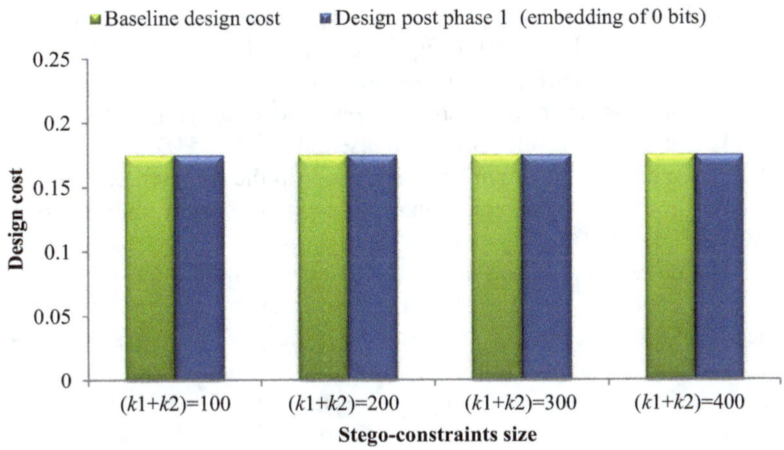

Figure 3.27 Comparison of design cost between baseline and post embedding '0' bits for varying size of total stego-constraints for 9A, 9M (Sengupta and Rathor, 2020)

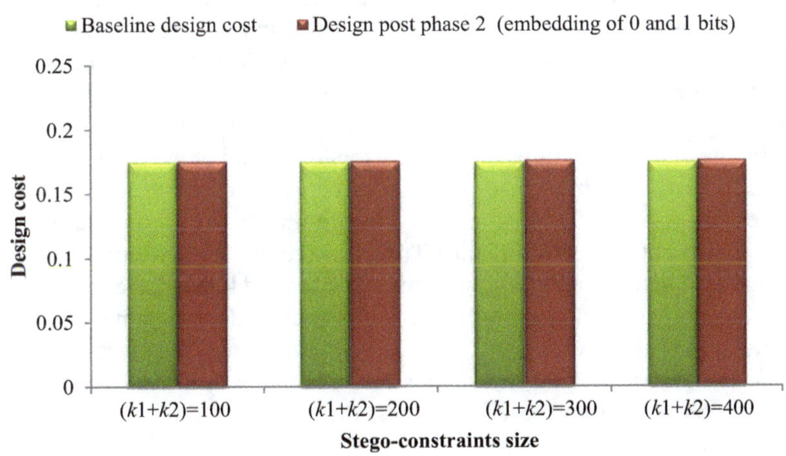

Figure 3.28 Comparison of design cost between baseline and post embedding '0' and '1' bits for varying size of total stego-constraints for 9A, 9M (Sengupta and Rathor, 2020)

with respect to the baseline cost. This slight increment may incur because of more allocation of the vendor which has a higher area and latency of resources than the other vendor, during embedding of '1' bits (or resource reallocation).

In addition, the design cost post embedding steganography is also dependent on the chosen design solution. This can be observed from Figures 3.23–3.28. On increasing the design solution from (3*A*, 3*M*) to (5*A*, 5*M*), the design cost post steganography decreases for all constraint sizes. The underlying reason behind this decrement is that as design solution is increased up to (5*A*, 5*M*), the design latency substantially decreases with only slight increment in the design area. This, in turn, leads to overall design cost reduction. However, as design solution is further increased from (5*A*, 5*M*) to (11*A*, 11*M*), the design latency does not reduce substantially. The underlying reason is that the resources (multipliers and adders) are not efficiently exploited in scheduling for the design solutions greater than (5*A*, 5*M*). Therefore, the increment in the area due to increased design solution is more dominant than the decrement in the design latency. Hence, design cost begins to increase as design solution is further increased from (5*A*, 5*M*) onwards. More information on the analysis of design cost for other design solutions such as (3*A*, 5*M*), (7*A*, 9*M*) and (11*A*, 11*M*) can be found in Sengupta and Rathor (2020).

3.7 Conclusion

The use of multimedia hardware accelerators such as JPEG codec processors in medical imaging systems is well acknowledged for medical image compression to enable low capacity storage and low bandwidth transmission for remote diagnosis. Simultaneously, Trojan-infected or fake processor designs are required to be avoided from being integrated into medical imaging modalities such as CT scanner. This demands security of JPEG codec processor designs against Trojan insertion, counterfeiting and cloning threats to disable the likelihood of wrong diagnosis. This chapter discussed a double line of defence mechanism to secure JPEG codec processor designs against aforementioned hardware threats. The first line of defence enables the preventive control against Trojan insertion, counterfeiting and cloning threats, using structural obfuscation. Further, the second line of defence enables the detective control against counterfeiting and cloning threats, using crypto-based hardware steganography. The integration of a secure and authentic JPEG codec processor in medical imaging modalities rebuilds the trust in diagnosis decisions. Further, as observed from the case studies, the employed double line of defence mechanism is capable of offering enhanced security at marginal design overhead.

Contents of this chapter build a readers understanding over following concepts:

1. Need of multimedia hardware accelerators such as a JPEG codec processor in medical imaging systems.
2. The need to secure a JPEG codec processor used in medical imaging system, against Trojan insertion, counterfeiting and cloning threats.
3. Various hardware threats and protection scenarios to secure JPEG codec processors.
4. Structural-obfuscation-based first line of defence.
5. Crypto-steganography-based second line of defence.

6. Stego-encoder for generating stego-constraints and stego-decoder for detecting steganography embedded into JPEG codec processor designs.
7. Details of crypto-steganography for 8-point DCT core used underneath the JPEG compression process.
8. A background on JPEG compression processor.
9. The entire process of employing a double line of defence to secure JPEG compression processor design.
10. Case studies in terms of security and design cost analysis for different design solutions of JPEG compression processors.
11. Case studies in terms of security and design cost analysis for varying sizes of stego-constraints.

3.8 Questions and exercise

1. Why do modern medical systems depend upon electronics and internet technology?
2. Give an example of why medical scan images are prohibitively large.
3. What is hybrid compression of medical images?
4. Why are secure JPEG codec processors used in medical imaging systems?
5. What are the different hardware threats that can attack an image compression chip?
6. What is the role of structural obfuscation?
7. Explain scheduling, allocation and binding in high level synthesis.
8. What are the first line of defence and second line of defence in the context of securing a JPEG codec processor?
9. Why is crypto-hardware steganography employed?
10. What are the components inside crypto-based steganography encoder? What is the role of this encoder?
11. What is the cover design data in a hardware steganography system?
12. How is secret design data extracted from CIG of a JPEG compression algorithm?
13. What are the roles of five different stego-keys in crypto-stego process?
14. How many times Trifid-cipher-based encryption is performed in crypto-based steganography process?
15. How is steganography detection performed?
16. Design the state matrix of JPEG compression process for a sample stego-key1 to stego-key5.
17. Explain the role of Rijndael's Galois (finite) field arithmetic in a double line of defence of JPEG codec hardware accelerator.
18. Describe the block diagram of a JPEG compression hardware accelerator.
19. What is the role of an 8-point DCT in JPEG-based image compression?
20. What is the role of quantization in JPEG codec and how is it applied?
21. Explain run-length encoding algorithm.

22. The total size of the stego-key used in JPEG compression is 775 bits. Please show the break-up of the stego-key bits.
23. Define MSE and PSNR.
24. Write the equations MSE and PSNR. How is a desirable value for each obtained?
25. How many registers are required to implement a JPEG compression hardware accelerator using 4+ and 4*?
26. How many registers are required to implement a JPEG compression hardware accelerator using 5+ and 5*?
27. Explain reverse engineering. How is Trojan insertion prevented by thwarting reverse engineering?
28. What is the bitstream truncation in crypto-based steganography from the security perspective?
29. Explain the security property of mix-column diffusion.
30. Explain the security property of row diffusion.
31. Explain the role of S-box substitution in forward AES from the security perspective.
32. Why is each digit represented in hexadecimal notation in crypto-based steganography process?

References

R. Agarwal, C. S. Salimath and K. Alam (2019), 'Multiple image compression in medical imaging techniques using wavelets for speedy transmission and optimal storage' *Biomed. Pharmacol. J.*, vol. 12(1).

Y.-Y. Chen and S.-C. Ti (2004), 'Embedded medical image compression using DCT based subband decomposition and modified SPIHT data organization,' *Proc. 4th IEEE Symposium on Bioinformatics and Bioengineering*, Taichung, Taiwan, pp. 167–174.

Y.-Y. Chen (2007), 'Medical image compression using DCT-based subband decomposition and modified SPIHT data organization,' *Int. J. Med. Inf.*, vol. 76(10), pp. 717–725.

S. B. Gokturk (2001), "Region of Interest Based Medical Image Compression," Stanford AI Lab, http://ai.stanford.edu/~gokturkb/Compression/FinalReport.htm.

S. B. Gokturk, C. Tomasi, B. Girod and C. Beaulieu (2001), 'Medical image compression based on region of interest, with application to colon CT images,' *Proc. of the 23rd Annual International Conference of the IEEE Engineering in Medicine and Biology Society*, Istanbul, Turkey, 3, pp. 2453–2456.

D. A. Koff and H. Shulman (2006), 'An overview of digital compression of medical images: can we use lossy image compression in radiology?,' *Can. Assoc. Radiol. J.*, vol. 57(4), pp. 211–217.

F. Koushanfar, I. Hong, and M. Potkonjak (2005), 'Behavioral synthesis techniques for intellectual property protection,' *ACM Trans. Des. Autom. Electron. Syst.*, vol. 10(3), pp. 523–545.

Y. Lao and K. K. Parhi (2015), 'Obfuscating DSP circuits via high-level transformations,' *IEEE Trans. Very Large Scale Integr. (VLSI) Syst.*, vol. 23(5), pp. 819–830.

B. Le Gal and L. Bossuet (2012), 'Automatic low-cost IP watermarking technique based on output mark insertions,' *Des. Autom. Embedded Syst.*, vol. 16(2), pp. 71-92.

H. R. Mahdiany, A. Hormati and S. M. Fakhraie (2001), 'A hardware accelerator for DSP system design,' *Proc. ICM*, pp. 141–144.

C. Pilato, S. Garg, K. Wu, R. Karri and F. Regazzoni (2018), 'Securing hardware accelerators: a new challenge for high-level synthesis,' *IEEE Embedded Syst. Lett.*, vol. 10(3), pp. 77–80.

A. Sengupta, S. Bhadauria and S. P. Mohanty (2017b), 'Low-cost security aware HLS methodology,' *IET Comput. Digital Tech.*, vol. 11(2), pp. 68–79.

A. Sengupta, S. Bhadauria, and S. P. Mohanty (2017c), 'TL-HLS: methodology for low cost hardware Trojan security aware scheduling with optimal loop unrolling factor during high level synthesis,' *IEEE Trans. CAD Integr. Circuits Syst.*, vol. 36(4), pp. 655–668.

A. Sengupta and S. Bhadauria (2016), 'Exploring low cost optimal watermark for reusable IP cores during high level synthesis,' *IEEE Access*, vol. 4, pp. 2198–2215.

A. Sengupta, E. R. Kumar and N. P. Chandra (2019), 'Embedding digital signature using encrypted-hashing for protection of DSP cores in CE,' *IEEE Trans. Consum. Electron.* vol. (3), pp. 398–407.

A. Sengupta and S. P. Mohanty (2019a), 'IP core and integrated circuit protection using robust watermarking', *IP Core Protection and Hardware-Assisted Security for Consumer Electronics*, e-ISBN: 9781785618000, pp. 123–170.

A. Sengupta and S. P. Mohanty (2019b), 'IP core protection and hardware-assisted security for consumer electronics', The Institute of Engineering and Technology (IET), Book ISBN: 978-1-78561-799-7, e-ISBN: 978-1-78561-800-0.

A. Sengupta and M. Rathor (2019a), 'IP core steganography for protecting DSP kernels used in CE systems,' *IEEE Trans. Consum. Electron.* vol. 65(4), pp. 506–515.

A. Sengupta and M. Rathor (2019b), 'Crypto-based dual-phase hardware steganography for securing IP cores,' *Lett. IEEE Comput. Soc.*, vol. 2(4), pp. 32–35.

A. Sengupta and M. Rathor (2020), 'Structural obfuscation and crypto-steganography-based secured JPEG compression hardware for medical imaging systems,' *IEEE Access*, vol. 8, pp. 6543–6565.

A. Sengupta and M. Rathor (2019c), 'Protecting DSP kernels using robust hologram-based obfuscation,' *IEEE Trans. Consum. Electron.*, vol. 65(1), pp. 99–108.

A. Sengupta, D. Roy, S. P. Mohanty and P. Corcoran (2017a), 'DSP design protection in CE through algorithmic transformation based structural obfuscation,' *IEEE Trans. Consum. Electron.*, vol. 63(4), pp. 467–476.

A. Sengupta, D. Roy, S. P. Mohanty and P. Corcoran (2018), 'Low-cost obfuscated JPEG CODEC IP core for secure CE hardware,' *IEEE Trans. Consum. Electron.*, vol. 64(3), pp. 365–374.

A. Sengupta (2016), 'Intellectual property cores: protection designs for CE products,' *IEEE Consum. Electron. Mag.*, vol. 5(1), pp. 83–88.

A. Sengupta (2017), 'Hardware security of CE devices [hardware matters],' *IEEE Consum. Electron. Mag.*, vol. 6(1), pp. 130–133.

A. Sengupta and D. Roy (2017), 'Antipiracy-aware IP chipset design for CE devices: a robust watermarking approach [hardware matters],' *IEEE Consum. Electron. Mag.*, vol. 6(2), pp. 118–124.

A. Sengupta, R. Sedaghat and Z. Zeng (2010), 'A high level synthesis design flow with a novel approach for efficient design space exploration in case of multi-parametric optimization objective,' *Microelectron. Reliab.*, vol. 50(3), pp. 424–437.

A. Sengupta (2020), 'Frontiers in securing IP cores – forensic detective control and obfuscation techniques', The Institute of Engineering and Technology (IET), ISBN-10: 1-83953-031-6, ISBN-13: 978-1-83953-031-9.

X. Zhang and M. Tehranipoor (2011), 'Case study: detecting hardware Trojans in third-party digital IP cores,' *IEEE International Symposium on Hardware-Oriented Security and Trust*, San Diego, CA, pp. 67–70.

Chapter 4

Integrating multi-key-based structural obfuscation and low-level watermarking for double line of defence of DSP hardware accelerators

Anirban Sengupta[1]

The chapter describes a double line of defence mechanism for securing hardware accelerators using key-based structural obfuscation (SO) and physical-level watermarking. The presented approach discussed in this chapter is capable of securing against combined threat models of reverse engineering (leading to Trojan insertion) and intellectual property (IP) piracy as preventive and detective control.

The chapter is organized as follows: Section 4.1 discusses about the background of the chapter; Section 4.2 presents the salient features of the chapter; Section 4.3 shows some practical applications applicable for this approach; Section 4.4 explains some contemporary approaches of this domain; Section 4.5 explains the details of the double line of defence process; Section 4.6 highlights the low-cost optimized multi-key-based SO process; Section 4.7 presents the KSO-PW tool of the presented approach; Section 4.8 discusses the case studies on digital signal processing (DSP) applications and Section 4.9 concludes the chapter.

4.1 Introduction

In this era of consumer electronics, DSP hardware accelerators have begun to dominate because of its vital role in image processing, audio processing, video processing and so forth applications. Today, nobody wants to compromise with the rate of video streaming and quality of videos. Moreover, high-quality audio such as 8-dimensional (D) audio are fascinating the consumers today. Various kinds of image filters have set their role in applications such as robotics vision, biometric fingerprinting and medical imagery. Therefore, the proliferating demand of high-definition video, high-quality audio and various kinds of image filtering is the key reason for blooming of DSP hardware accelerators in modern consumer electronics era. Apart from consumer's applications, the utility of DSP hardware accelerators is

[1]Computer Science and Engineering, Indian Institute of Technology Indore, Indore, India

well pronounced in several critical applications such as military, banking and healthcare (Schneiderman, 2010; Sengupta, 2020).

So far, we have discussed only the application side of the DSP hardware accelerators. However, another side of DSP hardware accelerators is its design-for-security (DFS) that is also being given strong attention by industry and researchers today. The DFS perspective of a DSP hardware accelerator is vital for its usage in both critical and non-critical applications. This is because ensuring a secured design of DSP hardware accelerators builds up trust in hardware. This chapter focuses on DFS perspective of DSP hardware accelerators. Now the question arises as to why the DFS is prevailing in modern system-on-chip (SoC) design technology. The key reason is the distribution of design chain across the globe. In other words, various offshore entities (fabless design houses, SoC integrators and foundries) participate in the journey of an electronic system from its idea to physical existence (Castillo *et al.*, 2007; Plaza and Markov, 2015; Sengupta, 2016, 2017). During this journey, a DSP hardware accelerator design can be infected with malicious logic insertion by any untrustworthy design house involved in the design process (Zhang and Tehranipoor, 2011). For example, (i) a dishonest third-party IP (3PIP) vendor may covertly insert a Trojan horse at a safer place in the design and send Trojan-infected IPs to the SoC integrator, (ii) a dishonest SoC integrator may covertly insert a Trojan horse before sending the design to the fabrication unit or foundry and (iii) an adversary in fabrication unit may insert the Trojan in the mask or by altering the dopant level. Thereby, a DSP hardware accelerator design can be compromised by an attacker using Trojan horse attack. This may lead to the failure of hardware accelerators deployed in critical systems. Apart from Trojan threat, other threats such as counterfeiting and cloning are also becoming a challenge for trustworthy hardware designs (Sengupta *et al.*, 2019; Sengupta and Rathor, 2019a). This is because the economic temptation and intents of sabotaging the genuine vendor's reputation and revenue may push an adversary (untrusted entity in the design chain) towards counterfeiting and cloning of hardware designs (Sengupta and Roy, 2017; Sengupta and Mohanty, 2019).

The earlier discussion is the underlying reason of the ramification of the very large scale integration (VLSI) design process in the form of DFS. The DFS can be performed at various phases of design process viz. high (behavioural) level, register transfer level (RTL), gate level and physical/layout level. This chapter highlights on how a high level and the low level (physical level) can simultaneously be exploited to employ security against Trojan insertion, counterfeiting and cloning threats (Sengupta and Rathor, 2020). The DFS at both high and low levels strengthens the trust in hardware designs. Sengupta and Rathor (2020) performed the high-level DFS by employing multi-key-based SO during high-level synthesis (HLS). Further, low-level DFS has been performed by embedding watermarking at physical level. Thereby, Sengupta and Rathor (2020) integrated the SO at high level and watermarking at low level to provide a double line of defence for securing DSP hardware accelerators against Trojan insertion, counterfeiting and cloning threats. The SO-based (Sengupta *et al.*, 2017) security during the high-level design process ensures preventive control against Trojan insertion, counterfeiting and cloning threats. This is because to insert a Trojan horse or to counterfeit and clone the designs, an adversary performs reverse engineering (RE) to deduce the original structure and functionality of the design. On successful RE, an adversary becomes competent to insert malicious Trojan horse or counterfeit/clone the design. However, the SO alters the design structure in such a way that the RE becomes highly obscure for the attacker. Thus, the SO technique obstructs the RE performed by an adversary and provides security against Trojan insertion, counterfeiting and cloning threats. Further, the watermarking-based security during the low-level (physical-level) design process ensures detective control against counterfeiting and cloning threats. This is because a robust watermark embedded into the designs enables the detection of counterfeiting and cloning.

4.2 Salient features of the chapter

The chapter discusses the security of DSP hardware accelerators based on the following key-points (Sengupta and Rathor, 2020):

- Discussion on DFS technique to generate highly secured DSP hardware accelerators using a double line of defence against Trojan insertion, counterfeiting and cloning threats.
- Discussion on the first line of defence using multi-key-driven robust SO, as preventive control against aforementioned hardware threats.
- Discussion on obfuscation process using the following high-level structural transformations which are executed sequentially: (i) key-driven loop unrolling (LU), (ii) key-driven partitioning, (iii) key-driven redundant operation elimination (ROE), (iv) key-driven tree height transformation (THT) and (v) key-driven folding-knob-based transformation.
- Discussion on the second line of defence using physical-level watermarking during early floorplanning of obfuscated DSP design. The watermark depends on vendor's signature comprising multiple variables, where each variable carries a robust mapping for conversion into respective watermarking constraints. The watermark insertion is overhead free regardless of the size of the DSP design.

4.3 Some practical applications of DSP hardware accelerators for modern electronic systems

In modern electronic systems such as television, digital camera, tablets, headsets, cell phones and laptops, the DSP hardware accelerators have numerous practical applications. These applications of the DSP hardware accelerators include filtering of digital data, attenuation, compression and decompression, audio and video encoding/decoding, speech recognition and so forth (Sengupta and Rathor, 2020). To facilitate these applications, DSP algorithms such as finite impulse response (FIR) filter, infinite impulse response (IIR) filter, discrete Fourier transform, fast Fourier transform (FFT), discrete wavelet transform (DWT), autoregressive filter (ARF), discrete cosine transform (DCT) and inverse DCT (IDCT) are executed as core functions. Because of data-intensive computations involved in theses DSP algorithms, it is efficient to realize them using hardware. Thereby, the DSP hardware accelerators are designed as dedicated processors such as application-specific integrated circuits (ASICs) or reconfigurable logics in programmable devices such as field-programmable gate array to execute data-intensive applications so that high performance can be achieved. Each kind of the DSP hardware accelerator performs a specific function such as FIR filter core is used for signal attenuation and image processing, DCT core is used for transforming data from spatial to frequency domain during image compression, IDCT is used for transforming data from frequency to spatial domain during image decompression, FFT core is used for fast transformation of data from time/spatial to frequency domain for image enhancement (e.g. biometric fingerprint image enhancement), DWT core is used for de-noising, data compression and feature extraction. Thus, because of the wide utility of the DSP hardware accelerators, their demand in consumer electronic applications is proliferating.

4.4 Overview of contemporary approaches

This section discusses an overview of contemporary approaches in two parts. The first part of the discussion includes SO-based contemporary approaches, whereas the second part includes watermarking-based ones. The discussions on both the parts are as follows (Sengupta and Rathor, 2020):

The SO-based approaches have been proposed for securing both sequential and combinational kinds of circuits. For securing sequential circuits against piracy, structural-transformation-based obfuscation has been proposed by Li and Zhou (2013). The authors performed the following four operations to achieve the best possible obfuscation: (i) retiming, (ii) re-synthesis, (iii) sweep and (iv) conditional stuttering. Further, Chakraborty and Bhunia (2009, 2011) proposed obfuscation techniques to ensure the security of designs against Trojan horse insertion. However, these approaches have not been proposed for securing larger designs such as the DSP hardware accelerators. Their application is limited to small combinational and sequential circuits. Since their target hardware is not designed using HLS framework, therefore, the DFS at high level is not possible. However, there are some other approaches which perform SO for performing DFS at high level. For example, Lao and Parhi (2015) performed SO-based DFS by applying folding transformation on the iterative data-flow graph (DFG) of digital filters such as FIR. Further, Sengupta *et al.* (2017) also performed SO-based DFS technique at high level using the following transformations: LU, logic transformation, THT, ROE and loop-invariant code motion. This technique (Sengupta *et al.*, 2017) targets the security of DSP cores. Further, Sengupta *et al.* (2018) targeted the security of multimedia hardware accelerators such as joint photographic experts group compression processor for securing using THT-based SO. Further Sengupta and Rathor (2019b) performed hologram-motivated SO by concealing one DSP architecture into another. In contrast to these contemporary approaches, the SO-based DFS approach to be discussed in this chapter has the following enhancements (Sengupta and Rathor, 2020): (i) multiple techniques of structural transformations have been performed, (ii) all the applied techniques of structural transformations are driven through a designer's chosen key value, therefore, resulting into higher control over the extent to which obfuscation can be applied, (iii) an attacker requires to know both the correct keys and the employed multiple techniques of obfuscation to expose the true functionality of the design, hence resulting into higher security against RE attack, (iv) applicability on both iterative and non-iterative algorithms of DSP and (v) key-driven partitioning and key-driven folding-knob-based transformation along with key-driven LU, key-driven ROE and key-driven THT-based structural transformations.

To enable the detection of counterfeiting and cloning, some watermarking-based contemporary approaches have been proposed for DSP hardware accelerators. For example, Hong and Potkonjak (1999) proposed watermarking technique based on binary encoding of author's signature. Further, Le Gal and Bossuet (2012) proposed an in-synthesis watermarking technique which embeds author's signature as output marks. Furthermore, Sengupta and Bhadauria (2016) proposed watermarking technique based on four-variables author signature, whereas Sengupta *et al.* (2018) proposed watermarking technique based on seven-variables author signature. Roy and Sengupta (2019) proposed a watermark which is embedded at multiple levels of design abstraction such as high level and RTL. However, in contrast to theses contemporary approaches, the low-level watermarking to be discussed in this chapter has the following differences (Sengupta and Rathor, 2020): (i) the low-level watermarking proposed by Sengupta and Rathor (2020) is embedded during floorplanning at physical level, (ii) the author signature comprises three distinct variables α, β and γ, where each variable has a robust mapping into watermarking constraints, (iii) the embedding of watermark does not result into design cost overhead and (iv) the physical-level watermark to be discussed in this chapter has been embedded as a second line of defence, where the first line of defence is employed using SO. The physical-level watermark as a second line of defence offers detective control in case the first line of defence is compromised. More explicitly, if an adversary nullifies the SO by deducing the original functionality of an obfuscated design through RE then the physical-level watermark acts as a second line of defence. Therefore, if the attacker counterfeits or clones the design post compromising the SO-based security, then the physical-level watermark-based security helps in detecting counterfeiting and cloning.

4.5 Double line of defence using structural obfuscation and physical-level watermarking

Sengupta and Rathor (2020) integrated SO and physical-level watermarking together to secure DSP hardware accelerators using a double line of defence.

4.5.1 Top down perspective of the approach

An abstract view of the SO and physical-level watermarking-based double line of defence process is shown in Figure 4.1. As shown in the figure, the following inputs are required to generate a watermark-implanted obfuscated DSP hardware accelerator as output (Sengupta and Rathor, 2020):

1. algorithmic description of a DSP application in the form of C/C++ or transfer function or mathematical relationship of inputs and output;
2. resource constraints;
3. module library;
4. SO secret keys (SO-key1, SO-key2, SO-key3, SO-key4, SO-key5) and
5. vendor's multivariable signature comprising three variables α, β and γ.

As shown in Figure 4.1, SO-based first line of defence is employed during HLS process and requires algorithmic description of the DSP application, resource constraints, module library and SO keys, as inputs. After employing the first line of defence, a structurally obfuscated RTL design is generated post-HLS. Thus, obtained structurally obfuscated RTL design is fed as input along with the vendor's signature to the second line of defence algorithm. The second line of defence algorithm is

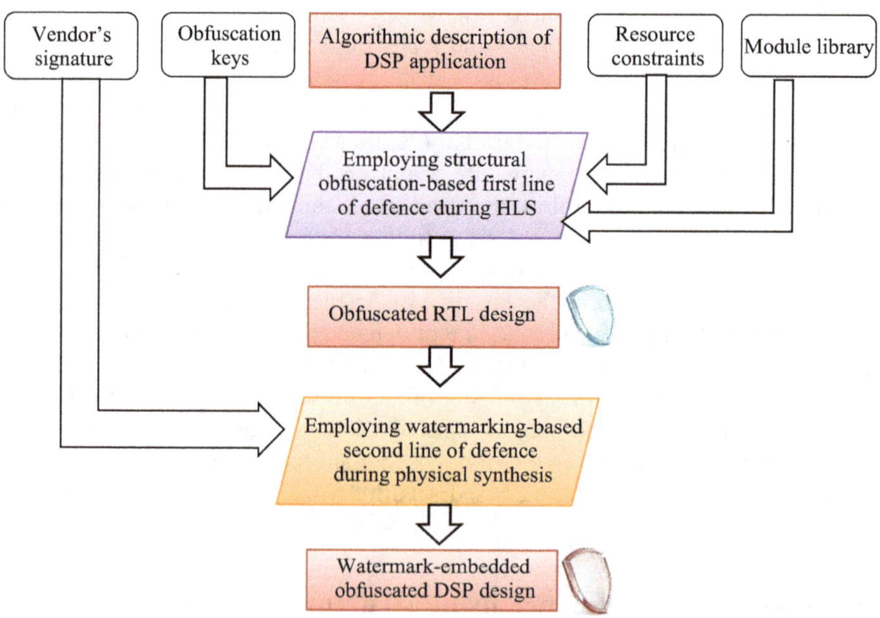

Figure 4.1 An overview of multi-key-based structural obfuscation and physical-level watermarking-based double line of defence (Sengupta and Rathor, 2020)

performed using watermarking-based physical-level synthesis. Post-watermarking, a watermark-implanted obfuscated design is generated which is secured using the double line of defence. The motivation of employing the double line of defence is to secure DSP hardware accelerators against the following threat scenarios: (i) first, the Trojan insertion threat (resulting from RE) infecting 3PIP cores, which, in turn, compromises the security of SoC design. The first line of defence using SO provides security against Trojan which can possibly be inserted in an untrustworthy regime such as foundry, (ii) second, counterfeiting/cloning threats that result into the integration of fake designs or IP cores into SoC, hence compromising the security and reliability of an electronic system. The first and second lines of defence ensure security against counterfeiting/cloning, where multiple SO-key-based SO provides preventive control, while physical-level watermarking offers detective control. The second line of defence is not directly contextual unless the first line of defence is overtaken by an adversary. However, somehow if an attacker de-obfuscates the obfuscated design and finds the correct functionality, only then doors are open for him/her for realizing malicious objectives of counterfeiting/cloning. In such a threat scenario, watermarking-based second line of defence secures the designs by enabling detective control over counterfeiting/cloning.

A more informative secure design flow of the SO and watermarking-based double line of defence technique is shown in Figure 4.2. As shown in the figure, the algorithmic description of DSP application is first represented in the form of control DFG (CDFG). Further, the CDFG is subjected to multiple secret SO-keys-based SO technique which performs the following high-level structural transformations: (i) key-driven LU, (ii) key-driven partitioning of CDFG, (iii) key-driven ROE, (iv) key-driven THT and (v) key-driven folding-knob-based transformation. Post employing these multiple key-driven techniques, the CDFG is transformed in an obfuscated form. This transformed CDFG is subjected to scheduling, allocation and binding phases of HLS to generate an obfuscated design in the form of scheduled and allocated CDFG. Further, data path and controller are synthesized to generate an obfuscated RTL circuit as shown in Figure 4.2. Thus, multiple secret SO-keys-based SO-based first line of defence is employed. Now let us discuss how the SO prevents RE which may result into Trojan insertion, counterfeiting/cloning.

When an SoC or a stand-alone IC design is sent to an offshore foundry for fabrication, the design to be fabricated can be infected with Trojan (malicious logic) or it can be counterfeited/cloned. In order to insert a Trojan, or counterfeit/clone a design, an adversary first tries to interpret the true functionality and structure of design. To do so, the adversary performs RE. Once she/he successfully interprets the true functionality through RE, she/he can easily insert Trojan at safer places inside the design. The Trojan is inserted such that they remain dormant until the payload is activated by the trigger logic. The triggering is designed to occur only at rare events so that the Trojan logic cannot be detected typically during pre and post silicon simulation/validation. Therefore, in order to evade the detection of Trojan during validation, they are inserted at safe places in the design by an adversary. The insertion of Trojan at safe places in the design is only possible when an adversary successfully interprets the original functionality/structure of the

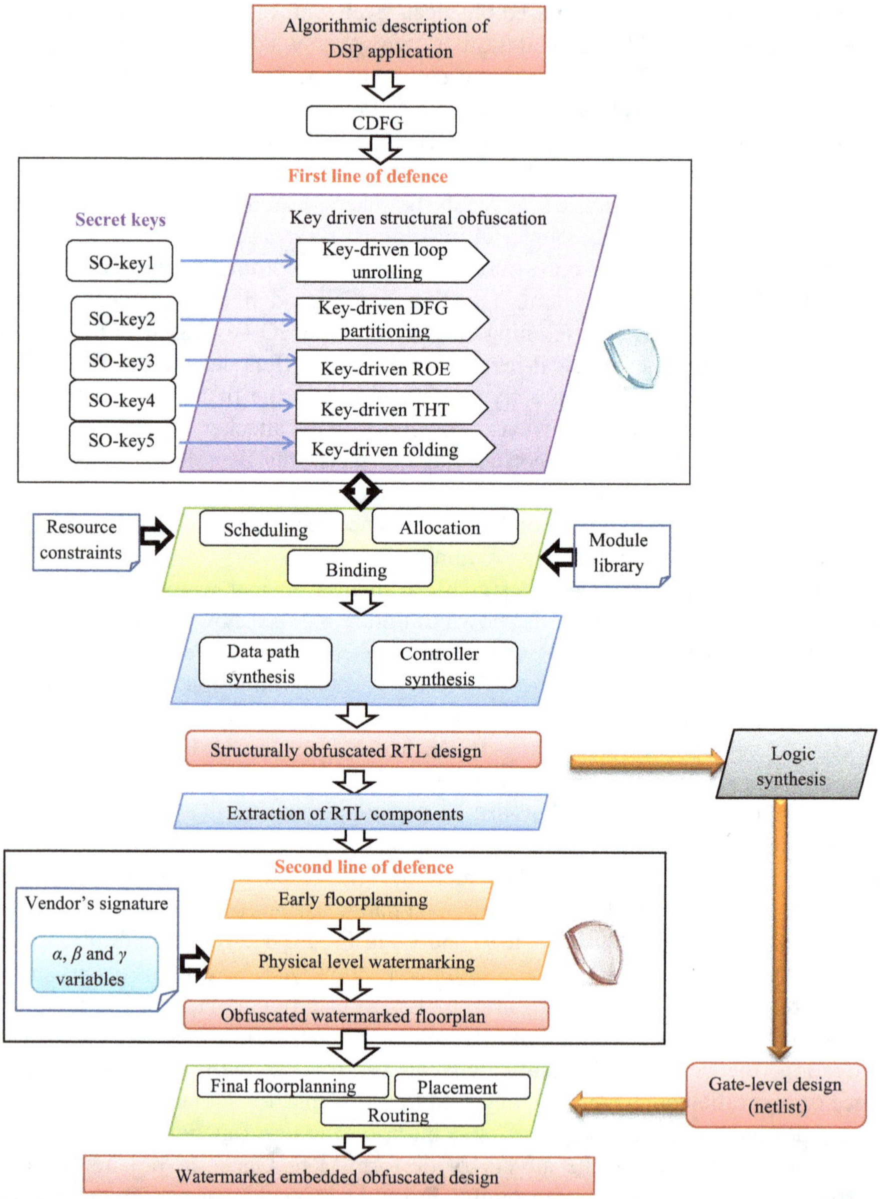

Figure 4.2 Secure design flow based on a double line of defence approach (Sengupta and Rathor, 2020)

design. Further, once the functionality of design is known to the adversary, counterfeiting or cloning can also be executed. The SO falls under the preventive control-based DFS technique against Trojan insertion and piracy. This is because

SO aims to modify the design to such an extent that RE becomes arduous for an attacker to perform. Hence, the SO technique thwarts RE and provides preventive control against Trojan insertion and piracy. The multiple SO-key-based SO incurs a very high amount of obscurity into the generated RTL/gate-level design structure (post-HLS) in terms of following modifications, without affecting functionality:

1. changes in the number of functional unit (FU) resources (such as multipliers, adders and subtractors) post obfuscation;
2. changes in the interconnect-hardware (such as multiplexers and demultiplexers) in terms of their size and total count;
3. changes in the total count of storage resources such as registers and latches and
4. changes in the interconnectivity of hardware resources.

So far, we have seen how SO-based first line of defence thwarts Trojan attack and piracy. Now, let us move ahead in the double line of defence-based secure design flow of hardware accelerators as shown in Figure 4.2. As shown in the figure, a structurally obfuscated RTL data path is generated post employing the first line of defence. Further, in order to employ the second line of defence, first a set of RTL components is extracted from the structurally obfuscated RTL data path. This set of RTL components is used to perform physical-level watermarking. The watermarking is employed by performing a physical-level design step referred to as *early floorplanning* (proposed by Sengupta and Rathor, 2020). The early floorplanning is performed using the set of extracted RTL components. In other words, an early floorplan of RTL components is prepared. Further, this early floorplan is subjected to watermarking based on vendor's signature. The vendor's signature is a combination of three unique variables (α, β and γ), where mapping rules of each variable convert the signature into respective watermarking constraints or hardware security constraints. The watermarking constraints are implanted into the early floorplan of the design, thus resulting into an obfuscated watermarked floorplan. Further, the obfuscated watermarked floorplan is subjected to final floorplanning, placement and routing phases of physical synthesis to obtain an obfuscated and watermarked layout file. The final floorplanning, placement and routing phases require gate-level netlist which is generated from logic synthesis of obfuscated RTL design. The embedded watermark in the early floorplan step (proposed by Sengupta and Rathor, 2020) of physical design flow enables detection against piracy/fake designs hence acts as detective control as a double line of defence.

4.5.2 Details of a double line of defence

As discussed earlier, a double line of defence for the DSP hardware accelerators has been deployed by integrating multiple SO-key-based SO as a first line of defence and physical-level watermarking as a second line of defence. This subsection discusses the double line of defence mechanism in details. The flow chart of the double line of defence process is shown in Figure 4.3. As shown in the flow chart, the entire flow has been divided into two portions where the first portion depicts the flow of employing multiple SO-key-based SO, as a first line of defence, while the

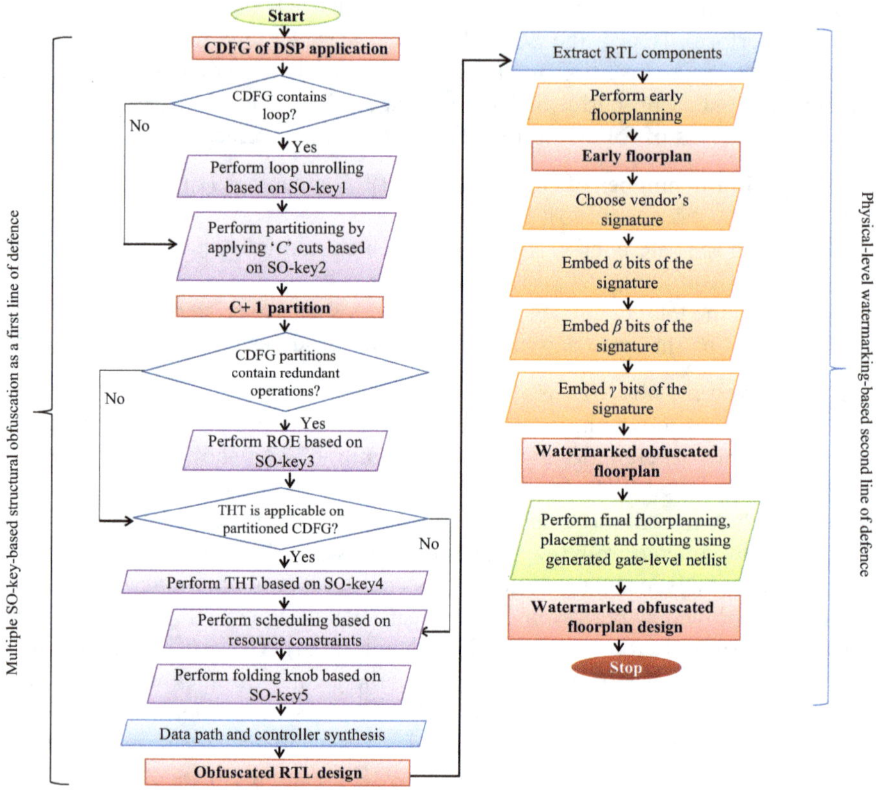

Figure 4.3 Flow chart of the structural obfuscation and watermarking-based double line of defence approach for securing DSP hardware accelerators (Sengupta and Rathor, 2020)

second portion shows the physical-level watermarking, as a second line of defence. The detailed discussion of the approach with demonstration on DSP cores is given as follows (Sengupta and Rathor, 2020):

4.5.2.1 Multiple SO-key-driven structural-transformation-based obfuscation – the first line of defence

The multiple SO-key-driven structural-transformation-based obfuscation requires the following inputs: algorithmic description of DSP application, multiple SO secret keys (SO-key1 to SO-key5) and module library and designer's specified resource constraints. As shown in the flow chart in Figure 4.3, the process starts with the conversion of algorithmic description of DSP application into its CDFG representation. Further, multiple high-level transformations are performed on CDFG in order to obtain a structurally transformed design leading to its equivalent structurally obfuscated design. Each high-level transformation technique is driven

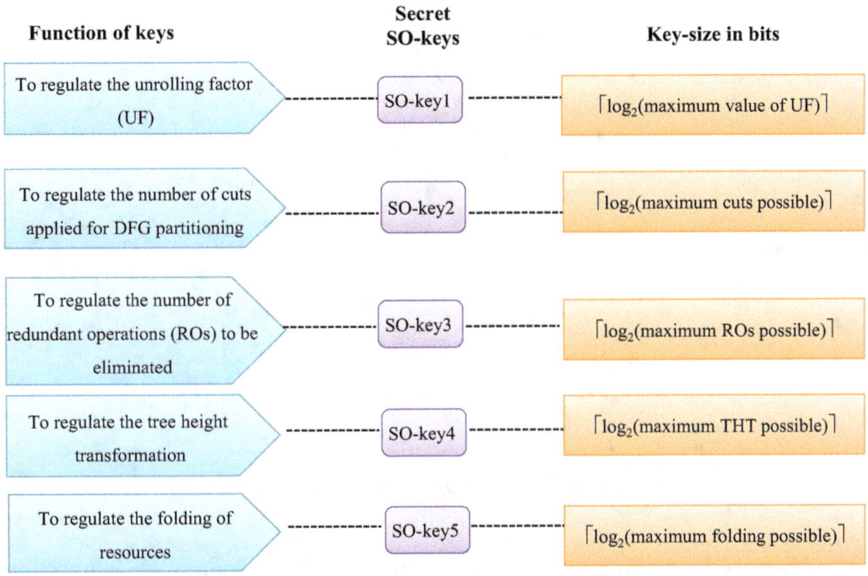

Figure 4.4 Functions and size of each SO-key used in structural-obfuscation-based first line of defence (Sengupta and Rathor, 2020)

through a designer's chosen secret SO-key which tailors the extent to which obfuscation has to be performed. Moreover, the involvement of multiple secret keys in the SO process renders the back engineering of the obfuscated design more complicated by an adversary who is assumed to be associated with an untrustworthy foundry. The function of five secret SO-keys and their sizes are shown in Figure 4.4. Now, let us see about each key-driven high-level transformation-based structural transformation/obfuscation technique one-by-one.

Key-driven loop-unrolling-based structural transformation

LU is a high-level transformation technique where loop body of an iterative DSP application is unrolled in order to incorporate parallelism in execution. The unrolling of loop body can be exploited to result into a structural-transformation-based obfuscation design. This is because upon LU, at circuit implementation level (RTL/gate level), the FU resource count, the interconnect hardware resource (such as multiplexers and demultiplexers) count, storage elements (latches and registers) resource count change drastically. This leads to huge variations in the structure without changing the functionality. Hence, this transformation impedes the deduction of true structure and functionality of the design through RE by an attacker. The extent to which the LU has to be performed is regulated by a loop unrolling factor (UF).

The LU-based structural transformation employed by Sengupta and Rathor (2020) is driven by secret SO-key1 which acts as selected UF. Therefore, the SO-key1 size depends on the maximum value of UF. The maximum value of UF is

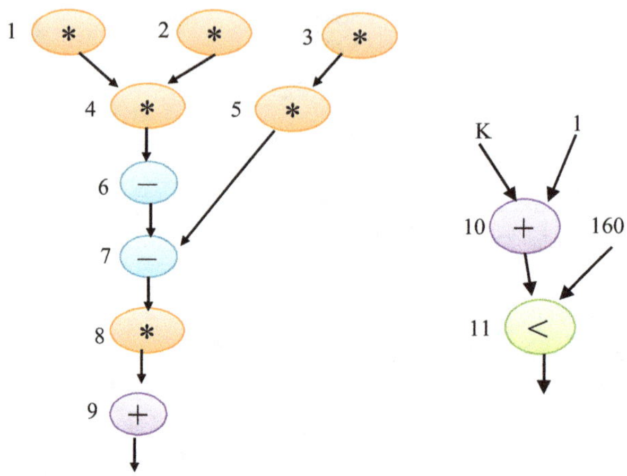

Figure 4.5 DFG of sample application

chosen same as the maximum number of iterations (K) in a loop of DSP algorithm. For example, the maximum number of iterations in a 160-tap FIR filter is K=160. The SO-key1 size in bits is shown in Figure 4.4.

A demonstration of LU-based transformation is shown using a loop-based DSP application. The DFG is shown in Figure 4.5, where the total number of iterations is assumed to be 160. The LU-based transformation is performed on the basis of designer's chosen secret value of SO-key1. Since K=160, therefore, maximum UF value is also equal to 160. Hence, the SO-key1 size is calculated to be $\lceil \log_2(160) \rceil$ 68 bits. For a designer's chosen secret value of SO-key1='00000100', the UF value to be used for LU is 4 (which is a decimal equivalent of '00000100'). For UF value equal to 4, the loop unrolled graph is shown in Figure 4.6. As shown in the figure, the loop body of the algorithm has been unrolled four times. This increases the total number of operations in the application nearly four times. This leads to more utilization of FU resources, multiplexers, demultiplexers and registers. Hence, a drastic variation in RTL structure is incurred which results into SO. This is how the first technique of structural-transformation-based obfuscation is performed.

Key-driven CDFG-partitioning-based structural transformation

CDFG partitioning is a high-level transformation technique which can be exploited in such a way that an unrolled CDFG (in the case of iterative DSP application) is partitioned in order to yield structural-transformation-based obfuscation. The impact of CDFG partitioning for SO manifests in the RTL data path in the form of modified interconnectivity of FU resources and modified size and number of multiplexers and demultiplexers. However, these modifications in the design do not impact any change in the functionality. Each partition of CDFG is further subjected to remaining structural transformation techniques in order to enhance obscurity.

The CDFG partitioning-based structural transformation employed by Sengupta and Rathor (2020) is driven through designer's chosen secret SO-key2. The value

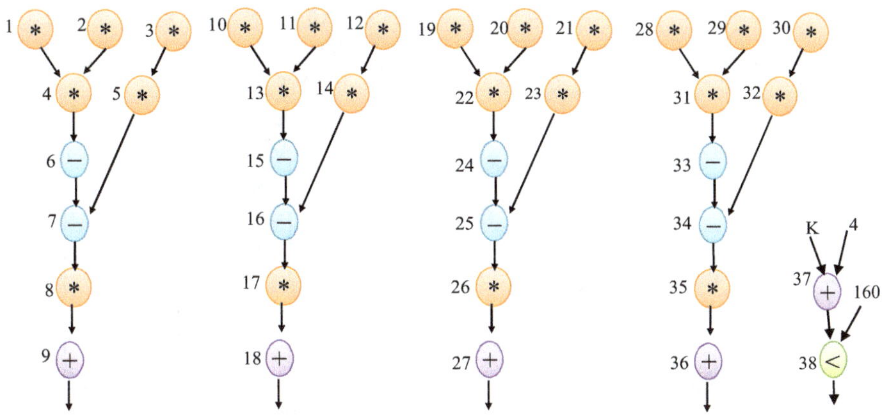

Figure 4.6 Loop-unrolled DFG after unrolling with UF=4

of the chosen key determines the number of cuts to be performed on the CDFG in order to partition the graph; upon performing C number of cuts, the target CDFG is partitioned into $C+1$ number of partitions. The way of applying cuts is again designer's selected. One of the ways that was used by Sengupta and Rathor (2020) is that the applied cuts should lead to such partitions where the smallest one should comprise minimum two nodes/operations bearing a data dependency (even if partitions are of unequal size). Thus, the size of secret SO-key2 depends on the maximum number of cuts that can be performed on CDFG. The function and size of SO-key2 are highlighted in Figure 4.4.

A demonstration of CDFG partitioning-based transformation is shown using a DSP application named an '8-point DCT'. The generic equation of an 8-point DCT is as follows:

$$X[0] = K1 * x[0] + K2 * x[1] + K3 * x[2] + K4 * x[3] + K5 * x[4] + K6 * x[5] + K7 * x[6] + K8 * x[7] \tag{4.1}$$

where $K1$–$K8$ denote DCT coefficients, $x[0]$–$x[7]$ indicate input values and $X[0]$ indicates the first output sample of an 8-point DCT. The corresponding DFG representation of an 8-point DCT core is shown in Figure 4.7. The DFG partitioning-based transformation is performed on the basis of designer's chosen secret value of SO-key2. Maximum six cuts can be applied to generate the smallest possible partitions complying with the condition that there should be minimum two operations with data dependency in each partition. Therefore, the size of SO-key2 is calculated to be $\lceil \log_2(6) \rceil = 3$ bits. For a designer's chosen value of SO-key2='010', the number of cuts to be applied is 2 (which is a decimal equivalent of '010'). The DFG of an 8-point DCT on applying C=2 cuts is shown in Figure 4.7.

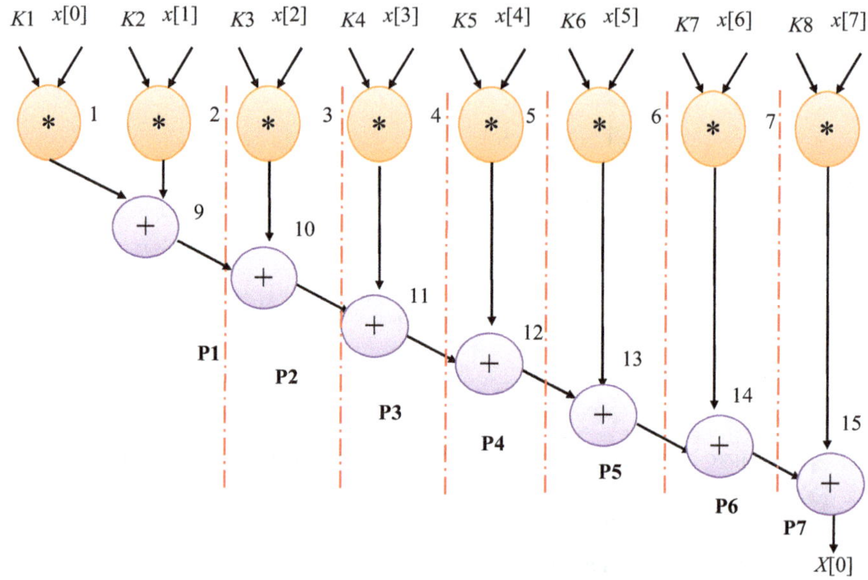

Figure 4.7 DFG of an 8-point DCT core showing that maximum six cuts can be possible with minimum two nodes in each partition. Due to six cuts, seven partitions (P1–P7) could be yielded

Post applying two cuts, resultant three partitions P1, P2 and P3 are shown in Figure 4.8. As shown in the figure, partitioning alters the interconnectivity of resources which leads to change in the size and number of interconnect hardware resources post behavioural synthesis, without affecting functionality.

Key-driven ROE-based structural transformation

ROE is a high-level transformation technique that can be exploited to achieve SO. Here, such nodes or operations which generate the same output because of having the same inputs and the same operation type are eliminated from the CDFG, in order to achieve structural transformation which, in turn, yields obfuscated design. To do so, first those operations are identified as redundant ones which have their inputs and operation type equivalent with another available node in the CDFG. Next, the identified redundant nodes are removed without affecting the design function. The ROE-based high-level transformation technique contributes in SO by incurring the following modifications in the RTL data path while preserving functionality: modification in the interconnectivity of FU resources and size of multiplexers and demultiplexers.

The ROE-based structural transformation employed by Sengupta and Rathor (2020) is driven through designer's chosen secret SO-key3. The value of the chosen key determines the number of redundant nodes that have to be removed from the partitioned CDFG. The size of SO-key3 depends on the maximum number of redundant nodes that have been identified across all the CDFG partitions. The function and size of SO-key3 are highlighted in Figure 4.4.

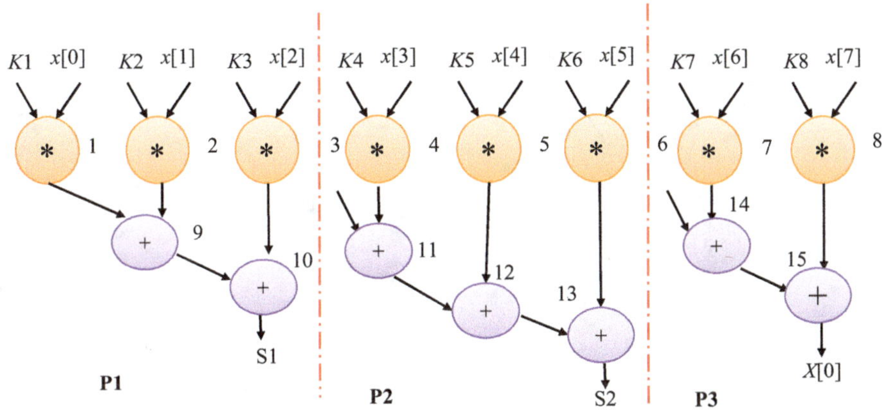

Figure 4.8 Partitioned DFG of an 8-point DCT core. Note*: Red dotted lines show the cuts applied in partitioning*

The ROE-based transformation cannot be applied on an 8-point DCT core as redundant nodes are not present in the corresponding DFG partitions. Therefore, a demonstration of ROE-based transformation is shown using autoregressive filter (ARF) application. The DFG representation of ARF is shown in Figure 4.9. The applied three cuts to generate its four partitions are also shown in Figure 4.9. The resultant partitions P1, P2, P3 and P4 are shown in Figure 4.10. The ROE-based structural transformation is performed on the partitioned CDFG on the basis of designer's chosen value of SO-key3. As shown in the partitioned DFG of ARF, there are total maximum four redundant nodes across the partitions. For a designer's chosen value of SO-key3='100', the number of operations to be eliminated is 4 (which is a decimal equivalent of '100'). As shown in Figure 4.10, operation numbers 11 and 12 are redundant in partition P2 as their inputs and operation types are the same as operation numbers 10 and 9, respectively. Similarly, operation numbers 21 and 22 are redundant in partition P3 as their inputs and operation types are the same as operation numbers 20 and 19, respectively. These redundant operations (number 11 and 12 in partition P2 and number 21 and 22 in partition P3) are eliminated. Post applying ROE-based high-level transformation, the transformed DFG of ARF is shown in Figure 4.11. As shown in the transformed DFG post ROE, inputs of operation 14 change from (S5, S6) to (S3, S4) and inputs of operation 26 change from (S9, S10) to (S7, S8). Thus, the elimination of operations and changes in the interconnectivity of operations incur alterations in the RTL structure in terms of change in FU resource interconnectivity and size of interconnect hardware resources such as multiplexers and demultiplexers.

Key-driven THT-based structural transformation

THT is a high-level transformation technique where the height of a CDFG is subjected to alteration by breaking the sequential data dependency in the graph and forcing parallel dependency of some operations without affecting the design

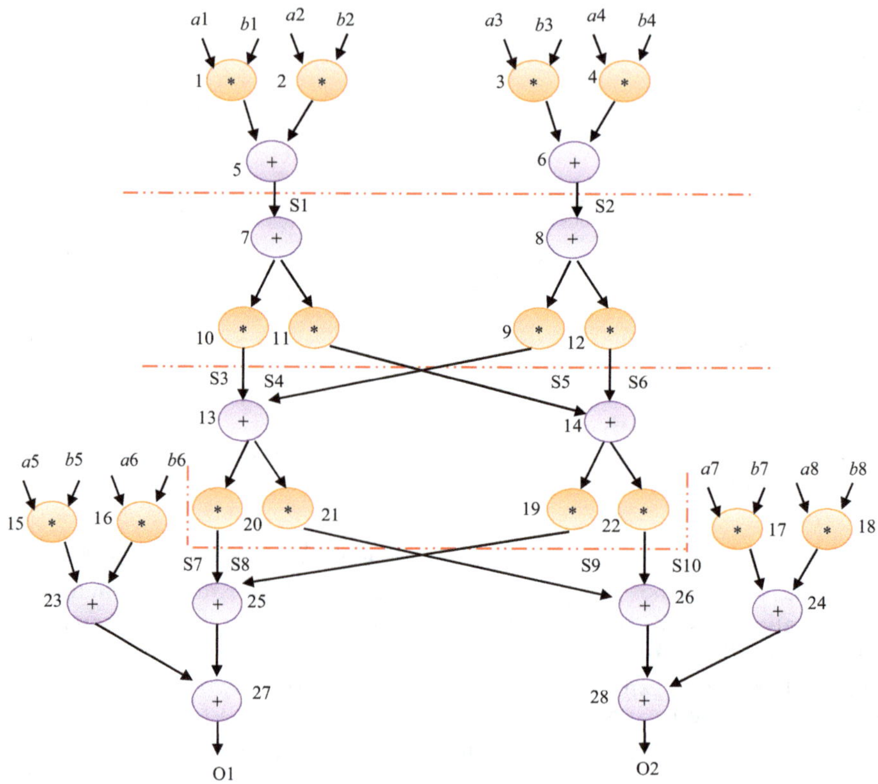

Figure 4.9 DFG of ARF application

functionality. The alteration in the height of graph by modifying the data dependency of nodes incurs structural-transformation-based obfuscation in the RTL structure post HLS. The THT-based high-level transformation technique contributes in SO by incurring the following modifications in the RTL data path: modification in the interconnectivity of FU resources and size of multiplexers and demultiplexers.

The THT-based structural transformation employed by Sengupta and Rathor (2020) is driven through designer's chosen secret SO-key4. The value of the chosen secret key determines the number of partitions that have to be subjected to the THT. The size of SO-key4 depends on the maximum number of partitions that can be subjected to the THT. The function and size of SO-key4 are highlighted in Figure 4.4.

A demonstration of THT-based transformation is shown using an 8-point DCT application. Earlier, it has been demonstrated that how partitions of DFG of an 8-point DCT are generated (shown in Figure 4.8). Since ROE is not possible in the partitioned DFG of an 8-point DCT, therefore, it is directly subjected to THT-based transformation. The THT-based structural transformation is performed on the partitioned CDFG on the basis of designer's chosen value of SO-key4. As shown in the partitioned DFG of DCT, there is maximum one partition (i.e. P2) which can be

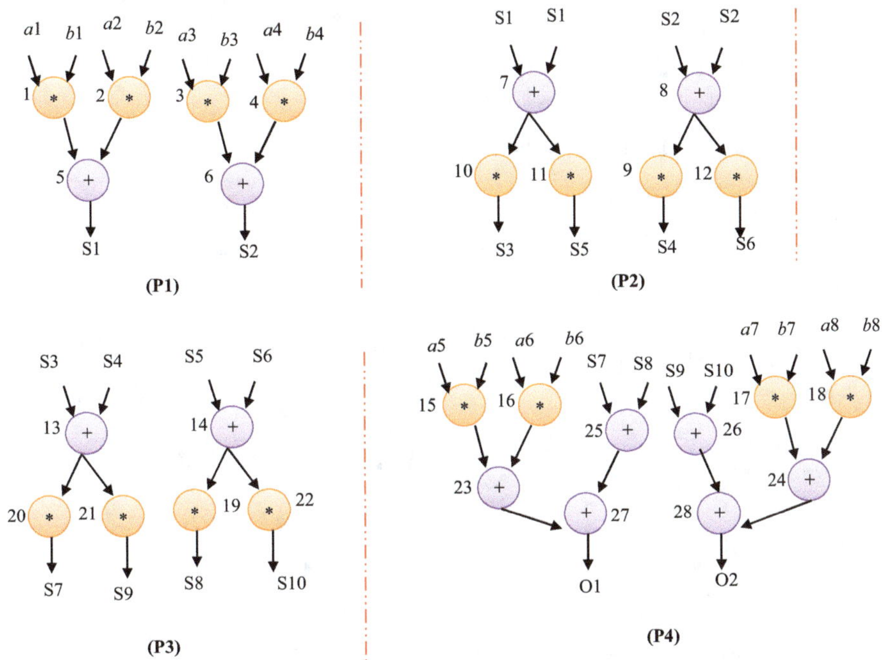

Figure 4.10 Partitioned DFG of ARF application

subjected to the THT. Therefore, the size of SO-key4 is only 1 bit. For a designer's chosen value of SO-key4='1', the partition P2 is subjected to the THT-based structural transformation. The THT-transformed partitioned DFG of DCT is shown in Figure 4.12. As shown in the figure, the sequential data dependency of operation 12 in partition P2 has been converted to the parallel dependency with respect to the operation number 11. This leads to alteration in the height of sub-graph in partition P2. Thus, THT-based structural transformation is performed to further enhance SO in the controller design.

Key-driven folding-knob-based structural transformation

Post performing LU, partitioning, ROE-based and THT-based structural transformations, all operations of the obtained obfuscated partitioned CDFG are scheduled in control steps. The folding-knob-based structural transformation is performed post-scheduling of the obfuscated partitioned DFG to further enhance the obscurity in the design structure. Folding is a high-level transformation technique which can be exploited to achieve SO, such that a common resource is shared among a set of operations that have the same operation type but are present in different control steps. *The number of operations in a set sharing the resource is called the 'folding factor'*. Thus, the folding of FU resource units depends on the secret value of chosen 'folding factor'. Further, the number of folding transformations is controlled by the designer chosen secret value of SO-key5. *The value of the key5 denotes the*

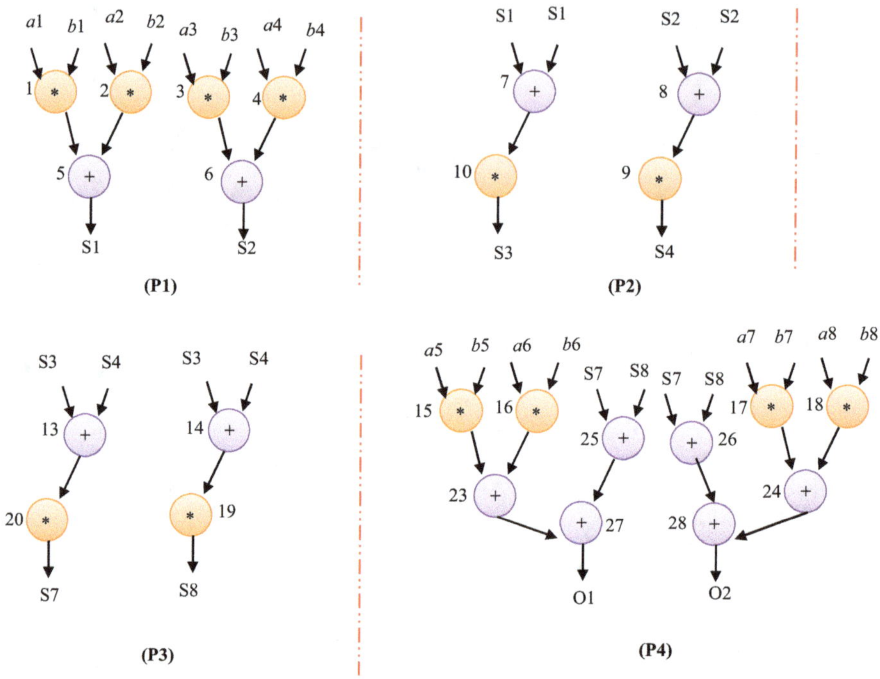

Figure 4.11 Partitioned DFG after ROE-based transformation

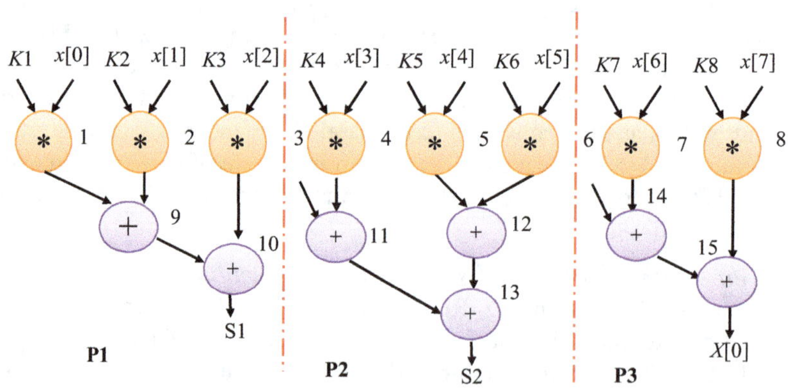

Figure 4.12 THT-based obfuscation technique on partitioned DFG of an 8-point DCT

number of folding transformations to be applied which is referred to as the 'folding knob'. The maximum size of SO-key5 depends on the maximum number of occurrences of folding transformation (i.e. maximum value of folding knob) across the scheduled DFG. The folding-knob-based transformation affects the number of

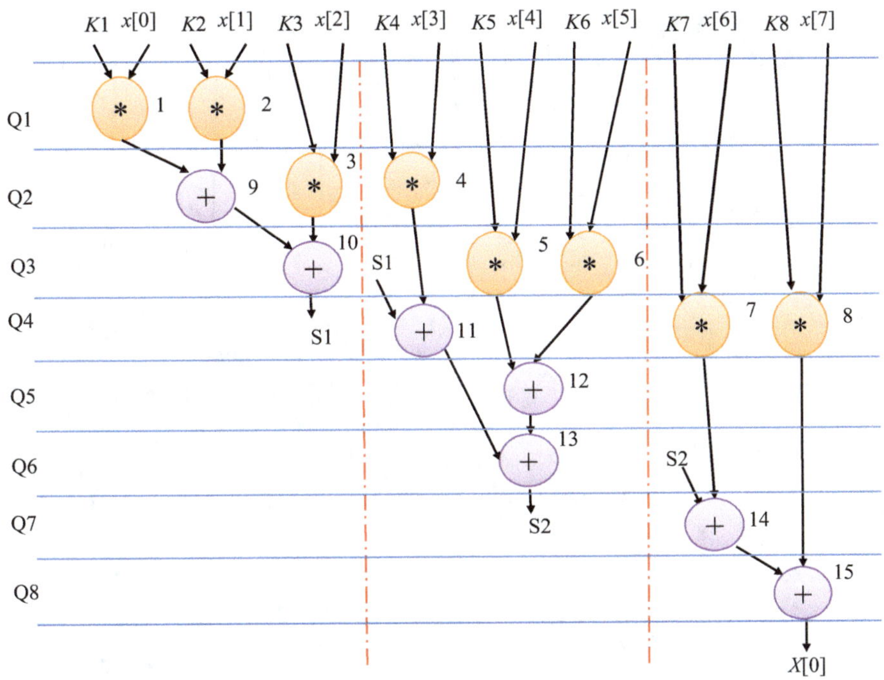

Figure 4.13 Scheduling of obfuscated partitioned DFG of DCT application using two multipliers and one adder

FU resources and delay elements (also known as registers) post HLS, thereby structurally obfuscating the design without causing any change to the functionality.

The folding knob transformation has been demonstrated using an 8-point DCT core. Sengupta and Rathor (2020) considered folding factor=2 for applying folding-knob-based transformation. So far, THT-obfuscated partitions of DFG of an 8-point DCT have been explained in previous step. In order to perform folding knob transformation, the THT-obfuscated partitions of DFG are scheduled as shown in Figure 4.13. It is evident from the figure that there are maximum five occurrences of folding transformation based on folding factor=2. First folding transformation can be performed in scheduled partition P1 using either (operations 1 and 2) or (operations 1 and 3) because of having the same operation type and different control steps. Second folding transformation can be performed in scheduled partition P1 using (operations 9 and 10). Third folding transformation can be performed in scheduled partition P2 using either (operations 4 and 5) or (operations 4 and 6). Fourth folding transformation can be performed in scheduled partition P2 using any of the following set of operations: (operations 11 and 12), or (operations 11 and 13) or (operations 12 and 13). Fifth folding transformation can be performed in scheduled partition P3 using (operations 14 and 15). For designer-selected value of So-key5='011', the value of folding knob is 3, i.e. folding transformation is to

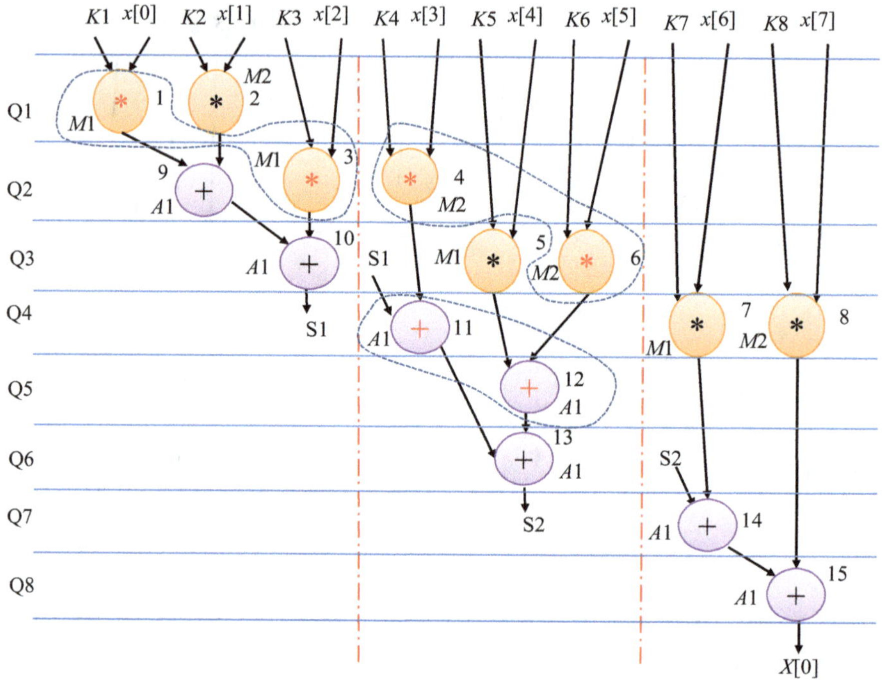

Figure 4.14 Folding-knob-based transformation on obfuscated partitioned DFG of an 8-point DCT

be applied at three occurrences in the scheduled graph. First folding transformation is applied using operations 1 and 3 in partition P1; hence, they are essentially executed through the same FU resource instance (i.e. *M*1) as shown in Figure 4.14. The second folding transformation is applied using operations 4 and 6 in partition P2; hence, they are essentially executed through the same FU resource instance (i.e. *M*2) as shown in the figure. The third folding transformation is applied using operations 11 and 12 in partition P2; hence, they are essentially executed through the same FU resource instance (i.e. *A*1) as shown in the same figure. The operations that are subjected to folding are encircled in loops as shown in Figure 4.14. The allocation of other operations to the FU resources does not comply with the folding transformation. A possible allocation of remaining operations is shown in Figure 4.14. As shown in the figure, the obfuscated partitioned DFG of DCT application uses two multipliers and one adder for FU resource allocation.

Once multiple key-driven techniques of SO are performed, the individual obfuscated scheduled partitions of the graph are synthesized into individual obfuscated RTL data path. The individual obfuscated RTL data path of three scheduled DFG partitions (refer Figure 4.14) of an 8-point DCT are shown in Figure 4.15. Further, these individual obfuscated RTL data paths are integrated into

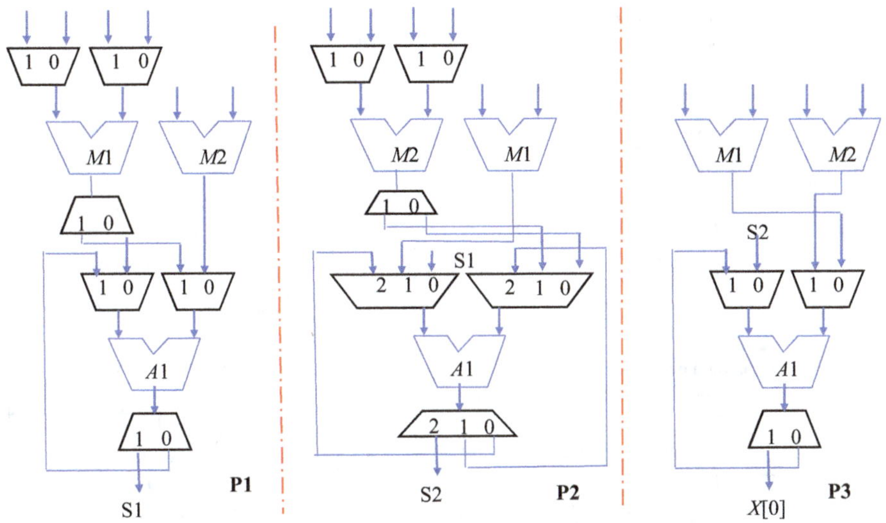

Figure 4.15 Obfuscated RTL data path of individual scheduled DFG partitions of an 8-point DCT

a single obfuscated RTL data path. Figure 4.16 shows the integration of individual obfuscated data path of scheduled DFG partitions into a single obfuscated RTL data path. This obtained structurally obfuscated RTL architecture is harder to reverse engineer by an adversary; hence, it becomes secured against Trojan insertion and piracy threats.

4.5.2.2 Multiple variables signature-based physical-level watermarking – the second line of defence

The watermarking based second line of defence is performed on the top of structurally obfuscated design. In this approach of the second line of defence (Sengupta and Rathor, 2020), watermarking has been performed at physical level by embedding three-variables vendor's signature. To do so, physical-level design step in the form of early floorplanning has been introduced by Sengupta and Rathor (2020) in the physical design flow. In the early floorplanning step of physical design flow, an early floorplan of RTL components is prepared which is subjected to watermarking constraints. The overall process of accomplishing physical-level watermark is illustrated using the following steps:

Formation of a set of RTL components

First, RTL components are extracted from the structurally obfuscated data path. A set *R* of RTL components such as FUs (adders, multipliers, subtractors and comparator) and interconnect hardware resources (multiplexers and demultiplexers of various sizes) is formed. *Different types of RTL components in the set R are arranged in the decreasing order of their size, and different instances of the same RTL components are arranged in the set in increasing order of their*

RTL components of data path of scheduled partition P1

FU resources	*Associated interconnect resources*
M1	*2:1 Mux1_M1, 2:1 Mux2_M1* *1:2 Demux1_M1*
M2	–
A1	*2:1 Mux1_A1, 2:1 Mux2_A1* *1:2 Demux1_A1*

+

RTL components of data path of scheduled partition P2

FU resources	*Associated interconnect resources*
M1	–
M2	*2:1 Mux1_M2, 2:1 Mux2_M2* *1:2 Demux1_M2*
A1	*4:1 Mux1_A1, 4:1 Mux2_A1* *1:4 Demux1_A1*

+

RTL components of data path of scheduled partition P3

FU resources	*Associated interconnect resources*
M1	–
M2	–
A1	*2:1 Mux3_A1, 2:1 Mux4_A1* *1:2 Demux2_A1*

RTL components of integrated data path

FU resources	*Associated interconnect resources*
M1	*4:1 Mux1_M1 (x1), 4:1 Mux2_M1 (x2)* *1:4 Demux1_M1 (d1)* *2:1 Mux1_M1 (x9), 2:1 Mux2_M1 (x10)* *1:2 Demux1_M1 (d5)*
M2	*4:1 Mux1_M2 (x3), 4:1 Mux2_M2 (x4)* *1:4 Demux1_M2 (d2)* *2:1 Mux1_M2 (x11), 2:1 Mux2_M2 (x12)* *1:2 Demux1_M2 (d6)*
A1	*4:1 Mux1_A1 (x5), 4:1 Mux2_A1 (x6)* *1:4 Demux1_A1 (d3)* *4:1 Mux3_A1 (x7), 4:1 Mux4_A1 (x8)* *1:4 Demux2_A1 (d4)* *2:1 Mux1_A1 (x13), 2:1 Mux2_A1 (x14)* *1:2 Demux1_A1 (d7)* *2:1 Mux3_A1 (x15), 2:1 Mux4_A1 (x16)* *1:2 Demux2_A1 (d8)*

Figure 4.16 Integration of all individual obfuscated data paths into a single obfuscated RTL data path of DCT. Note: *'x' and 'd' represent multiplexers and demultiplexers, respectively. Numbering of components has been done in decreasing order of their size and increasing order of instance number*

instance number. For example, a set of RTL components obtained from obfuscated RTL data path of an 8-point DCT (shown in Figure 4.16) is given in the following:

$$R = \{M1, M2, d1, d2, d3, d4, x1, x2, x3, x4, x5, x6, x7, x8, A1, d5, d6, d7, d8, x9, x10, x11, x12, x13, x14, x15, x16\} \quad (4.2)$$

The RTL components shown in the set R have been ordered on the basis of their size and instance number. Note: these components have been highlighted using red colour in the integrated obfuscated RTL data path shown in Figure 4.16.

Thus, a set of RTL components is formed, which is used in the further steps of embedding physical-level watermarking.

Preparing early floorplan

Once a set of RTL components is formed, it is utilized in preparing early floorplan. To prepare the early floorplan, the elements (i.e. RTL components) in the set R are traversed from left to right. While traversing the set, each component of different types is selected according to their order of presence (in the set R) and placed in the floorplan in order to make it grow diagonally, i.e. d appears before x. *Note: however, selection within the components of the same size can be made randomly.* To facilitate the growth of floorplan, it is partitioned into equal-sized blocks. The size of each block is given by the following:

$$\text{Block size} = (\text{largest component height in the set } R) * (\text{largest component width in the set } R) \tag{4.3}$$

For the set of components of obfuscated RTL data path of an 8-point DCT core, the largest component height is of multiplier and the largest component width is also of multiplier. For the height=8 units and the width=4 units of multiplier, the size of each block is 8×4.

Once first block is filled post the placement of RTL components, further placement of components is performed in the second block which is placed just right to the first block in the floorplan. Further, the third block is placed on the top of the first block in order to make diagonal growth of floorplan and maintain a rectangular envelop. Thus, early floorplan is prepared using RTL components as shown in Figure 4.17. The second line of defence in the form of physical-level watermarking is performed on this generated early floorplan of the obfuscated design.

Selection of vendor's signature and conversion into watermarking constraints

In order to embed vendor's watermark into the early floorplan, a secret signature comprising three unique variables α, β and γ is selected. To enable the embedding of signature into the floorplan, the mapping of signature variables into watermarking constraints is shown in Table 4.1. Based on these mapping rules, each digit of the signature is decoded and embedded into the floorplan one-by-one.

For embedding watermark into the floorplan of the obfuscated DCT design, let us assume that the vendor's secret signature is

'$\alpha\ \alpha\ \beta\ \beta\ \gamma\ \beta\ \gamma\ \beta$'.

Embedding of α digits into floorplan

As evident from the mapping rule of α variable given in Table 4.1, the embedding of α digits of the signature is performed using FU resource components only.

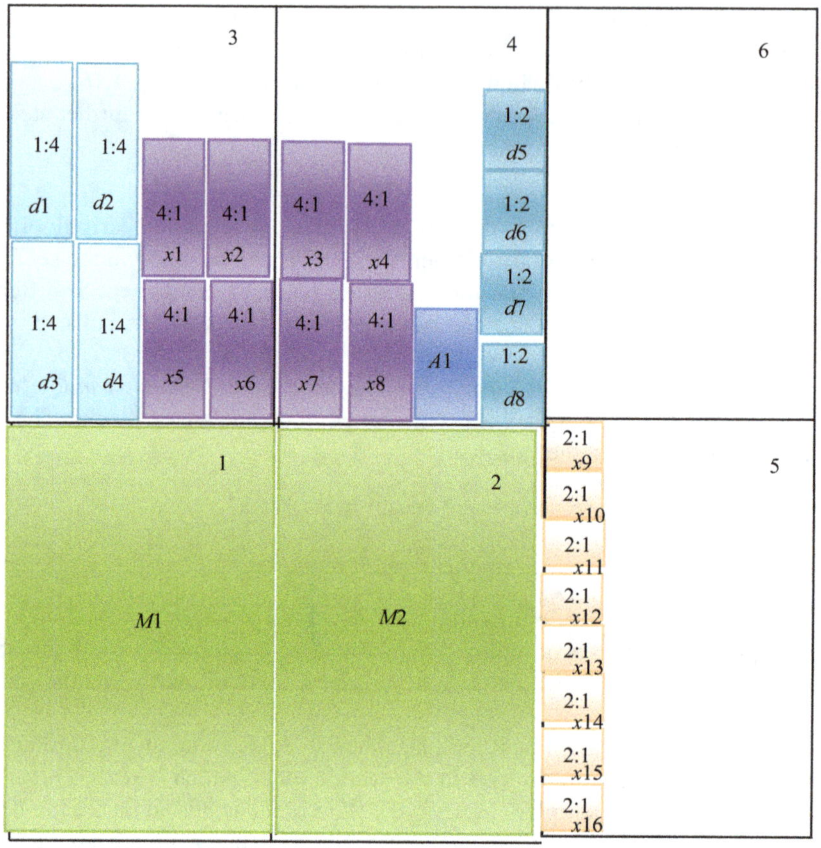

Figure 4.17 Early floorplan of obfuscated DCT design before embedding watermark

Table 4.1 Mapping of signature variables into watermarking constraints

Variables in author's signature	Mapping rules to map into watermarking constraints
α	Odd FU resource needs to be placed on the top of even FU by swapping between two FU resources of the same type
β	Odd multiplexer (Mux) needs to be placed on the top of even multiplexer by swapping between two multiplexers of the same size
γ	Odd demultiplexers (Demux) need to be placed to the right of even demultiplexer by swapping between two demultiplexers of the same size

Therefore, in order to embed α digits, first a subset $R1$ containing only FU components (multipliers, adders, etc.) is extracted from the set R, where the arrangement of FU resources follows the decreasing order of size and the increasing order of instance number. Further, the elements in subset $R1$ are traversed using two pointers p and q in order to embed each α digit. Initially, pointer p indexes the first FU component in the subset $R1$, whereas pointer q indexes the second FU component. For each α digit, pth FU component is swapped with the qth component in the subset $R1$. If swapping leads to the placement of an odd FU component on the top of the even FU component in the floorplan, then one α digit is successfully embedded. Now, those FU components which participated in the embedding of α digit are eliminated from the subset $R1$. However, if the swapping does not place an odd FU component on the top of the even FU component (i.e. embedding of α digit is not possible) then the swapping is undone (previous placement of components in the floorplan is restored). Further, the pointer q is subjected to increment by one. Again, pth FU component is swapped with qth component for embedding of the α digit. For each swapping, q is subjected to increment by one until it reaches to its maximum value (m) which is represented by the total number of FU components of the same type in the subset $R1$. When the maximum value is reached by pointer q, the pointer p is incremented by one and pointer q is set to $p+1$. The pointer p can go upto $m-1$. When the traversal of all components of a particular FU type is accomplished, pointers p and q are made to index the first two components of another FU type in the subset $R1$. This process is followed until the embedding of all α digits in the signature is accomplished.

In the case of an 8-point DCT design, the subset $R1$ extracted from set R is as follows:

$$R1 = \{M1, M2, A1\} \tag{4.4}$$

For embedding first α digit of the signature, the subset $R1$ is traversed to perform swapping between components of the same size. Initially, components $M1$ and $M2$ are indexed by pointers p and q, respectively. Therefore, components $M1$ and $M2$ are subjected to swapping. However, this swapping does not place an odd FU component on the top of even FU component (according to the embedding rule of α variable). Therefore, this swapping is undone. Further, the swapping of components by increasing the pointer q and/or p in the subset $R1$ is not possible because of very limited number of FU components in the subset. Therefore, embedding of any α digit is not possible in the floorplan of obfuscated 8-point DCT design.

Embedding of β digits into floorplan

As evident from the mapping rule of β variable given in Table 4.1, the embedding of β digits of the signature is performed using Mux components only. Therefore, in order to embed β digits, first a subset $R2$ containing only Mux components of different sizes is extracted from the set R, where the arrangement of Mux components follows the decreasing order of size and the increasing order of instance number. Likewise, the embedding process of α digits, the elements in subset $R2$ are traversed using two pointers p and q in order to embed each β digit of the signature.

Initially, pointers p and q index the first and second Mux components of the same size. For each β digit, pth Mux component is swapped with the qth Mux component of the same size in the subset $R2$. If swapping leads to the placement of an odd Mux component on the top of the even Mux component in the floorplan, then one β digit is successfully embedded. *Now, those Mux components which participated in the embedding of β digit are eliminated from the subset R2. However, if the swapping does not make embedding of β digit possible, then the swapping is undone.* During the embedding of β digits, the pointers p and q are incremented in the same manner as was discussed for embedding of α digits.

The embedding of β digits is demonstrated using early floorplan (shown in Figure 4.17) of obfuscated DCT design. In order to do so, the subset $R2$ is extracted from set R which is given as follows:

$$R2 = \{x1, x2, x3, x4, x5, x6, x7, x8, x9, x10, x11, x12, x13, x14, x15, x16\} \tag{4.5}$$

To embed first β digit of the signature, the subset $R2$ containing all Mux components is traversed and swapping is performed between the components of the same size. Initially, Mux components $x1$(4:1) and $x2$(4:1) are indexed by pointers p and q, respectively, hence subjected to swapping for embedding first β digit. The **swapping of $x1$ and $x2$** will result into embedding of first β digit (as per the embedding/mapping rule of β). This is because, it will place $x1$ (odd component) over $x6$ (even component). Since the embedding of first β digit has utilized Muxes $x1$ and $x6$, therefore, they are removed from the subset $R2$. The modified subset $R2$ is as follows: $R2=\{x2, x3, x4, x5, x7, x8, x9, x10, x11, x12, x13, x14, x15, x16\}$. Now for embedding next β digit, pointers p and q are made to index again the first two components, i.e. $x2$(4:1) and $x3$(4:1), respectively, in the subset $R2$. Now **swapping of $x2$ and $x3$** in the floorplan is performed which will not result into the embedding of second β digit. This is because the swapping will cause the placement of $x2$ (even) on the top of $x7$ (odd) and $x3$ (odd) on the top of $x5$ (odd). Here, neither of the cases follows the embedding/mapping rule of β. Therefore, this swapping is undone. Now, pointer q is incremented by one and it starts indexing $x4$ in the subset $R2$. Further, **swapping of $x2$ and $x4$** in the floorplan is performed which also will not result into the embedding of second β digit. This is because the swapping will cause the placement of $x2$ (even) on the top of $x8$ (even) and $x4$ (even) on the top of $x5$ (odd). Here again, neither of the cases follows the embedding/mapping rule of β. Therefore, this swapping is also undone. Now, pointer q is again incremented by one and it indexes $x5$ in the subset $R2$. Now, the **swapping of $x2$ and $x5$** will result into the embedding of second β digit (as per the embedding/mapping rule of β). This is because, it will place $x5$ (odd component) over $x2$ (even component). Since the embedding of the β digit has utilized Muxes $x2$ and $x5$, therefore, they are also removed from the subset $R2$. The modified subset $R2$ is as follows: $R2=\{x3, x4, x7, x8, x9, x10, x11, x12, x13, x14, x15, x16\}$. Now for embedding next β digit, pointers p and q are made to index again the first two components, i.e. $x3$(4:1) and $x4$(4:1), respectively, in the subset $R2$. The **swapping**

of *x*3 and *x*4 will result into the embedding of the third β digit. This is because, it will place *x*3 (odd component) over *x*8 (even component). Since the embedding of the β digit has utilized Muxes *x*3 and *x*8, therefore, they are also removed from the subset *R*2. The modified subset *R*2 is as follows: *R*2={*x*4, *x*7, *x*9, *x*10, *x*11, *x*12, *x*13, *x*14, *x*15, *x*16}. Further, the **swapping of *x*4 and *x*7** will result into the embedding of the fourth β digit. This is because, it will place *x*7 (odd component) over *x*4 (even component). This is how all the four β digits of the signature are embedded into the floorplan. The intermediate floorplan post embedding all the β digits is shown in Figure 4.18. As shown in the figure, the embedded β digits (i.e. odd Muxes on the top of even Muxes) have been highlighted using yellow colour.

Embedding of γ digits into floorplan
As evident from the mapping rule of γ variable given in Table 4.1, the embedding of γ digits of the signature is performed using Demux components only. Therefore, in order to embed γ digits, first a subset *R*3 containing only Demux components of

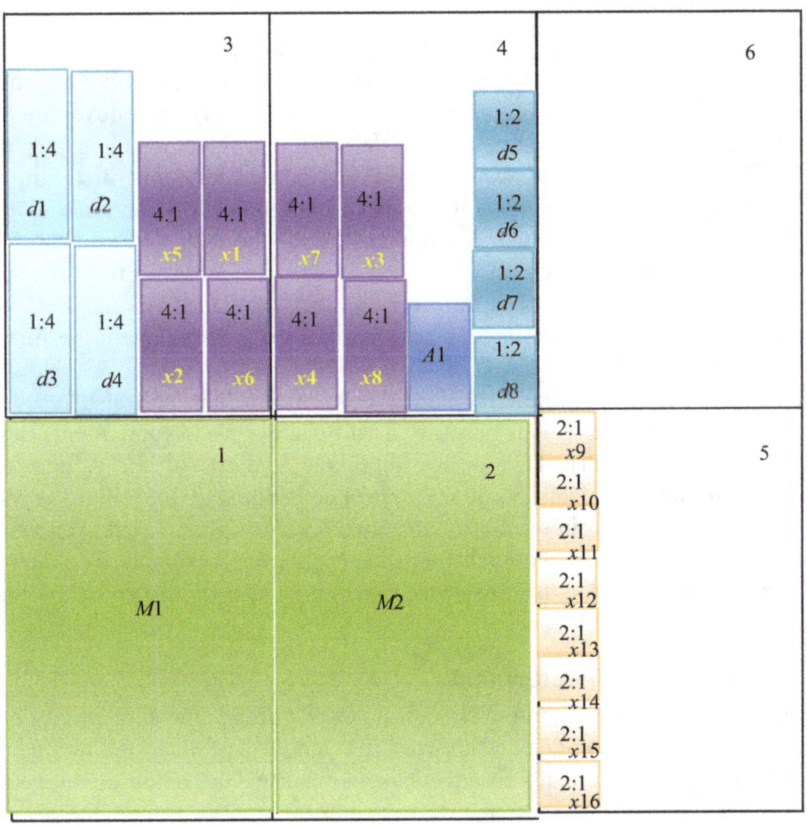

Figure 4.18 Intermediate watermarked floorplan post embedding β digits of the signature

different sizes is extracted from the set R, where the arrangement of Demux components follows the decreasing order of size and the increasing order of instance number. Likewise, the embedding rule of α digits, the elements in subset $R3$ are traversed using two pointers p and q in order to embed each γ digit of the signature. Initially, pointers p and q index the first and second Demux components of the same size. For each γ digit, pth Demux component is swapped with the qth component of the same size in the subset $R3$. If swapping leads to the placement of an odd Demux component to the right of an even Demux component in the floorplan, then one γ digit is successfully embedded. Now, those Demux components which participated in the embedding of γ digit are removed from the subset $R3$. However, if the swapping does not make the embedding of γ digit possible, then the swapping is undone. During the embedding of γ digits, the pointers p and q are incremented in the same manner as was discussed for embedding of α and β digits.

The embedding of γ digits in the floorplan of the obfuscated DCT design is discussed as follows. First, the subset $R3$ is extracted from set R as given next.

$$R3 = \{d1, d2, d3, d4, d5, d6, d7, d8\} \tag{4.6}$$

To embed the first γ digit of the signature, the subset $R3$ containing all Demux components is traversed and swapping is performed between components of the same size. Initially, Demux components $d1(1{:}4)$ and $d2(1{:}4)$ are indexed by pointers p and q, respectively hence subjected to swapping for embedding the first γ digit. The swapping of $d1$ and $d2$ will result into the embedding of first γ digit (as per the embedding/mapping rule of γ). This is because, it will place $d1$ (odd component) right to the $d2$ (even component). Since the embedding of the first γ digit has utilized Demuxes $d1$ and $d2$, therefore, they are removed from the subset $R3$. The modified subset $R3$ is as follows: $R3=\{d3, d4, d5, d6, d7, d8\}$. Now, for embedding the next γ digit, pointers p and q are made to index again the first two components, i.e. $d3(1{:}4)$ and $x4(1{:}4)$, respectively, in the subset $R3$. Now, the swapping of $d3$ and $d4$ in the floorplan is performed which will also result into the embedding of γ digit. This is because, it will place $d3$ (odd component) right to the $d4$ (even component). This is how both the γ digits of the signature are embedded into the floorplan. The floorplan post embedding all the γ digits is shown in Figure 4.19. As shown in the figure, the embedded γ digits (i.e. odd Demuxes at the right to the even Demuxes) have been highlighted using red colour. The floorplan in Figure 4.19 is the final watermark-embedded floorplan of the obfuscated RTL design of DCT core.

4.5.2.3 Detection of watermark

The detection of watermark into a design under-test proves the authenticity of the design. The absence of vendor's watermark into a design of the brand of a genuine vendor indicates that probably an adversary is misusing the brand of the genuine vendor to illegally sell counterfeited designs into the market. Hence, the detection of the absence of watermark identifies the counterfeiting. Only those designs of vendor's brand would be proved authentic which contain the watermark (vendor's

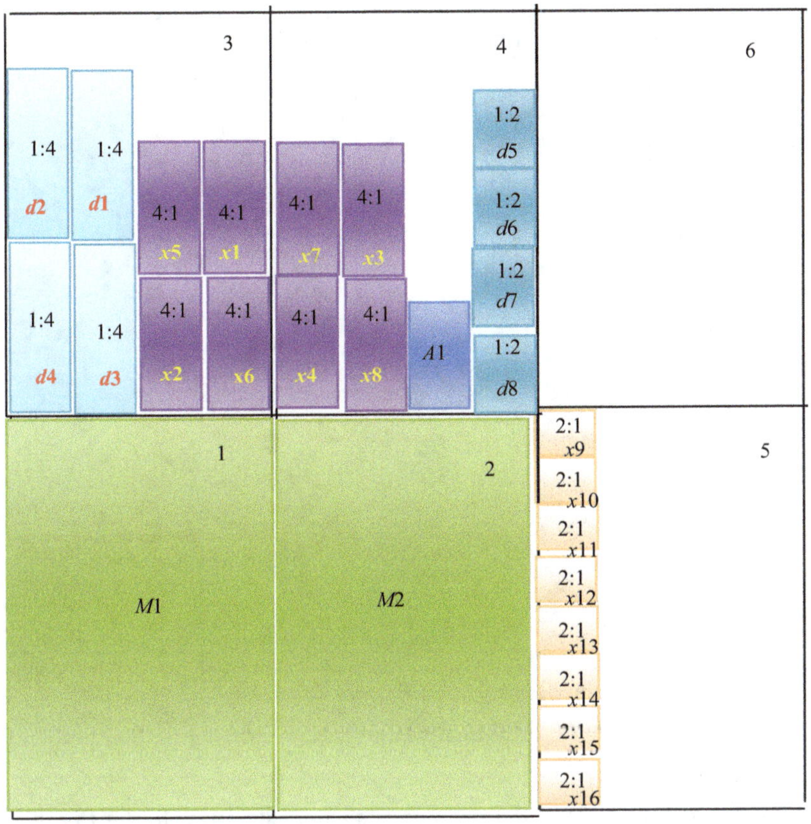

Figure 4.19 Final watermarked floorplan post embedding β and γ digits of the signature

signature) of genuine owner. The detection of author's signature to identify his/her watermark is performed in the following steps as shown in Figure 4.20.

Watermark constraints regeneration for verification

In this step, the watermarking constraints are regenerated from the vendor's signature comprising α, β and γ variables. The regeneration of watermarking constraints is performed by applying vendor's/designer's specified mapping rules to the signature digits. Further, the presence of regenerated watermarking constraints is verified into the design.

Inspection of watermark into the design

The final floorplanned design file obtained from the physical design tool is subjected to watermark detection. The following inputs are required to generate the final floorplanned design file during physical design process: (i) Verilog file of the DSP core design netlist, (ii) configuration file, (iii) watermarked early floorplan file of obfuscated design (this file contains the vendor's signature which represents the

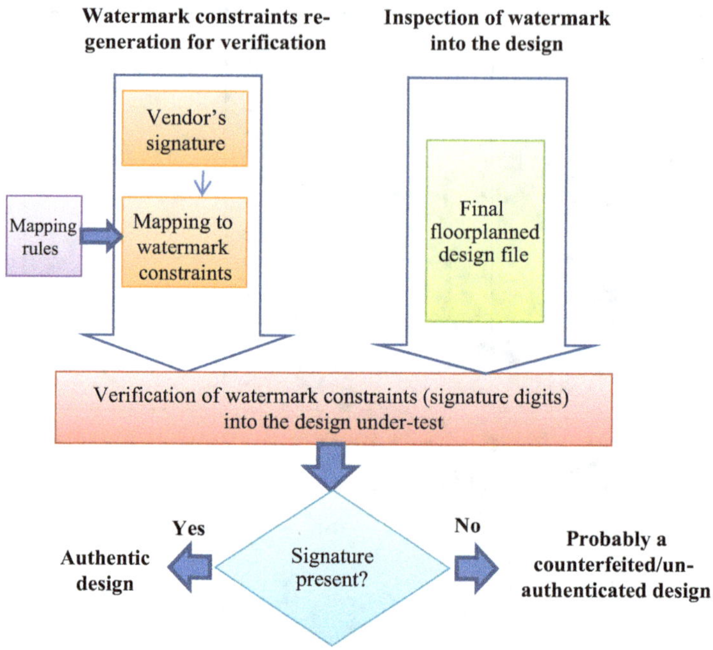

Figure 4.20 Detection of signature digits embedded during physical-level watermarking (Sengupta and Rathor, 2020)

watermark), (iv) library files and (v) any other relevant files. In the floorplanned design file, orientations of RTL components are inspected to detect the watermark.

Verification of watermark constraints (signature digits)

If orientation of RTL components in the floorplanned design file is according to the watermark constraints (obtained post mapping of signature digits in first step), then the presence of original vendor's (author's) signature is verified in the design. Detection of author's signature (representing watermark) in the design of the brand of genuine author proves that the design is authentic. However, if the author's signature is missing in the design of the brand of genuine author, then probably those designs are counterfeited/not authentic.

4.5.3 Key size analysis of the structural obfuscation

The total SO-key size (KS_{total}) of multi-structural-transformation-based obfuscation process is calculated as follows:

$$KS_{total} = \{\lceil(\log_2(U_m))\rceil + \lceil(\log_2(C_m))\rceil + \lceil(\log_2(R_m))\rceil + \lceil(\log_2(T_m))\rceil + \lceil(\log_2(F_m))\rceil\} \text{ bits} \tag{4.7}$$

where U_m indicates the maximum value of UF, C_m indicates the maximum cuts possible, R_m indicates the maximum ROs possible, T_m indicates the maximum THT possible and F_m indicates the maximum folding possible. Driving the multiple techniques of structural transformations viz. LU, partitioning, ROE, THT and folding knob through individual SO-keys enhances the security level against RE (which results into Trojan insertion and piracy). Because of the involvement of multilayered secret SO-keys in the structural-transformation-based obfuscation, the difficulty level of deducing true functionality and structure through RE by an attacker is augmented. This is because, now an attacker has to know both the multiple secret SO-keys and the structural transformation techniques to back-engineer the original functionality and structure. Further, finding true key combination among the exhaustive key combinations is difficult for the attacker. Hence, deobfuscation becomes arduous for an attacker as the secret SO-keys, which are used to perform a high-level transformation-based obfuscation, are not known to the attacker.

4.6 Low-cost optimized multi-key-based structural obfuscation

This section presents a discussion on why a low cost and optimized version of multi-key-based SO is useful and how it is achieved. The discussion has been divided into the following subsections:

4.6.1 Motivation for low-cost optimized structural obfuscation

Selection of an optimal (resulting into low-cost) design from a vast number of functionally equivalent architecture designs requires extensive analysis of the design space. Exploring low-cost solution using design space exploration (DSE) framework, while simultaneously performing multiple structural-transformation-driven obfuscation based security is vital in two ways: (i) it minimizes the overhead impact caused due to employing the SO-based security and (ii) it ensures that the designer's area, power and delay budget are satisfied. This motivates the integration of DSE framework with the multiple high-level transformation-driven SO-based security technique. This section discusses how the particle swarm optimization (PSO) process (Mishra and Sengupta, 2014) is exploited to explore the design space in order to yield a low-cost obfuscated design without hampering the functionality of the DSP core. Later some approaches are mentioned where even though obfuscation has been used for security, however, the exploration of low-cost solution has not been addressed.

A brief discussion on some approaches that secure DSP cores using obfuscation technique is as follows: approaches proposed by Chakraborty and Bhunia (2009, 2011) applied obfuscation-based IP core security against Trojan insertion attacks. However, HLS framework has not been leveraged by these approaches to perform obfuscation. Further, these approaches (Chakraborty and Bhunia, 2009, 2011) do not target DSP cores and do not explore low-cost solution. In approaches proposed by Lao and Parhi (2015) and Sengupta and Roy (2017), the SO has been applied on DSP cores. However, these approaches do not explore the design space to ensure the low design cost. Therefore, this section presents a discussion on how the PSO-DSE is integrated with the key-based transformation-driven SO for DSP cores to explore a low-cost obfuscated design.

4.6.2 High-level perspective

High-level view of the methodology of generating low-cost obfuscated designs is shown in Figure 4.21. It comprises three major modules: key-driven SO module, PSO module and fitness/cost evaluation module. The SO module employed in this approach accepts a CDFG (of the corresponding DSP application) and secret keys as inputs and produces a structurally modified but functionally equivalent CDFG design as output. The output of the SO module serves as the input to the PSO-DSE module which determines the optimal resource configuration that would incur minimal design cost. PSO-DSE process uses the fitness/cost evaluation module for computing cost and finally a low-cost highly secured structurally obfuscated design is generated.

4.6.3 Details of methodology

Figure 4.22 shows a detailed description of the low-cost multi-key-based SO approach. The inputs are a CDFG representing a DSP application, module library, PSO control parameters (iteration count, swarm size, acceleration coefficient, etc.) and secret multi-key values for obfuscation as shown in the figure. The first transformation, i.e. key-based LU is applied to the input CDFG depending on its applicability; thereafter, the CDFG is partitioned into as many parts as per the specification of the designer's secret key. Further, redundant nodes elimination process is performed from the CDFG based on the designer's secret key, and then finally key-based THT is applied to the CDFG. Subsequently, PSO-DSE is applied to determine optimal resource constraints that would yield minimal design cost. Based on the optimal resource constraints, final scheduling is performed on the CDFG. The final obfuscation technique, i.e. folding is performed on scheduled CDFG which results into structurally obfuscated scheduled CDFG of DSP core. Further, upon performing data path and controller generation of HLS, structurally obfuscated RTL design is generated.

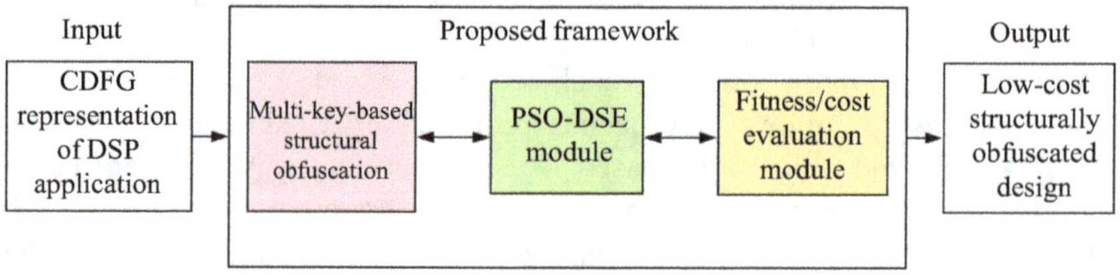

Figure 4.21 Overview of low-cost key-based structural obfuscation approach

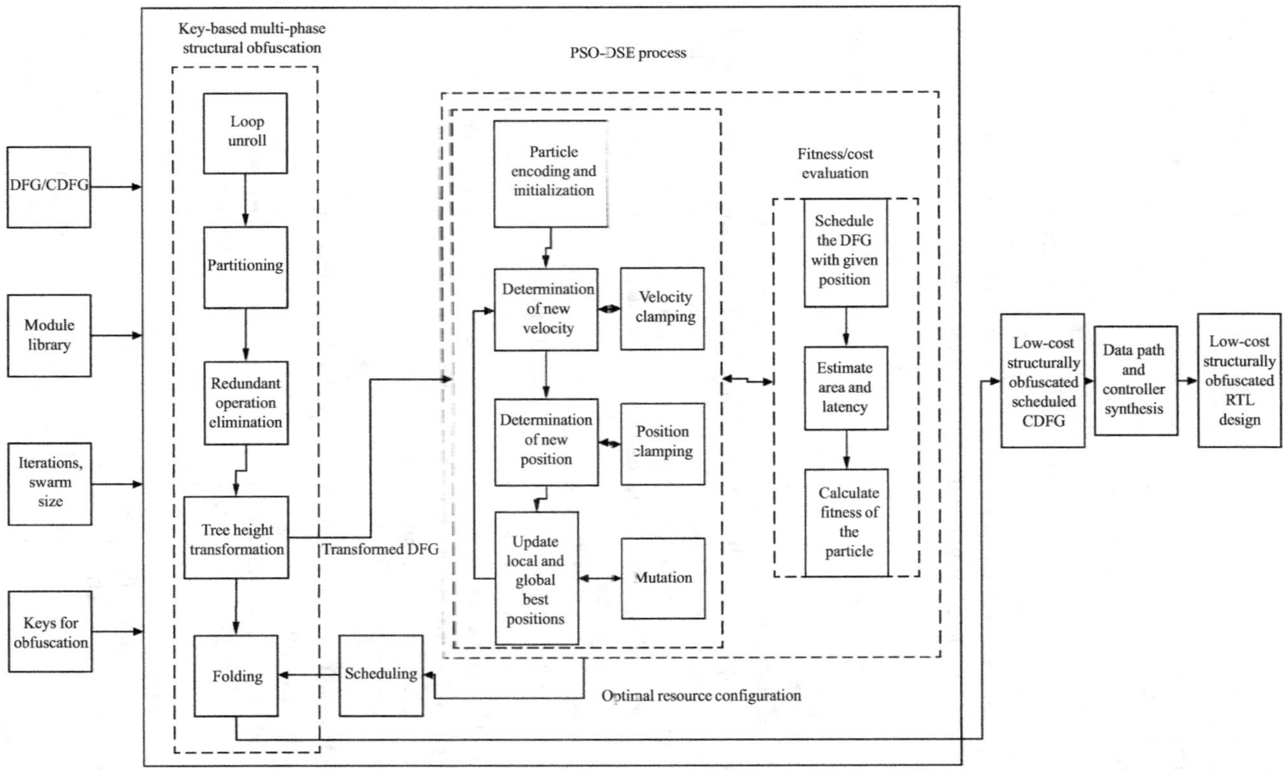

Figure 4.22 Details of low-cost key-based structural obfuscation approach

4.6.3.1 Summary of key-driven structural-transformation-based obfuscation and their effect on gate count/design cost

1. **LU** (Sengupta and Rathor, 2020): It is a high-level transformation technique used to obfuscate the loop-based DSP core by untwining the loop based on the secret SO-key1. The LU affects a number of control steps in the controller and increases logic density, thereby causing a change in the gate count (and structure) without changing functionality.
2. **Partitioning** (Sengupta and Rathor, 2020): This high-level transformation technique is performed to obfuscate the CDFG by dividing it into a number of partitions depending on the secret SO-key2. Partitioning affects the count of multiplexers and demultiplexers in the final RTL, thereby changing the gate count (and structure) without causing any change to the functionality of the CDFG.
3. **ROE** (Sengupta and Rathor, 2020): This high-level transformation technique is performed to obfuscate the CDFG by removing redundant nodes from the graph based on the secret SO-key3. The elimination of redundant nodes affects the logic density thereby decreasing cost and also decreasing the gate count (and structure) without causing any change to the functionality.
4. **THT** (Sengupta and Rathor, 2020): This high-level transformation technique is performed to obfuscate the CDFG by transforming the height of partitioned graph based on secret SO-key4. It leads to a decrease in the number of control steps, thereby decreasing gate complexity as well as count of the controller design and design cost.
5. **Folding** (Sengupta and Rathor, 2020): It is a high-level transformation technique in which a common resource is shared among a set of nodes that have the same operation type but are present in successive control steps in the scheduled graph. The value of the secret SO-key5 specified by the user denotes the number of folding transformations to be applied which is referred to as the folding knob. Sharing of resources enables the maximum utilization of resources, thereby decreasing the cost.

4.6.3.2 PSO-DSE-based framework for generating low-cost design solution

A generic PSO has been mapped to DSE process in HLS framework and is integrated with key-based SO to generate low-cost structurally obfuscated DSP design. The PSO-DSE process comprises five steps: (i) particle initialization, (ii) velocity computation, (iii) position computation, (iv) fitness evaluation and (v) finding global and local best positions. Steps (ii)–(v) are repeated for each particle over the iterations, until either terminating criterion is met (e.g. 100 iterations are completed) or there is no change in the global best position for some successive iterations (e.g. 10). The detailed explanation of each step is as follows (Mishra and Sengupta, 2014):

1. *Particle initialization:* Dimension of particle position equals the number of types of resources (e.g. in this approach, it is a 4D vector because there are four types of resources for the case studies tested: adder, subtractor, multiplier and comparator). Position of the first particle from the population is initialized with maximum resource constraints, i.e. R_{max}. Position of the second particle is initialized with minimum resource constraints, i.e. R_{min}. Position of the third particle is initialized to average of the positions of the first two particles, i.e. $(R_{max}+R_{min})/2$. Then the position of the remaining particles is initialized with random values between minimum and maximum resource constraints.
2. *Velocity computation:* Velocity of a particle is computed using the following equation (Mishra and Sengupta, 2014):

$$V_{new} = \omega * V_{old} + b_1 * r_1 * (R_{lb} - R_{curr}) + b_2 * r_2 * \left(R_{gb} - R_{curr}\right) \tag{4.8}$$

where ω is inertia weight which linearly decreases from 0.9 to 0.4 (to obtain global best), b_1 and b_2 are acceleration coefficients. b_1 linearly decreases from 3 to 2, b_2 linearly increases from 2 to 3 over the iterations of PSO. Further, r_1 and r_2 are random values between 0 and 1, R_{curr}, R_{lb} are the current and best position of the current particle and R_{gb} is the best position considering all the particles so far. Initial velocity for every particle is assigned to zero. Velocity computation is followed by velocity clamping using the following (Mishra and Sengupta, 2014):

$$V^i_{new} = \begin{cases} V^i_{max}, & \text{if} V^i_{new} > V^i_{max} \\ -V^i_{max}, & \text{if } V^i_{new} < -V^i_{max} \\ V^i_{new}, & \text{otherwise} \end{cases} \tag{4.9}$$

where V^i means the velocity of ith dimension, and V^i_{max} is given by $(R^i_{max}-R^i_{min})/2$.
3. *Position computation*: Position is computed by moving in the direction of velocity. Position is calculated using the following (Mishra and Sengupta, 2014):

$$R_{new} = R_{old} + V_{new} \tag{4.10}$$

If the new position calculated exceeds the maximum resource constraints then clamping is applied by selecting a random value between minimum and maximum resource constraints for each dimension.
4. *Fitness evaluation:* Fitness value for each particle in the population is computed using the following cost function (Mishra and Sengupta, 2014):

$$\text{Cost} = w_1 * \left(\frac{A}{A_{max}}\right) + w_2 * \left(\frac{L}{L_{max}}\right) \tag{4.11}$$

where A and L are the area and latency of the design with constrains of the particle and A_{max}, L_{max} are the maximum possible area and latency. Further, w_1 and w_2 are the designer-specified weights (each taken as 0.5).

5. *Determining local and global best*: Local best is the best position (i.e. the position at which the cost value for the particle is minimum) achieved by a particle so far. Global best is the position at which the cost value is minimum for the entire population (i.e. minimum cost of all the local bests).
6. *Mutation:* Mutation is applied in order to avoid getting stuck in local minima. It is applied on every fifth iteration with designer's choice of probability of mutation. And, it is applied on local best position of every particle. Further, mutation is applied by performing a left-shift operation on every particle with even index, followed by position clamping. On every particle with odd index, mutation is applied by adding a random value (between minimum and maximum resource constraints) to the values at even index of the particle position vector and subtracting the same value from the values at odd index of the particle position vector, followed by position clamping. Then, fitness/cost is evaluated at this new position obtained after mutation. If the cost value is lesser than value computed using the local best, then the local best is updated with the new position.

Thus, low-cost structurally obfuscated design is generated post-performing PSO-DSE.

4.7 Structural obfuscation and physical-level watermarking tool for securing hardware accelerators

Authors have developed a ***KSO-PW tool*** (key-driven SO and physical-level watermarking tool) to simulate and analyse the functionality of SO and physical-level watermarking approach for securing DSP hardware accelerators, publicly available for free download: http://www.anirban-sengupta.com/Hardware_Security_Tools.php.

A snapshot of the graphical user interface (GUI) of the tool is shown in Figure 4.23, for FIR application input. The left portion of the tool shows the panel for providing required inputs to the tool, right portion shows the panel with output buttons to see the intermediate and final outputs of the SO and physical-level watermarking-based double line of defence approach. The panel in the middle shows the buttons for entering secret SO-keys (for multiple structural transformation techniques viz. LU, partitioning, ROE, THT and folding knob) and vendor's signature for watermarking. Since ROE is not applicable for FIR application, therefore, the button for entering the ROE key input remains invisible as shown in Figure 4.23. It is noteworthy in the GUI of KSO-PW tool that for a DSP application, only those secret key-buttons are activated for which the corresponding structural transformation techniques are applicable. The KSO-PW tool accepts the DSP application input in the form of CDFG along with module library file and resource constraints file. The tool shows outputs of all intermediate steps of SO and watermarking approach and finally generated watermarked floorplan of the obfuscated design. Further, it also shows design cost before performing SO and watermarking (i.e. baseline design cost), design cost after intermediate steps and the final design cost.

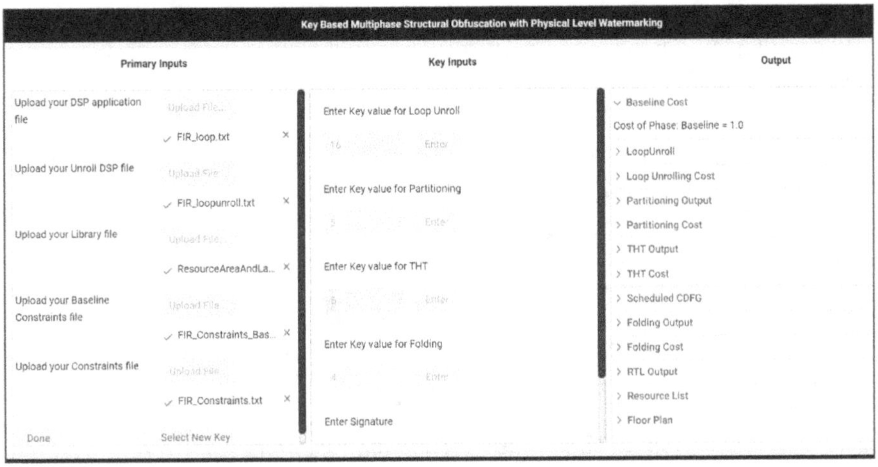

Figure 4.23 A snapshot of GUI of KSO-PW tool

Let us generate all the intermediate and final outputs of SO and watermarking-based double line of defence approach for 160-tap FIR filter core using the KSO-PW tool. Post feeding CDFG of 160-tap FIR filter as input, the secret SO-keys for LU, partitioning, THT and folding are fed to the tool as shown in Figure 4.24. As sown in the figure, the entered SO-key1 for LU is 16 (i.e. UF=16). Therefore, the loop body of 160-tap FIR filter application is unrolled 16 times. The output of LU is shown at the output panel in Figure 4.24. The LU output shown onto the GUI is interpreted as follows: each operation has been denoted by four tuples: (operation type, input1 operation #, input2 operation # and current operation #). The design cost post performing LU is shown in Figure 4.25. Further as shown in Figure 4.26, the entered secret SO-key2 for partitioning is 5. Therefore, the loop unrolled FIR filter application is divided into five partitions. The output of partitioning is shown on to the output panel in Figure 4.26. The design cost post performing partitioning is shown in Figure 4.27. As discussed earlier, ROE is not applicable on partitioned DFG of FIR applications as redundant nodes/operations are not present. Further as shown in Figure 4.28, the entered secret SO-key4 for THT is 5. Therefore, the THT is applied on all the five partitions. The output of THT is shown onto the output panel in Figure 4.28. As shown in the figure, partitions have modified post applying THT. The design cost post performing THT is shown in Figure 4.29. Further, the CDFG of obfuscated partitioned FIR application is scheduled into 17 control steps. Scheduled partitioned CDFG is shown in Figures 4.30–4.32. The scheduling of operations in control steps 1–6 is shown in Figure 4.30, scheduling in control steps 7–13 is shown in Figure 4.31 and scheduling in control steps 14–17 is shown in Figure 4.32. Further as shown in Figure 4.33, the entered secret SO-key5 for folding is 4. Therefore, folding is performed on four folding occurrences in the scheduled partitioned CDFG of FIR filter application. The output of folding is shown onto the output panel in Figure 4.33. As shown in the figure, operation/node

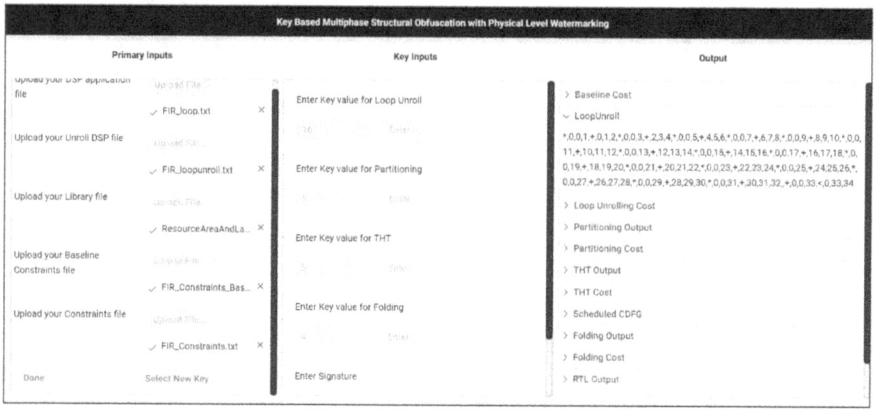

Figure 4.24 Different SO-key values fed and output of loop unrolling of FIR filter application shown onto the tool

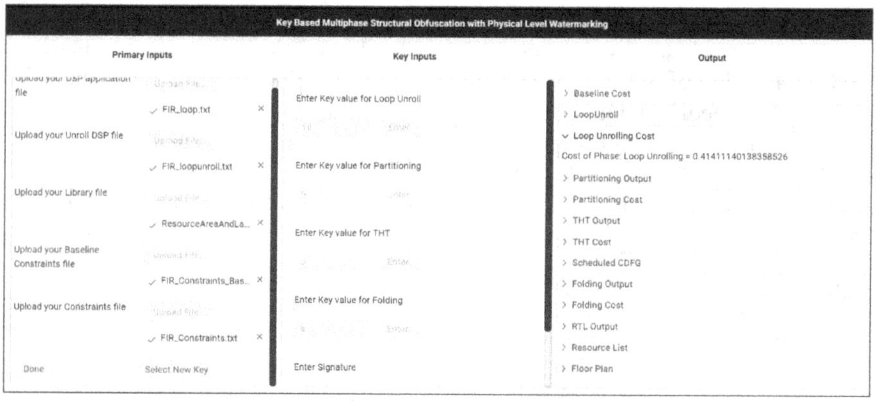

Figure 4.25 Design cost post loop unrolling of FIR filter application

pairs (4, 8), (14, 12), (20, 18) and (26, 24) have been used for folding based on folding factor of 2. Therefore, both the operations of the same type in each pair are essentially allocated to the same FU resource unit. The design cost post performing folding is shown in Figure 4.34. Further, Figures 4.35–4.38 show the RTL output post-HLS of each scheduled partition. The RTL output of partition 1 is shown in Figure 4.35. The RTL output of partition 2 and an excerpt of partition 3 are shown in Figure 4.36. The RTL output of remaining excerpt of partition 3 and an excerpt of partition 4 are shown in Figure 4.37. The RTL output of remaining excerpt of partition 4 and RTL output of partition 5 are shown in Figure 4.38. The interconnection of different components in the RTL output shown onto the GUI (in Figures 4.35–4.38) is interpreted as follows:

For M (multiplier): Input1 operation #, input2 operation #, (output operation #)

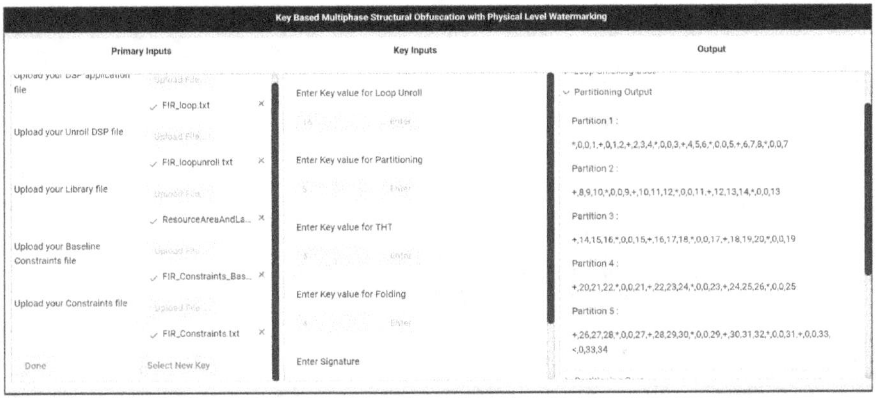

Figure 4.26 Output of partitioning of loop unrolled FIR filter application shown onto the tool

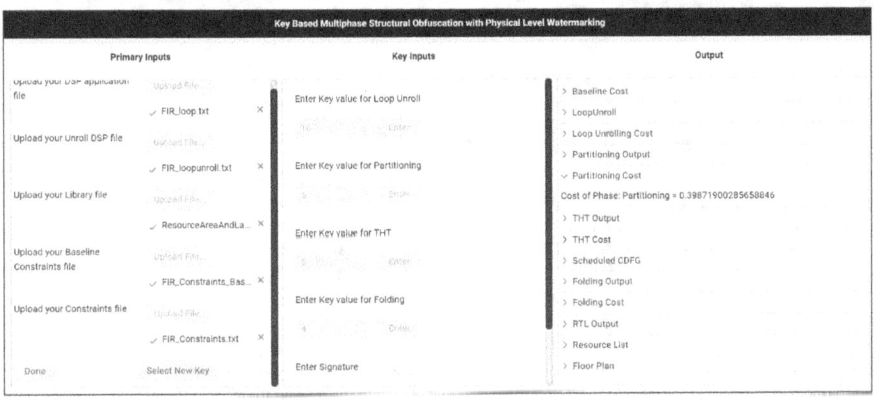

Figure 4.27 Design cost post partitioning shown onto the tool

('0' denotes *M* gets primary input, i.e. no intermediate input from an output of an operation)

For *A* (adder): Input component 1, input component 2, output component

For Mux-x (4:1): First, second, third and fourth inputs of multiplexer, respectively, accepting input from operation #

For Demux-d (1:4): First, second, third and fourth output of demultiplexer, respectively, providing input to operation #

For C1 (comparator): Input operation #, output operation #

Further, the set of RTL components extracted from obfuscated RTL data path of FIR filter for preparing an early floorplan is shown in Figure 4.39. As shown in the figure, there are 4 instances of multiplier (i.e. M->1, M->2, M->3 and M->4), 1 instance of Demux 1:8 (i.e. d8->1), 2 instances of Mux 8:1 (i.e. x8->1, x8->2),

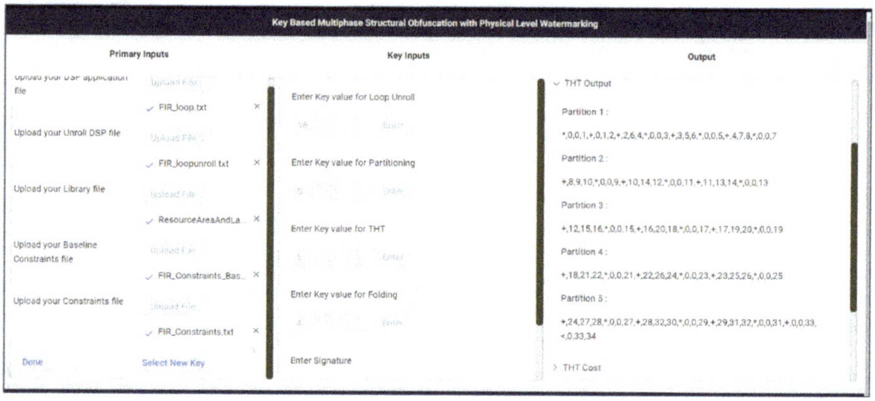

Figure 4.28 Output of THT shown onto the tool

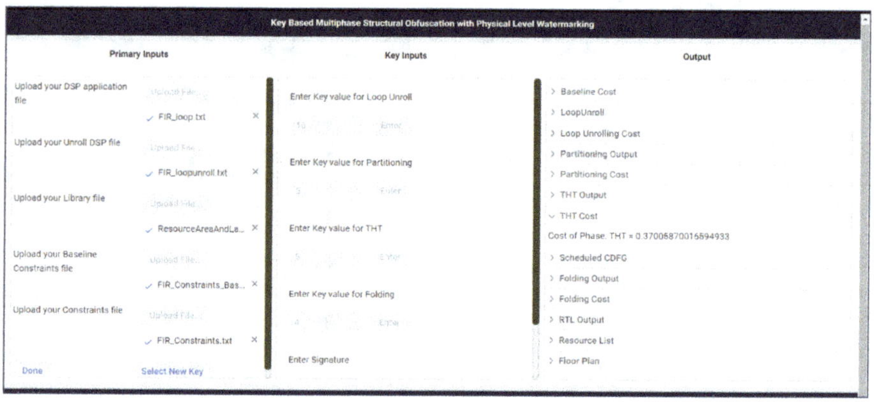

Figure 4.29 Design cost post THT shown onto the tool

1 instance of comparator (i.e. C->1), 9 instances of Demux 1:4 (i.e. d4->1, d4->2, d4->3, d4->4, d4->5, d4->6, d4->7, d4->8, d4->9), 18 instances of Mux 4:1 (i.e. x4->1, x4->2, x4->3, x4->4, x4->5, x4->6, x4->7, x4->8, x4->9, x4->10, x4->11, x4->12, x4->13, x4->14, x4->15, x4->16, x4->17, x4->18) and 1 instance of adder (A->1). These RTL components and their instances have been arranged (as shown in Figure 4.39) on the basis of decreasing order of their size and increasing order of their instance number. An excerpt of floorplan is shown in Figure 4.40, where the orientations of components in blocks 8 and 9 have been highlighted. To generate a watermarked floorplan, a six-digit author's signature '$\alpha\beta\alpha\beta\gamma\gamma$' is entered as shown in Figure 4.40. On clicking on the '*View Final Floorplan*' button shown at output panel in Figure 4.40, the final watermarked floorplan is generated in a new window. The final watermarked floorplan of FIR filter application is shown in Figure 4.41. As shown in the figure, odd FU instance

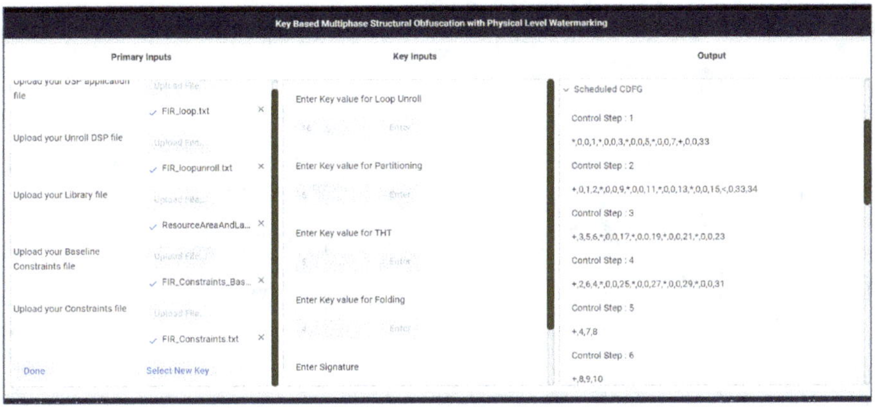

Figure 4.30 Output of scheduling shown onto the tool; excerpt-1: control steps 1–6

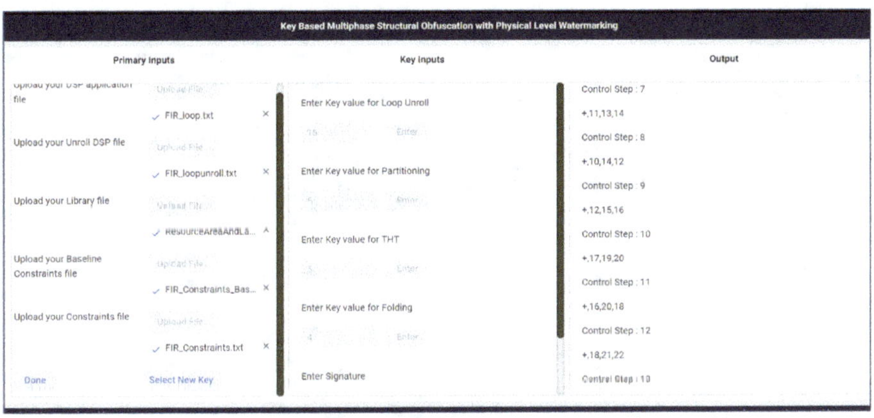

Figure 4.31 Output of scheduling shown onto the tool; excerpt-2: control steps 7–13

M->1 on the top of even FU instance M->4 shows the embedding of first α digit of the signature '$\alpha\beta\alpha\beta\gamma\gamma$'. Further, odd FU instance M->3 on the top of even FU instance M->2 shows the embedding of second α digit. Similarly, odd Mux instance x4->1 on the top of even Mux instance x4->2 shows the embedding of first β digit, whereas odd Mux instance x4->3 on the top of even Mux instance x4->4 shows the embedding of second β digit. Further, odd Demux instance d4->1 at the right to even Demux instance d4->4 shows the embedding of first γ digit, whereas odd Demux instance d4->3 at the right to even Demux d4->2 shows the embedding of second γ digit.

Thus, the SO and physical-level watermarking can be simulated and analysed using the KSO-PW tool developed by the authors. This tool is useful for various

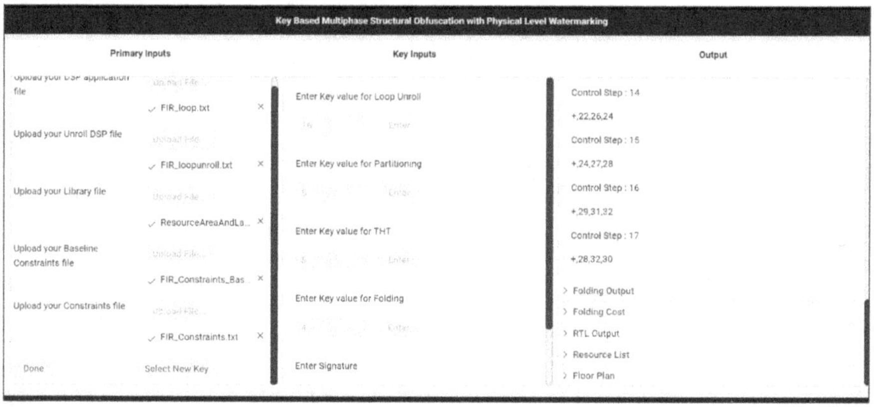

Figure 4.32 Output of scheduling shown onto the tool; excerpt-3: control steps 14–17

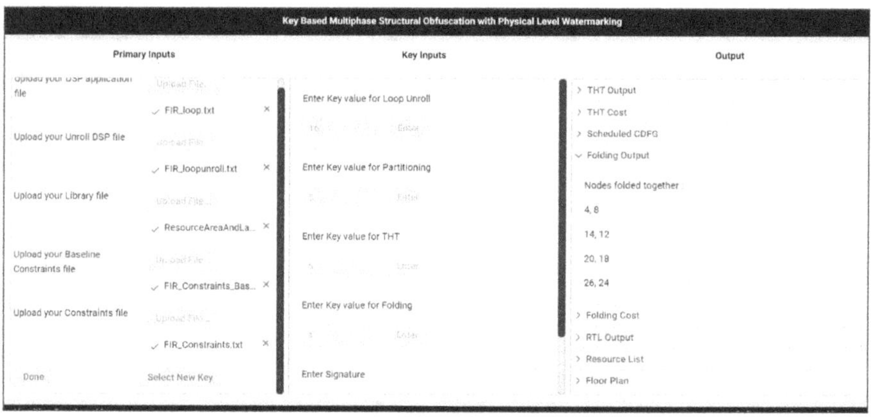

Figure 4.33 Nodes/operations subjected to folding shown onto the tool

kind of DSP hardware accelerator applications such as FIR filter, IIR filter, DCT, and autoregression filter (ARF). In addition, the tool evaluates and shows the design cost pre and post performing a double line of defence for hardware security.

4.8 Analysis of case studies

The double line of defence approach, proposed by Sengupta and Rathor (2020), offers security to DSP hardware accelerators without incurring any design cost overhead. This section discusses the security and design cost analysis of a double line of defence approach based on multi-key-driven SO and physical-level watermarking. The security analysis due to SO has been discussed in terms of difference

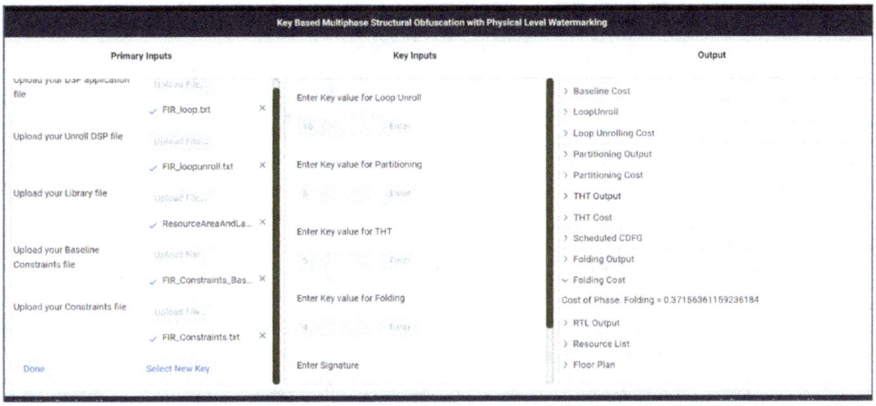

Figure 4.34 Design cost post folding shown onto the tool

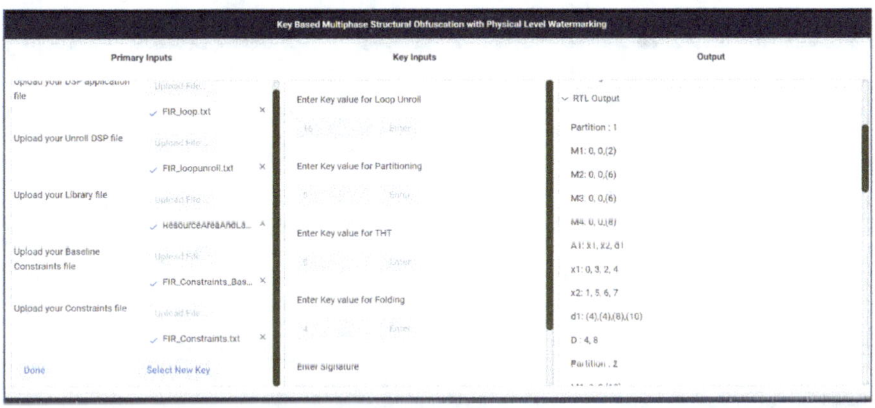

Figure 4.35 RTL output (of partition 1) of obfuscated FIR filter design shown onto the tool

in gate count (that creates obscurity) incurred due to obfuscation and SO-key size. Further, security analysis due to physical-level watermarking has been discussed in terms of probability of coincidence metric, tamper tolerance ability and brute-force attack analysis. Further, an analysis of total key size due to multi-key-driven SO and physical-level watermarking has been discussed. Furthermore, the security and design cost analysis for low-cost optimized multi-key-based SO have been discussed in a separate subsection. This case study of security and design cost analysis for various DSP applications gives a deeper insight about the robustness of multi-key-based SO and physical-level watermarking-based double line of defence approach and its impact on overall design cost. The discussions are presented as follows.

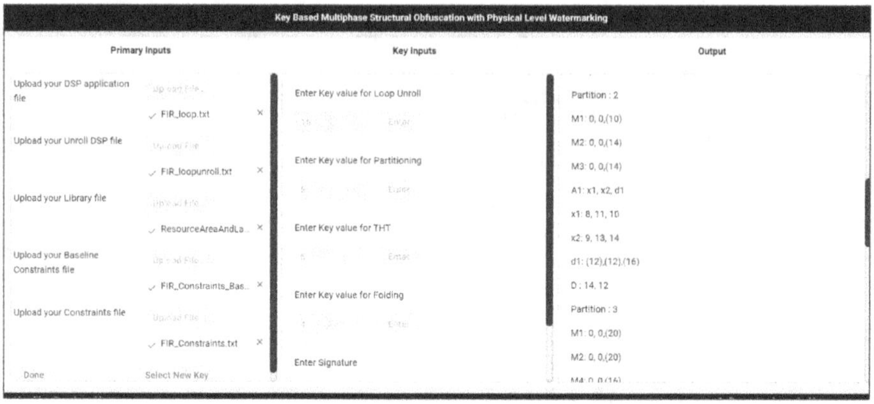

Figure 4.36 RTL output (of partitions 2 and 3) of obfuscated FIR filter design shown onto the tool

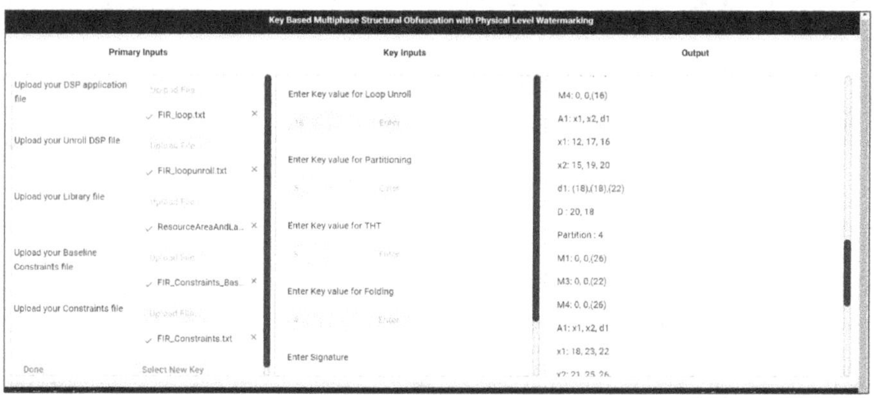

Figure 4.37 RTL output (of partitions 3 and 4) of obfuscated FIR filter design shown onto the tool

4.8.1 Analysis of case studies for a double line of defence – structural obfuscation and physical-level watermarking

4.8.1.1 Security analysis

The multiple SO-keys-based SO acts as a first line of defence and physical-level watermarking acts as a second line of defence to secure DSP hardware accelerators against RE, Trojan insertion and piracy threats. The security achieved using both the lines of defence has been discussed separately as follows (Sengupta and Rathor, 2020).

Security analysis of multi-key-based structural obfuscation

As discussed earlier, the multiple SO-keys-based SO provides security to DSP hardware accelerators in the form of first line of defence. Here, the SO-based first

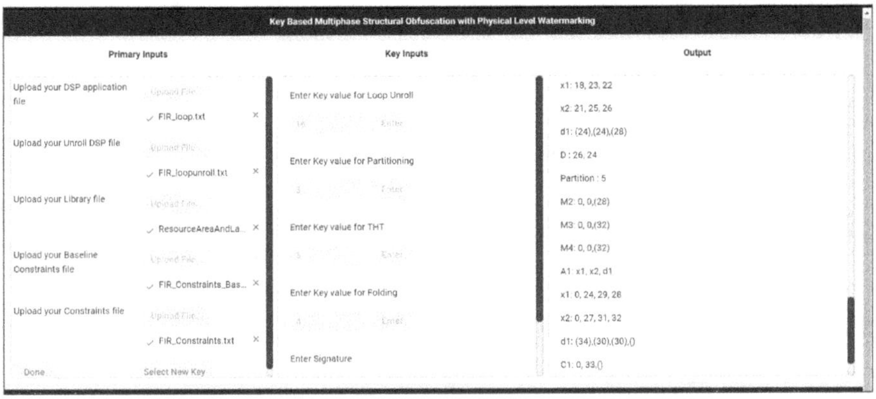

Figure 4.38 RTL output (of partitions 4 and 5) of obfuscated FIR design filter shown onto the tool

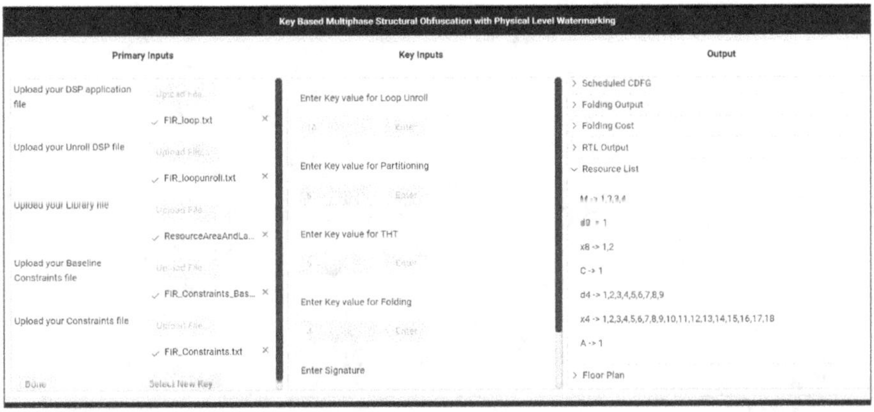

Figure 4.39 Extracted RTL components (resource list) shown on to the tool

line of defence acts as preventive control against Trojan (malicious logic) insertion and piracy threats. The preventive control is enabled because the employed multiple SO-keys-based SO incurs massive obscurity into the design structure, hence rendering the interpretation of true functionality and structure of the design highly complicated for an attacker. Therefore, an attacker, who aims to RE to infect the design using Trojan insertion or aims to pirate the design, fails/gets hindered and hence becomes unable to realize his malicious intents. The robustness of SO due to employing multiple structural transformations is measured in terms of difference in gate count pre-obfuscation and post-obfuscation. Figure 4.42 highlights the difference in gate count pre and post employing SO. The huge difference in gate count (that creates obscurity) shown in Figure 4.42 indicates the robustness of multiple SO-keys-based SO technique. The change in gate count (without affecting

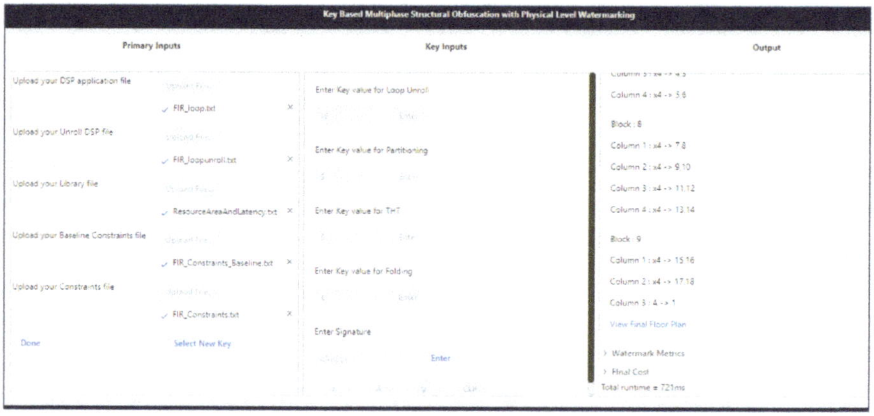

Figure 4.40 Feeding of author's signature into the tool to generate watermarked floorplan

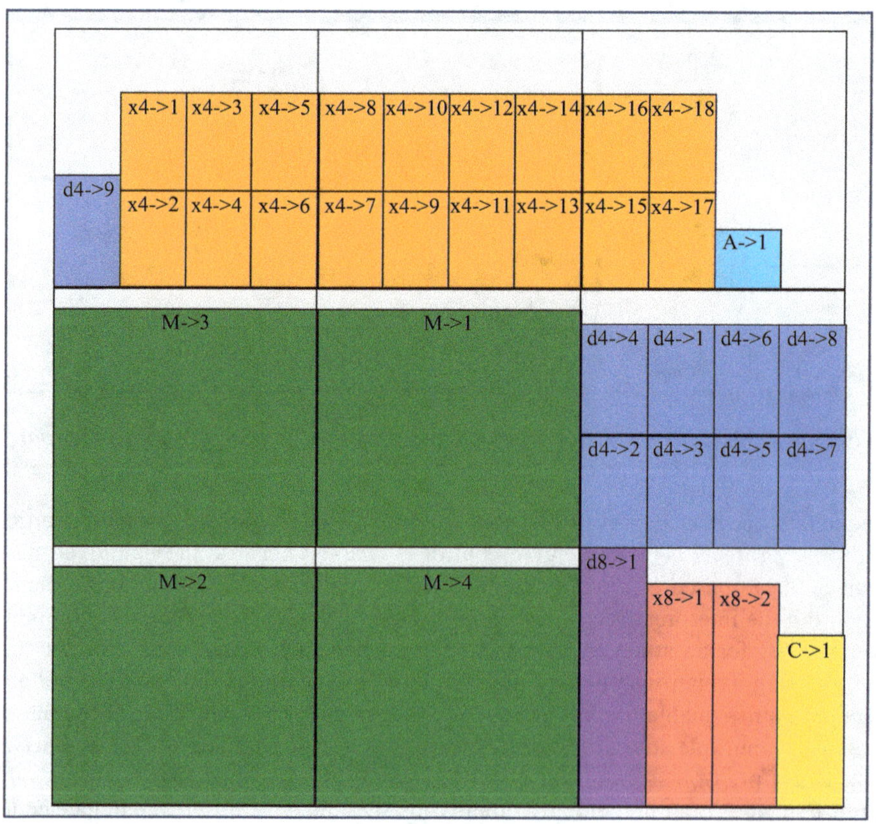

Figure 4.41 Final watermarked floorplan displayed by the tool

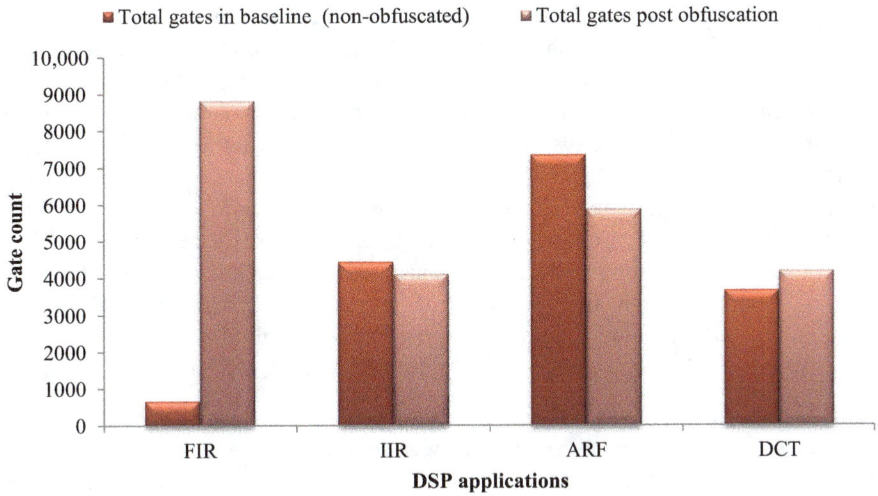

Figure 4.42 Strength of structural obfuscation in terms of difference in gate count

functionality) incurred due to SO is an indication of the robustness of obfuscation because of the following reasons: (i) the change in gate count post obfuscation does not follow any particular pattern (for inferring the original structure/functionality) but rather depends on the obfuscation techniques employed and designer's chosen SO-keys and (ii) the change in gate count not only makes a difference in the count but also results into alterations in the gates connectivity hence leads to an obfuscated netlist.

As shown in Figure 4.42, the amount of gates changed due to obfuscation depends on the type and size of the application (as evident from the figure, there is no fixed pattern in the gate count difference for different DSP application). This is because different applications have different operation count and data dependency. Thus, the nature of the applications determines whether a particular type of obfuscation technique (such as LU, partitioning, ROE, THT and folding knob) is applicable or not and further to what extent it is applicable. Thereby, for different DSP applications, the applied structural transformation techniques of SO creates different kinds of modifications in the resource interconnectivity, the number of resources, size and count of the Muxes/Demuxes and storage elements, which, in turn, modifies the overall gate count (creating obscurity while preserving functionality). For example, the gate count of IIR and ARF filters reduces post applying multiple SO-keys-based SO. The reason is that the applied obfuscation techniques do not result into larger size multiplexers and demultiplexers with respect to the non-obfuscated (baseline) counterpart. Therefore, the gate count post obfuscation reduces, whereas the gate count of FIR and DCT core is augmented post-obfuscation as shown in Figure 4.42 (Sengupta and Rathor, 2020).

In addition, the multiple secret SO-keys used to regulate the process of SO play a vital role in enhancing the robustness of obfuscation. The reasons are as follows: (i) each structural transformation technique is regulated using a specific secret key value which decides the extent to which the structural transformation has to be performed, (ii) only being aware of the applied transformations cannot help an adversary in performing RE. An attacker needs to be aware of both the applied structural transformation techniques and the secret SO-keys used and (iii) each individual secret SO-key contributes to augment the total size (space) of key; hence, it becomes challenging to find an exact correct key among exhaustive possibilities. Therefore, the incorporation of secret SO-keys in the SO process enhances the security level manifold. Figure 4.43(a) highlights the total SO-key size for different DSP applications (Sengupta and Rathor, 2020).

Security analysis of physical-level watermarking

As discussed earlier, the physical-level watermarking provides security to DSP hardware accelerators in the form of second line of defence. Here, the physical-level watermarking-based second line of defence acts as detective control against counterfeiting and cloning threats. The detective control is enabled by embedding vendor's secret signature into the early floorplan of the physical design process. The embedded watermark is detected to identify counterfeited and cloned designs. The robustness of the watermark embedded at physical level has been analysed by Sengupta and Rathor (2020) in terms of the following: (i) probability of coincidence, (ii) tamper tolerance, (iii) brute-force analysis and (iv) key bits required to represent the total space of vendor's secret signature. Let us see the discussion on each analysis one-by-one.

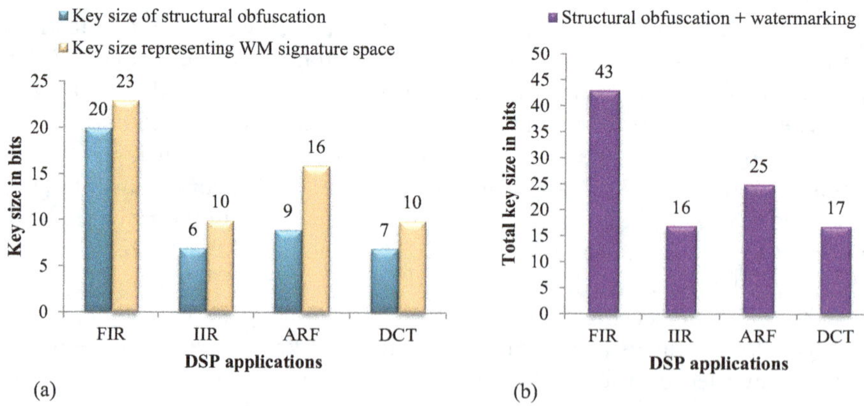

Figure 4.43 Key size analysis: (a) key size of structural obfuscation and key size of watermarking and (b) total key size of a double line of defence approach (Sengupta and Rathor, 2020)

The probability of coincidence (Pc) metric is measured as follows (Sengupta and Rathor, 2020):

$$Pc = \left(\prod_{i=1}^{\alpha} \frac{1}{\{\sum_{k1 \in F_p} (k1(k1-1)/2)\} - x++} \right) * \left(\prod_{j=1}^{\beta} \frac{1}{\{\sum_{k2 \in X_q} (k2(k2-1)/2)\} - y++} \right) * \left(\prod_{k=1}^{\gamma} \frac{1}{\{\sum_{k3 \in D_r} (k3(k3-1)/2)\} - z++} \right) \tag{4.12}$$

where $k1$ denotes the total number of FU resource components of type F_p, where p denotes the total types of FU resources; $k2$ denotes the number of multiplexers of size X_q, where q denotes different sizes of multiplexers in the design and $k3$ denotes the number of demultiplexers of size D_r, where r denotes different sizes of demultiplexers in the design. The ranges of variables x, y and z are as follows: 0arx $\leq \alpha - 1$, 0,$y \leq \beta - 1$, 0 $z \leq \gamma - 1$, where x, y and z variables are incremented with the embedding of each digit of signature variables α, β and γ, respectively. The interpretation of individual terms in the formula is as follows:

$\left\{\sum_{k1 \in F_p} (k1(k1-1)/2)\right\}$ indicates all swapping pairs corresponding to all the types of FU resource components in the set extracted from obfuscated RTL design (i.e. swapping pairs of all multiplier instances+swapping pairs of all adder instances+swapping pairs of all instances of pth FU resource).

$\left\{\sum_{k1 \in F_p} (k1(k1-1)/2)\right\} - x++$ indicates **remaining** swapping pairs after embedding an α digit.

$1/\left\{\sum_{k1 \in F_p} \left(\frac{k1(k1-1)}{2}\right)\right\} - x++$ indicates the probability of obtaining/detecting one pair of FU modules corresponding to an α digit.

$\left(\prod_{i=1}^{\alpha} \left(1/\left\{\sum_{k1 \in F_p} \left(\frac{k1(k1-1)}{2}\right)\right\} - x++\right)\right)$ indicates the probability of obtaining/detecting all pairs of FU modules corresponding to all embedded α digits in a non-watermarked design by an attacker.

$\left\{\sum_{k2 \in X_q} (k2(k2-1)/2)\right\}$ indicates all swapping pairs corresponding to all the sizes of Mux components in the set extracted from obfuscated RTL design (i.e. swapping pairs of all the instances of Mux of one size+swapping pairs of all the instances of Mux of next size+swapping pairs of all the instances of Mux of qth size).

$\left\{\sum_{k2\in X_q}(k2(k2-1)/2)\right\}-y++$ indicates **remaining** swapping pairs after embedding a β digit.

$\left(1/\left\{\sum_{k2\in X_q}\left(\frac{k2(k2-1)}{2}\right)\right\}-y++\right)$ indicates the probability of obtaining/detecting one pair of Mux modules corresponding to a β digit.

$\left(\prod_{j=1}^{\beta}1/\left\{\sum_{k2\in X_q}\left(\frac{k2(k2-1)}{2}\right)\right\}-y++\right)$ indicates the probability of obtaining/detecting all the pairs of Mux modules corresponding to all embedded β digits in a non-watermarked design by an attacker.

$\left\{\sum_{k3\in D_r}(k3(k3-1)/2)\right\}$ indicates all the swapping pairs corresponding to all the sizes of Demux components in the set extracted from obfuscated RTL design (i.e. swapping pairs of all the instances of Demux of one size+swapping pairs of all the instances of Demux of next size+swapping pairs of all the instances of Demux of rth size).

$\left\{\sum_{k3\in D_r}(k3(k3-1)/2)\right\}-z++$ indicates **remaining** swapping pairs after embedding a γ digit.

$1/\left\{\sum_{k3\in D_r}\left(\frac{k3(k3-1)}{2}\right)\right\}-z++$ indicates the probability of obtaining/detecting one pair of Demux modules corresponding to a γ digit.

$\left(\prod_{k=1}^{\gamma}\left(1/\left\{\sum_{k3\in D_r}\left(\frac{k3(k3-1)}{2}\right)\right\}-z++\right)\right)$ indicates the probability of obtaining/detecting all the pairs of Demux modules corresponding to all embedded γ digits in a non-watermarked design by an attacker.

The probability of coincidence captures the probability of coincidently finding the same signature in a non-watermarked design. If the probability of coincidence value is high, the watermarked is considered weak. Therefore, lower *Pc* value is desirable, which indicates that a large amount of digital evidence is embedded into the design, hence indicating high robustness of watermark. Figure 4.44 shows the probability of coincidence obtained using physical-level watermarking. The figure shows that significantly lower *Pc* is achieved for all DSP applications. The reason of obtaining lower *Pc* is that the signature digits corresponding to the multiple variables (i.e. α, β and γ) have been embedded using different types and size of RTL components in the early floorplan stage. Thus, a stronger detective-control-based security is achieved by embedding a robust physical-level watermark (Sengupta and Rathor, 2020).

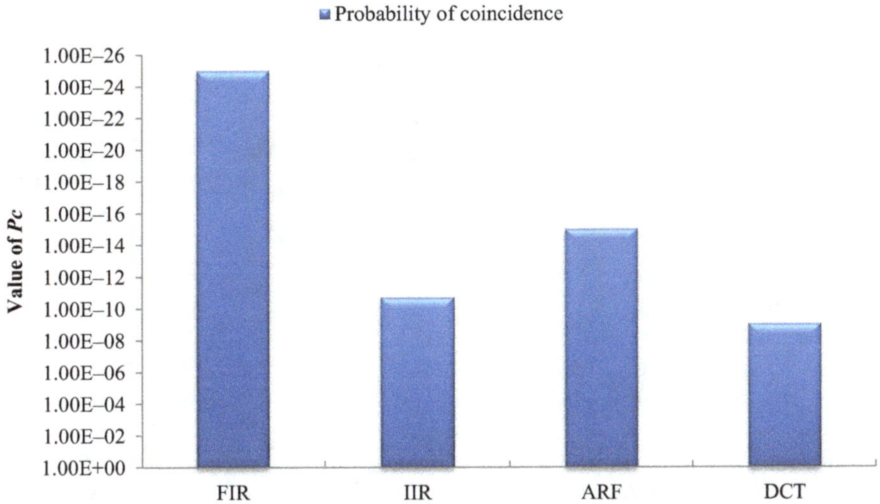

Figure 4.44 Robustness of physical-level watermarking in terms of Pc (Sengupta and Rathor, 2020)

The tamper tolerance capability of embedded physical-level watermark is measured in terms of total signature combinations representing the signature space of watermark. Because of embedding watermark corresponding to multiple signature variables (α, β and γ), the signature space of watermark is significantly high. Therefore, the attacker's effort of finding correct signature, to eliminate the signature digits by tampering watermarking constraints, becomes significantly high. Hence, ability to tolerate the tampering, caused by the adversary, is high. The formula to estimate the tamper tolerance capability (T^P) is as follows (Sengupta and Rathor, 2020):

$$T^P = Z^Q \tag{4.13}$$

where Z represents the number of signature variables used in the watermark and Q represents the size of the vendor's signature. The value of Z^Q represents the signature space of watermark (i.e. total possible combinations of signature), which, in turn, shows the tamper tolerance capability of the watermark. Figure 4.45(a) depicts the tamper tolerance capability (using (4.13)) of the physical-level watermark for vendor's chosen signature strength. As shown in the figure, very high value of tamper tolerance is achieved. This indicates that the physical-level watermark proposed by Sengupta and Rathor (2020) is strong against tampering.

Further, the security against removal attack is ensured using brute-force attack analysis of the signature. The security against removal attack on signature using brute-force analysis is measured in terms of probability of finding the valid

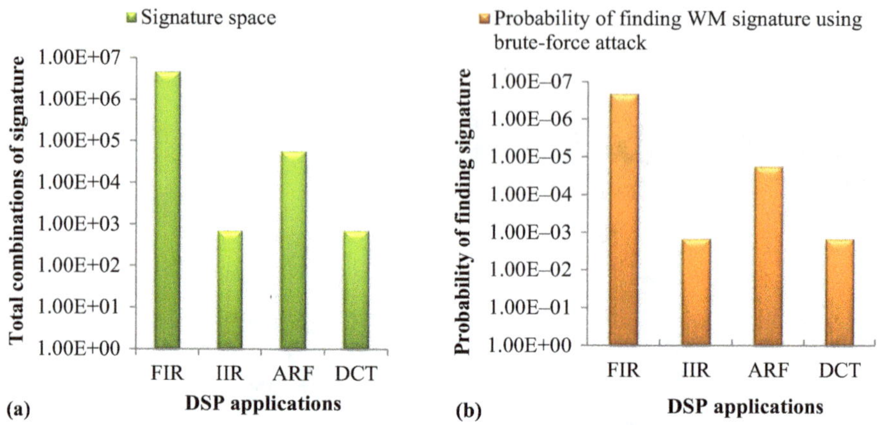

Figure 4.45 Security analysis of physical-level watermarking in terms of (a) tamper tolerance analysis and (b) brute-force attack analysis (Sengupta and Rathor, 2020)

signature within exhaustive signature combinations (signature space). Hence, the security is measured using the following formula (Sengupta and Rathor, 2020):

$$S^B = \frac{1}{Z^Q} \tag{4.14}$$

where S^B indicates the probability of finding correct signature by an attacker using brute-force analysis and Z^Q represents the signature space of watermark. The lower the value of S^B, the higher the security against removal attack on signature. Figure 4.45(b) depicts the brute-force analysis using the security metric given in (3.14). As shown in the figure, very low probability of finding correct signature using brute-force analysis is achieved. Hence, it indicates the strong security against removal attack on signature by an attacker (Sengupta and Rathor, 2020).

Further, the number of bits required to represent the total space of the signature embedded during physical-level watermarking is calculated using $\lceil \log_2 (Z^Q) \rceil$. This formula gives the key size in bits which captures the signature space of watermark. Figure 4.43(a) shows the key size required to represent the signature space. Further, Figure 4.43(b) shows the total key size of the double line of defence approach which sums up the key size of the SO and the required key size to capture the whole signature space of the physical-level watermark. It indicates the hardship of an attacker in terms of finding correct key of SO and the valid signature of physical-level watermark (Sengupta and Rathor, 2020).

4.8.1.2 Design cost analysis

It is important to ensure that the employed security mechanism should not incur excessive design overhead. A security mechanism that results into minimal or zero design overhead is considered effective and practical. Therefore, the design cost of the double line of defence approach needs to be analysed. The following equation is used to evaluate the design cost (Sengupta and Rathor, 2020):

$$C_d(U_i) = \rho_1 \frac{L_d}{L_m} + \rho_2 \frac{A_d}{A_m} \tag{4.15}$$

where $C_d(U_i)$ is the design cost calculated for resource constraints U_i, further L_d and L_m are the design latency at specified resource constraints and maximum design latency, respectively, A_d and A_m are the design area at specified resource constraints and maximum area, respectively, and ρ_1 and ρ_2 are the weights which are fixed at 0.5. The analysis of design cost post-employing each line of defence is discussed as follows.

Design cost analysis of multi-key-based structural obfuscation

For the evaluation of the design cost of multi-key-based SO, the design area and latency are calculated using 15-nm NanGate library (Sengupta *et al.*, 2020). The calculation of deign area is based on the area of resources in the obfuscated design, whereas the calculation of latency is based on the scheduling of structurally obfuscated design (Sengupta *et al.*, 2020). Figure 4.46 compares the design cost post-employing SO with respect to the baseline (un-obfuscated) counterpart. As shown in the figure, the obfuscation mechanism incurs zero design overhead for most of the applications. This is because the applied high-level transformations during structurally obfuscating the design also results into a sort of optimization in the structure as by-product. As shown in Figure 4.46, the design cost of FIR filter

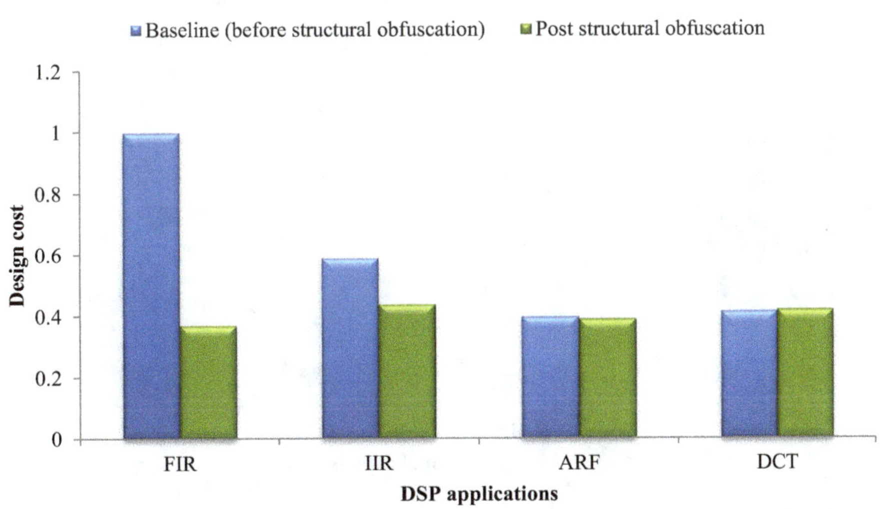

Figure 4.46 Design cost comparison pre and post structural obfuscation with respect to baseline (Sengupta et al., 2020)

application reduces significantly. The reason is the applicability of LU-based structural transformation which causes parallelism of operations. This leads to substantial reduction in the design latency, hence reducing the overall design cost. Further, design cost of some applications (such as DCT) slightly increases post-obfuscation. This is due to the nature of target application and applicability of different obfuscation techniques affecting the Mux/Demux size and their count.

Design cost analysis of structural obfuscation and physical-level watermarking

For the evaluation of design cost post-SO and physical-level watermarking, the design area is measured in terms of the area of the enveloping rectangle of the floorplan design (Sengupta and Rathor, 2020), whereas scheduling of design is exploited for determining the latency. The comparison of design cost of structurally obfuscated watermarked design with the baseline (un-obfuscated) design is shown in Figure 4.47. As shown in the figure, the SO and physical-level watermarking-based double line of defence mechanism incurs zero design cost overhead. Moreover, the design cost post employing a double line of defence reduces because of reduction in either design latency or floorplan area post employing SO. The impact of physical-level watermarking on design cost is nil. This is because, the physical-level watermark has been embedded into the floorplan by swapping the RTL components of the same type and the same size. Therefore, no design cost overhead incurs. Let us analyse the case study on FIR filter application. The cost of the obfuscated FIR filter design is significantly lesser than the baseline design. The underlying reason is the substantial reduction in the latency post-employing SO.

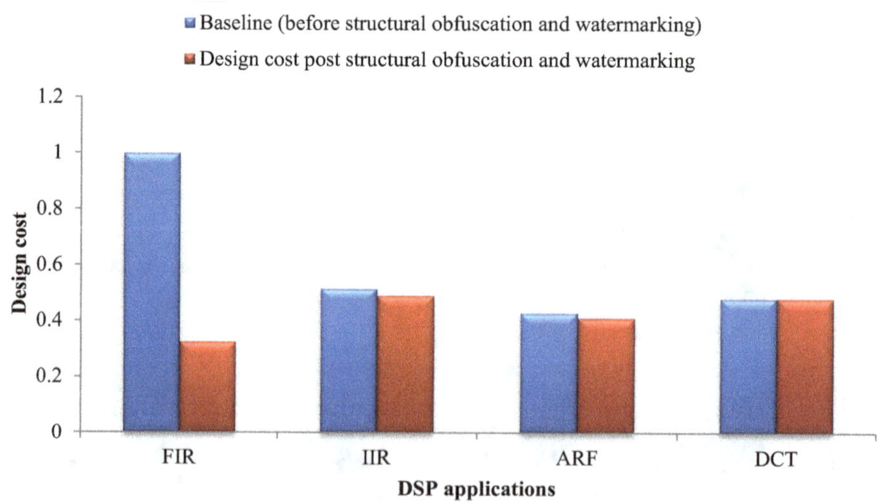

Figure 4.47 Design cost comparison pre and post structural obfuscation and physical-level watermarking with respect to baseline (Sengupta and Rathor, 2020)

The reduction in latency is achieved because of key-driven UF-based LU transformation which causes more parallelization of the operations (due to duplicate iterations of loop body) during scheduling, hence resulting into lesser delay. This kind of parallelization of operations is absent in the baseline (un-obfuscated) FIR filter design; therefore, it has more delay and hence larger design cost than the obfuscated version. Further, for other DSP applications shown in Figure 4.47, LU-based structural transformation is not applicable. Therefore, the design latency does not change post-SO. However, the SO results into a slight decrement in the area of enveloping rectangle of the structurally obfuscated floorplan. The area is reduced because of reduction in the sizes of Muxes and Demuxes post-obfuscation. In general, the type and size of DSP applications and the applicability of structural transformation together determine the increment/decrement in the interconnect hardware resources (size and number of the Muxes and Demuxes), storage resources and FU resources (Sengupta and Rathor, 2020).

4.8.2 Analysis of case studies for low-cost optimized multi-key-based structural obfuscation

The analysis of the low-cost multi-key-based SO process (presented in Section 4.6) has been discussed, for various DSP applications, in terms of design cost and security.

4.8.2.1 Security analysis

The multi-key-based SO approach secures a DSP design against RE by obfuscating it in terms of structural transformation resulting into affecting larger amount of gates structurally (while preserving functionality), such that it becomes unobvious to interpret to an attacker or outsider. The gates are affected in terms of change in gates interconnectivity and change in total gate count post-obfuscation. Figure 4.48 depicts the number of gates transformed (change in the gate count) due to this obfuscation process, with respect to its equivalent un-obfuscated design. As shown in the figure, significant change in gate count post-applying obfuscation is achieved thereby making it appear unobvious or non-meaningful during inspection. Higher the number of gates transformed (i.e. change in the gate count), more is the obfuscation expected in the design, thereby more difficult it is for an attacker to interpret it functionally, thus ensuring higher security. For example, a very high increase in the gate count of the obfuscated FIR filter design compared to the baseline design is observed owing to the LU transformation being applied on it along with other successive transformations.

4.8.2.2 Design cost analysis

The impact of PSO-based DSE on multi-key-based SO approach has been analysed in terms of design cost using (4.11). Figure 4.49 compares the cost of the obfuscation approach without PSO-DSE vs. the cost of the obfuscation approach with PSO-DSE, for different DSP cores. The design cost of multi-key-based structurally obfuscated design with PSO-DSE module is achieved to be lesser than that without

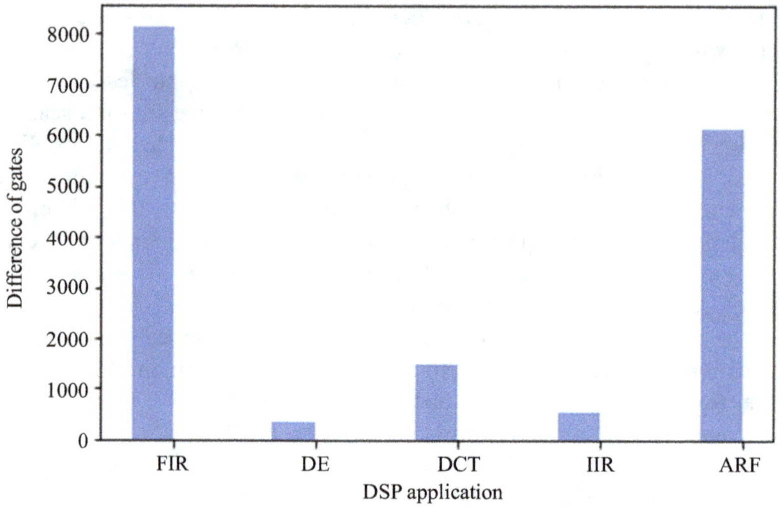

Figure 4.48 Security of multi-key-based structural obfuscation in terms of number of gates affected

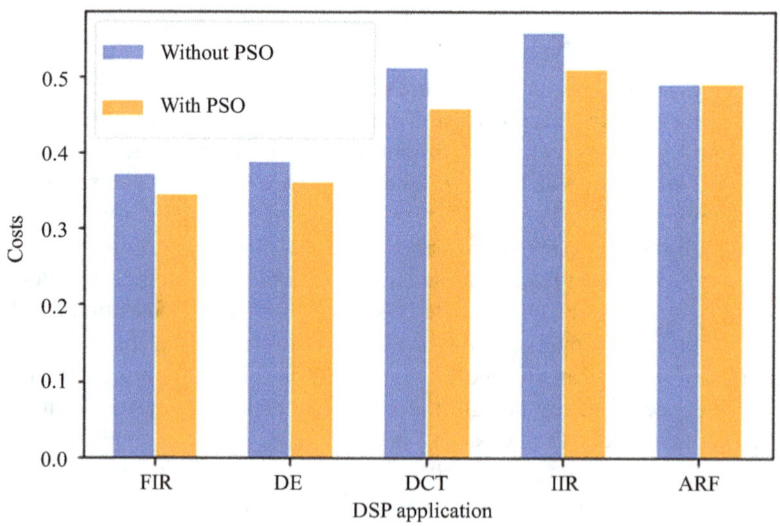

Figure 4.49 Design cost analysis of multi-key-based structural obfuscation approach with and without PSO-DSE

PSO-DSE module. This is because, PSO-based DSE produces an optimal architecture (resource configuration) which is used to schedule the structurally obfuscated design. On overage, 6.58% reduction is achieved upon integrating multi-key-based SO approach with PSO-DSE process.

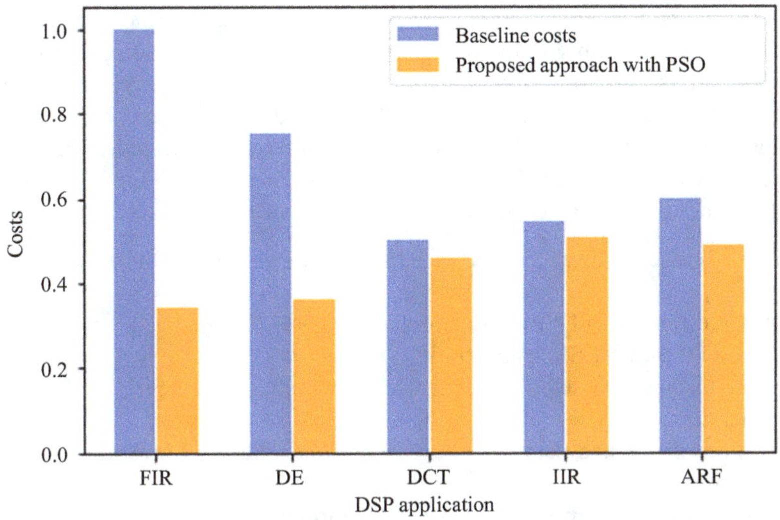

Figure 4.50 Design cost comparison of multi-key-based structural obfuscation approach with the baseline

Further, Figure 4.50 depicts the comparison of baseline cost with the cost of the multi-key-based SO approach with PSO-DSE module. Because of using optimal architecture obtained using PSO-DSE, lower design cost is achieved for the obfuscation approach, compared to baseline design cost. A drastic reduction in design cost is achieved in the case of FIR filter and sample DFG (DE) applications as shown in Figure 4.50. This is because in these applications, being loop-based, LU-based transformation was applied which contributed to the huge reduction in delay. Since LU increases parallelization, therefore, execution delay is decreased.

4.9 Conclusion

The DFS aspect in the VLSI design process has become very important because of potential hardware threats such as Trojan insertion and piracy. This chapter discusses DFS using a double line of defence technique to offer enhanced security to DSP hardware accelerators. The first line of defence based on multiple SO-key-driven SO provides preventive measure, whereas the second line of defence based on physical-level watermarking provides detective measure against the hardware threats. Employing multiple techniques of structural transformations, each driven through a key value, incurs huge obscurity in the design structure, hence resulting into a robust SO. Since the employed high-level transformation techniques also optimize the design structural, therefore, the probability of incurring design overhead due to SO is almost zero. Further, an author's watermark is embedded into early floorplan of obfuscated design during physical design process. The embedded watermark has a larger signature space because of comprising multiple variables.

Therefore, the watermark is highly robust because it results into very low probability of coincidence, high tamper tolerance ability and high security against the removal attack. In addition, the embedding rules of physical-level watermark are such that it does not result into design overhead. Additionally, PSO-based extensive DSE has been applied to yield a low-cost structurally obfuscated design with an average improvement of 6.58% in the design cost compared to the baseline.

At the end of this chapter, the following concepts are communicated to readers:

- importance and applications of DSP hardware accelerators in electronic systems;
- hardware threats to DSP hardware accelerators;
- need of DFS in VLSI design flow;
- a double line of defence-based DFS technique to secure DSP hardware accelerators;
- first line of defence using SO-key-driven multiple high-level transformation-based SO;
- physical-level watermark during early floorplanning;
- detection of physical-level watermarking;
- demonstration of the process of employing multi-key-based SO;
- demonstration of the process of generating a watermarked floorplan of an obfuscated design by embedding author signature;
- importance of PSO-DSE integration with the security algorithm;
- obtaining optimal architecture using PSO-DSE to apply SO-based security;
- security and design cost analysis of a double line of defence for various DSP applications and
- security and design cost analysis of the low-cost multi-key-based SO approach with respect to baseline version.

4.10 Questions and exercise

1. What is partitioning? What is the rule for partitioning?
2. How is key-driven loop unrolling performed?
3. What is key-driven folding-knob-based transformation?
4. What is a double line of defence employed for hardware accelerators?
5. What are the inputs required for designing watermark-implanted obfuscated DSP hardware accelerator?
6. How many secret keys are used for obfuscation and what is the role of each?
7. What impact does key-based structural obfuscation have on the RTL/gate-level design structure?
8. What is the concept of 'early floorplanning' and how is it useful?
9. In what sequence are the transformations in structural obfuscation performed? How would it vary if the transformation sequence is changed?
10. Explain the key size of each secret keys used.
11. Demonstrate key-based loop unrolling on 16-tap FIR.
12. Demonstrate key-based partitioning on 16-tap FIR.

13. Demonstrate key-based THT on 16-tap FIR.
14. Derive the generic expression of an 8-point DCT computation function.
15. How is the RTL circuit of key-based structurally obfuscation design generated?
16. What is the difference between folding factor and folding knob?
17. Explain encoding algorithm of physical-level watermarking.
18. Explain physical-level watermark detection algorithm. What are the inputs required for this process?
19. How do you evaluate the total key size of structurally obfuscated designs?
20. Explain the flow of low-cost optimized structural obfuscation process.
21. What is velocity clamping in PSO-DSE?
22. In the KSO-PW tool, what inputs are required? What is the output generated?
23. Give examples of DSP hardware accelerator applications where key-based loop unrolling is not applicable.
24. Give examples of DSP hardware accelerator applications where key-based THT is not applicable.
25. Give examples of DSP hardware accelerator applications where key-based ROE is not applicable.

References

E. Castillo, U. Meyer-Baese, A. Garcia, L. Parilla and A. Lloris (2007), 'IPP@HDL: efficient intellectual property protection scheme for IP cores,' *IEEE Trans. Very Large Scale Integr. VLSI Syst.*, vol. 15(5), pp. 578–590.

R. S. Chakraborty and S. Bhunia (2009), 'Security against hardware Trojan through a novel application of design obfuscation,' *Proc. of the International Conference on Computer-Aided Design*, ACM, pp. 113–116.

R. S. Chakraborty and S. Bhunia (2011), 'Security against hardware Trojan attacks using key-based design obfuscation,' *J. Electron. Test.*, vol. 27(6), pp. 767–785.

I. Hong and M. Potkonjak (1999), 'Behavioral synthesis techniques for intellectual property security,' *Proc. DAC*, pp. 849–854.

Y. Lao and K. K. Parhi (2015), 'Obfuscating DSP circuits via high-level transformations,' *IEEE Trans. Very Large Scale Integr. VLSI Syst.*, vol. 23(5), pp. 819–830.

B. Le Gal and L. Bossuet (2012), 'Automatic low-cost IP watermarking technique based on output mark insertions,' *Des. Autom. Embedded Syst.*, vol. 16(2), pp. 71–92.

L. Li and H. Zhou (2013), 'Structural transformation for best-possible obfuscation of sequential circuits,' *Proc. HOST*, Austin, TX, pp. 55–60.

V. K. Mishra and A. Sengupta (2014), 'MO-PSE: adaptive multi-objective particle swarm optimization based design space exploration in architectural synthesis for application specific processor design,' *Adv. Eng. Softw.*, vol. 67, pp. 111–124.

S. M. Plaza and I. L. Markov (2015), 'Solving the third-shift problem in IC piracy with test-aware logic locking,' *IEEE Trans. Comput. Aided Des. Integr. Circuits Syst.*, vol. 34(6), pp. 961–971.

D. Roy and A. Sengupta (2019), 'Multilevel watermark for protecting DSP kernel in CE systems [hardware matters],' *IEEE Consum. Electron. Mag.*, vol. 8(2), pp. 100–102.

R. Schneiderman (2010), 'DSPs evolving in consumer electronics applications,' *IEEE Signal Process. Mag.*, vol. 27(3), pp. 6–10.

A. Sengupta (2016), 'Intellectual property cores: protection designs for CE products,' *IEEE Consum. Electron. Mag.*, vol. 5(1), pp. 83–88.

A. Sengupta (2017), 'Hardware security of CE devices [hardware matters],' *IEEE Consum. Electron. Mag.*, vol. 6(1), pp. 130–133.

A. Sengupta and S. Bhadauria (2016), 'Exploring low cost optimal watermark for reusable IP cores during high level synthesis,' *IEEE Access*, vol. 4, pp. 2198–2215.

A. Sengupta, E. R. Kumar and N. P. Chandra (2019), 'Embedding digital signature using encrypted-hashing for protection of DSP cores in CE,' *IEEE Trans. Consum. Electron.*, vol. 65(3), pp. 398–407.

A. Sengupta and S. P. Mohanty (2019), 'IP core and integrated circuit protection using robust watermarking', IP Core Protection and Hardware-Assisted Security for Consumer Electronics', e-ISBN: 9781785618000, pp. 123–170.

A. Sengupta and M. Rathor (2019a), 'IP core steganography for protecting DSP kernels used in CE systems,' *IEEE Trans. Consum. Electron.*, vol. 65(4), pp. 506–515.

A. Sengupta and M. Rathor (2019b), 'Protecting DSP kernels using robust hologram-based obfuscation,' *IEEE Trans. Consum. Electron.*, vol. 65(1), pp. 99–108.

A. Sengupta and M. Rathor (2020), 'Enhanced security of DSP circuits using multi-key based structural obfuscation and physical-level watermarking for consumer electronics systems,' IEEE Trans. Consum. Electron., doi: 10.1109/TCE.2020.2972808.

A. Sengupta, M. Rathor, S. Patil and N. G. Harishchandra (2020), 'Securing hardware accelerators using multi-key based structural obfuscation,' *IEEE Lett. Comput. Soc.*, vol. 3(1), pp. 21–24.

A. Sengupta and D. Roy (2017), 'Antipiracy-aware IP chipset design for CE devices: a robust watermarking approach [hardware matters],' *IEEE Consum. Electron. Mag.*, vol. 6(2), pp. 118–124.

A. Sengupta and D. Roy (2017), 'Protecting an intellectual property core during architectural synthesis using high-level transformation based obfuscation,' *Electron. Lett.*, vol. 53(13), pp. 849–851.

A. Sengupta, D. Roy and S. P. Mohanty (2018), 'Triple-phase watermarking for reusable IP core protection during architecture synthesis,' *IEEE Trans. Comput. Aided Des. Integr. Circuits Syst.*, vol. 37(4), pp. 742–755.

A. Sengupta (2020), 'Frontiers in securing IP cores – forensic detective control and obfuscation techniques', *The Institute of Engineering and Technology (IET)*, ISBN-10: 1-83953-031-6, ISBN-13: 978-1-83953-031-9.

A. Sengupta, D. Roy, S. P. Mohanty and P. Corcoran (2017), 'DSP design security in CE through algorithmic transformation based structural obfuscation,' *IEEE Trans. Consum. Electron.*, vol. 63(4), pp. 467–476.

A. Sengupta, D. Roy, S. P. Mohanty and P. Corcoran (2018), 'Low-cost obfuscated JPEG CODEC IP core for secure CE hardware,' *IEEE Trans. Consum. Electron.*, vol. 64(3), pp. 365–374.

X. Zhang and M. Tehranipoor (2011), 'Case study: detecting hardware Trojans in third-party digital IP cores,' *IEEE International Symposium on Hardware-Oriented Security and Trust*, San Diego, CA, pp. 67–70.

Chapter 5

Multimodal hardware accelerators for image processing filters

Anirban Sengupta[1]

The chapter describes hardware accelerators for image processing filters, including design methodology and security technique employed for the following: blur filter, sharpening filter, embossment filter and Laplace edge–detection (ED) filter. The chapter is organized as follows: Section 5.1 discusses the reasons for using dedicated image processing filter hardware, Section 5.2 discusses the motivation for designing secure image processing filter hardware accelerators, Section 5.3 presents the salient features of this chapter, Section 5.4 discusses some selected contemporary approaches, Section 5.5 discusses the theory of 3 × 3 filter hardware accelerator, Section 5.6 presents designing of functionally reconfigurable obfuscated 3 × 3 filter hardware accelerator, Section 5.7 discusses the theory of 5 × 5 filter hardware accelerator, Section 5.8 presents designing of obfuscated 5 × 5 filter hardware accelerator, Section 5.9 presents designing of secured application specific filter hardware accelerators, Section 5.10 presents the equivalent MATLAB® codes for image processing filters, Section 5.11 presents additional information on image processing convolution filters, Section 5.12 presents analysis of case studies, Section 5.13 concludes the chapter and Section 5.14 presents some questions and exercise for the readers.

5.1 Introduction – why dedicated image processing filter hardware is needed?

Image processing functions such as image blurring, sharpening, embossment and ED are performed using specific 2D convolution filters, since images are 2D signals. The kind of filtering performed on images depends on corresponding filter kernel matrix. In the context of modern digital imaging technology, it is advantageous to realize the image processing filters using dedicated hardware. This is because an image filtering is a highly computationally/data intensive task since huge number of pixels are subjected to complex computations (processing) in order

[1]Computer Science and Engineering, Indian Institute of Technology Indore, Indore, India

to generate filtered images. Moreover, variables such as size of images, number of images to be processed per unit time (measured in frames per second) and the complexity of image processing/filtering algorithms are growing with the evolution of digital imaging technology. Therefore, an image filtering process consumes very high processing time and significant amount of power. However, the image filtering processing is expected to be performed under stringent time and power constraints. This is because image filtering tasks are an important application of portable consumer electronics (CE) systems in which a strong battery life and high performance are critical factors. By performing an image filtering function using dedicated hardware (processor), low power and high performance requirements can be satisfactorily fulfilled (Benda *et al.*, 2008; Dutta *et al.*, 2006; Ortega-Cisneros *et al.*, 2014).

Designing of image processing filter hardware accelerators using high-level synthesis (HLS) process that enables achieving the following benefits: (i) easier to model complex image processing filters at behavioural or high level. The high-level or behavioural description of an image processing filter is automatically converted into register transfer level (RTL) design using HLS process, thus reducing the design complexity and design time. (ii) The HLS process integrated with design space exploration helps in achieving target area, power and delay specifications (Mishra and Sengupta, 2014). Thus, designed low power, high performance and area efficient image processing filter hardware accelerators provide benefits of minimal power dissipation, longer battery life and enhanced user experience in terms of system performance (Mahdiany *et al.*, 2001; Sengupta and Mohanty, 2019; Sengupta, 2020).

This chapter discusses few powerful methodologies of designing image processing filter hardware accelerators for various image processing applications such as image blurring, sharpening, embossment and Laplace ED. Sengupta and Rathor (2020) proposed efficient hardware accelerator designs for image processing filters in two modes: (i) functionally reconfigurable processor mode where same hardware architecture can be reconfigured to act as different image processing filters and (ii) application specific processor mode where a fixed hardware architecture has been designed for a specific image processing/filtering application. In addition, Sengupta and Rathor (2020) proposed secured versions of image processing filter hardware accelerators.

5.2 Why secure image processing filter hardware accelerators?

Besides considering low power, high performance and area efficient design aspects of image processing filter hardware, a designer also needs to take care of security aspect during the design process (Sengupta, 2020). The security of image processing filter hardware accelerators needs to be ensured against a known hardware threat of hardware Trojan (Chakraborty and Bhunia, 2009; Zhang and Tehranipoor, 2011). A potential adversary in an untrusted design house may covertly insert

Trojan logic into the image processing filter hardware accelerator design by reverse engineering (RE) the design netlist (Sengupta, 2016,2017). The secretly inserted Trojan may lead to the following consequences: (i) leakage of consumer's secret data, (ii) performance degradation, (iii) excessive heat dissipation, (iv) battery explosion, (v) device failure, etc. Therefore, to ensure the systems reliability and consumer's safety, security against the Trojan attack is of paramount importance. Since the hardware Trojan is stealthy by nature and it triggers only upon certain rare events (conditions) in the circuit, they are not easily detectable during common pre-silicon or post-silicon validation (Chakraborty and Bhunia, 2011). Therefore, once the hardware Trojan is inserted into the design, their detection becomes arduous. In such a scenario, preventive control of Trojan insertion plays a crucial role (Sengupta and Roy, 2017; Sengupta and Rathor, 2020) for an image processing filter hardware accelerator design.

Image processing filter hardware accelerators can be secured against Trojan attack by employing structural-obfuscation-based preventive control mechanism (Lao and Parhi, 2015; Sengupta *et al.*, 2017,2018; Sengupta and Rathor, 2019b). The structural obfuscation process conceals the original design architecture through significant transformations in the structure without affecting the functionality, such that it becomes uninterpretable. Sengupta and Rathor (2020) employed loop unrolling, partitioning and tree-height-transformation (THT)-based transformations to structurally obfuscate the design of an image processing filter hardware accelerator. Thus the obtained structurally obfuscated design of image processing filters becomes challenging to be reverse engineered by an adversary, thus thwarting secret insertion of hardware Trojan. An abstract view of filtering process of images using structurally obfuscated (secure) image processing filter hardware accelerator is shown in Figure 5.1.

5.3 Salient features of the chapter

This chapter discusses the methodology of designing and securing image processing filter hardware accelerators, based on the following key features (Sengupta and Rathor, 2020):

- Discussion on HLS design methodology of designing hardware accelerators for image processing filters for two different kernel sizes, viz. 3 × 3 and 5 × 5.
- Discussion on multi-modal hardware accelerator architectures of image processing filters in the following two different modes: (i) functionally reconfigurable processor mode and (ii) application specific processor mode.
- Discussion on functionally reconfigurable hardware accelerator architecture which can be enabled to work as different image processing filters, viz. blur filter, sharpening filter, embossment filter and ED filter for 3 × 3 kernel size. The reconfiguration is achieved by using a selection vector that enables at a time a specific image processing filter function.

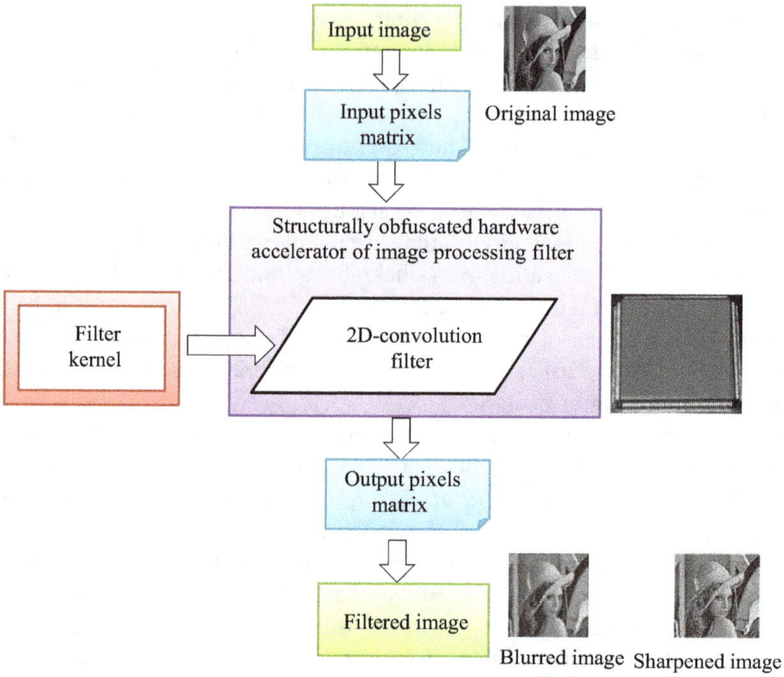

Figure 5.1 Abstract view of filtering of an image using structurally obfuscated hardware accelerator of image processing filters (Sengupta and Rathor, 2020)

- Discussion on application specific image processing filter hardware accelerators for various applications, viz. image blurring, sharpening, horizontal embossment (HE), vertical embossment (VE) and Laplace ED.
- Discussion on designing structurally obfuscated hardware accelerators for 3×3 and 5×5 kernel-size-based image processing filters, using following structural transformations, viz. loop unrolling, partitioning and THT.

5.4 Selected contemporary approaches

This section discusses some contemporary approaches related to hardware accelerators for image processing applications. Further, this section also highlights the key difference of contemporary approaches with the approach of designing multimodal and structurally obfuscated image processing filter hardware accelerators proposed by Sengupta and Rathor (2020). Let us first discuss the contemporary approaches. A semi-automatic mapping methodology has been proposed by Dutta *et al.* (2006) to produce hardware accelerators for a generic category of adaptive image filtering applications. Further, a co-processor for image median filter along with MicroBlaze processor for executing generic function has been proposed by Wu *et al.* (2009). A field-programmable-gate-array-based image processing hardware accelerator has been proposed by Tsiktsiris *et al.* (2018) and Vourvoulakis *et al.* (2012). Further, an image processing filter hardware accelerator which performs the filtering of the input data using Gabor functions has been proposed by Cappetta *et al.* (2017). Furthermore, hardware architecture for image processing filters has also been proposed by Azizabadi and Behrad (2013) and Ortega-Cisneros *et al.* (2014).

However, the approach of designing multi-modal and structurally obfuscated hardware accelerators for image processing filters (proposed by Sengupta and Rathor, 2020) differs from the contemporary approaches in the following ways: (i) Sengupta and Rathor (2020) introduced methodology of designing functionally reconfigurable processor for various image filtering applications. However, contemporary approaches did not propose such designs of reconfigurable functionality. (ii) Sengupta and Rathor (2020) introduced application specific processor designs for five different image processing filters, viz. blurring, sharpening, HE, VE and Laplace ED. However, contemporary approaches did not present application specific processor designs for various types of image processing filters. (iii) Sengupta and Rathor (2020) employed structural obfuscation mechanism to secure hardware accelerator designs of image processing filters against hardware Trojan threat. However, in contemporary approaches, discussion on hardware security against Trojan is not presented.

5.5 Theory of 3 × 3 filter hardware accelerator

This section presents discussion on the theory of 3 × 3 filter hardware accelerators for image processing applications (Sengupta and Rathor, 2020). More explicitly, we discuss how a computation function for generating output pixels of filtered image using 3 × 3 filter kernels is derived. There are a number of image processing applications that use 3 × 3 filter kernels, such as blurring, sharpening and Laplace ED. Let us start discussing by defining generic pixel matrices of input image and generic kernel of size 3 × 3. A pixel matrix of an input image of size $A \times B$ is defined using $[I]_{A\times B}$ as given in the following equation (Sengupta and Rathor, 2020):

$$I = \begin{bmatrix} X_{00} & X_{01} & \cdots & X_{0(B-1)} \\ X_{10} & X_{11} & \cdots & X_{1(B-1)} \\ \vdots & \vdots & \ddots & \vdots \\ X_{(A-1)0} & X_{(A-1)1} & \cdots & X_{(A-1)(B-1)} \end{bmatrix}_{A\times B} \tag{5.1}$$

where X_{ij} indicates pixels intensity value at the location of ith row and jth column in the input pixel matrix. The variables i and j vary from 0 to $(A-1)$ and 0 to $(B-1)$, respectively.

Next let us assume a generic kernel matrix of filter size $n \times m$ is represented using $[K]_{n\times m}$. For the chosen filter size of 3×3, the kernel matrix is defined using $[K]_{3\times 3}$ which has been shown in the following equation (Sengupta and Rathor, 2020):

$$K = \begin{bmatrix} K_{00} & K_{01} & K_{02} \\ K_{10} & K_{11} & K_{12} \\ K_{20} & K_{21} & K_{22} \end{bmatrix}_{3\times 3} \tag{5.2}$$

where K_{rs} indicates kernel value at the location of rth row and sth column in the kernel matrix. Here the variables r and s both vary from 0 to 2.

In order to generate filtered images using a 3×3 filter, the filter is applied on an input image by performing 2D-convoluation between input pixel matrix and the kernel matrix. The 2D convolution can be a 'valid convolution' or 'same convolution'. This chapter focuses discussion on image processing filters using same convolution. The key attribute of the same convolution is that it generates output matrix of the same size as input matrix, i.e. the size of a filtered (output) image remains same as input image. Now, let us understand how the same convolution is applied between input pixel matrix and kernel matrix to perform image processing/filtering. In order to do so, first a pre-processing is applied on input pixel matrix for extending its size. The pre-processing is based on a zero padding rule which is given as follows (Sengupta and Rathor, 2020):

$$L = \frac{(w-1)}{2} \tag{5.3}$$

where w denotes the size of filter kernel. Further, L denotes the value by which the size of input matrix is to be extended. More explicitly, L number of rows is added above and below of the input matrix and L number of columns is added to the left and right of the input matrix. The additional rows and columns padded in the input matrix are filled with zeros. Since $w = 3$ for 3×3 kernel size, L is computed to be 1. Hence, the number of rows and columns of input matrix are each increased by 2. Since the dimension of original matrix is $A \times B$, the modified dimension becomes $(A+2) \times (B+2)$. In general, the dimension of the modified input matrix post padding rows and columns is represented by $N \times M$. The modified matrix $[I]_{N\times M}$ of the input image is given in the following equation (Sengupta and Rathor, 2020):

$$I = \begin{bmatrix} 0 & 0 & 0 & \cdots & 0 & 0 \\ 0 & X_{00} & X_{01} & \cdots & X_{0(B-1)} & 0 \\ 0 & X_{10} & X_{11} & \cdots & X_{1(B-1)} & 0 \\ \vdots & \vdots & \vdots & \ddots & \vdots & \vdots \\ 0 & X_{(A-1)0} & X_{(A-1)1} & \cdots & X_{(A-1)(B-1)} & 0 \\ 0 & 0 & 0 & \cdots & 0 & 0 \end{bmatrix}_{N\times M} \tag{5.4}$$

The matrix [I] can be represented generically as follows (Sengupta and Rathor, 2020):

$$I = \begin{bmatrix} Y_{00} & Y_{01} & \cdots & Y_{0(M-1)} \\ Y_{10} & Y_{11} & \cdots & Y_{1(M-1)} \\ \vdots & \vdots & \ddots & \vdots \\ Y_{(N-1)0} & Y_{(N-1)1} & \cdots & Y_{(N-1)(M-1)} \end{bmatrix}_{N\times M} \tag{5.5}$$

where Y_{pq} indicates pixel values at the location of pth row and qth column in the modified matrix $[I]_{N\times M}$ of input image. The variables p and q vary from 0 to ($N-1$) and 0 to ($M-1$), respectively.

Once the size of input matrix $[I]_{A\times B}$ is extended based on padding rule shown in (5.3), the modified input matrix $[I]_{N\times M}$ is subjected to 2D convolution with the filter kernel matrix shown in (5.2). This type of convolution results into same convolution as the size of generated output matrix is same as input matrix $[I]_{A\times B}$. However, in general, the size of output matrix of same convolution is denoted by $(N-n+1) \times (M-m+1)$, where $N-n+1$ and $M-m+1$ are computed to be A and B, respectively. For example, if the size of input image is $A \times B = 512 \times 512$ and kernel matrix size $n \times m = 3 \times 3$, the size of modified input matrix post padding two rows and two columns (as $L = 2$) becomes $N \times M = 514 \times 514$. Subsequently, the size of output matrix is $(N-n+1) \times (M-m+1) = (514-3+1) \times (514-3+1) = 512 \times 512$, which is same as $A \times B$. The output matrix represents the pixels of the filtered/processed image. The generic representation of output matrix $[O]_{(N-n+1)\times(M-m+1)}$ is given as follows (Sengupta and Rathor, 2020):

$$O = \begin{bmatrix} O_{00} & O_{01} & \cdots & O_{0(M-m)} \\ O_{10} & O_{11} & \cdots & O_{1(M-m)} \\ \vdots & \vdots & \ddots & \vdots \\ O_{(N-n)0} & O_{(N-n)1} & \cdots & O_{(N-n)(M-m)} \end{bmatrix}_{(N-n+1)\times(M-m+1)} \tag{5.6}$$

where O_{ij} indicates output pixel values at the location of ith row and jth column in the output matrix [O]. The variables i and j vary from 0 to ($N-n$) and 0 to ($M-m$), respectively. Since the total number of pixels in the output matrix are $(N-n+1) \times (M-m+1)$, each output pixel value can generically be represented by O_V where V varies from 0 to $[(N-n+1) \times (M-m+1)-1]$. Using (5.2) and (5.5), the computation function used for computing output pixels O_V varying from O_0 to $O_{[(N-n+1)\times(M-m+1)-1]}$ is given as follows (Sengupta and Rathor, 2020):

$$\text{for } (V = 0; V<(N-n+1)\times(M-m+1); V++) \\ \left\{ O_V = \sum_{p,r=\text{min value}}^{p,r=\text{max value}} \left(\sum_{q,s=\text{min value}}^{q,s=\text{max value}} Y_{pq} \times K_{rs} \right) \right\} \tag{5.7}$$

In order to compute all output pixels, the 'for loop' shown earlier is executed $(N-n+1) \times (M-m+1)$ times. In the (5.7), Y_{pq} represents the pixel values in the modified input matrix $[I]_{N\times M}$ (given in (5.5)) and K_{rs} represents the values in the

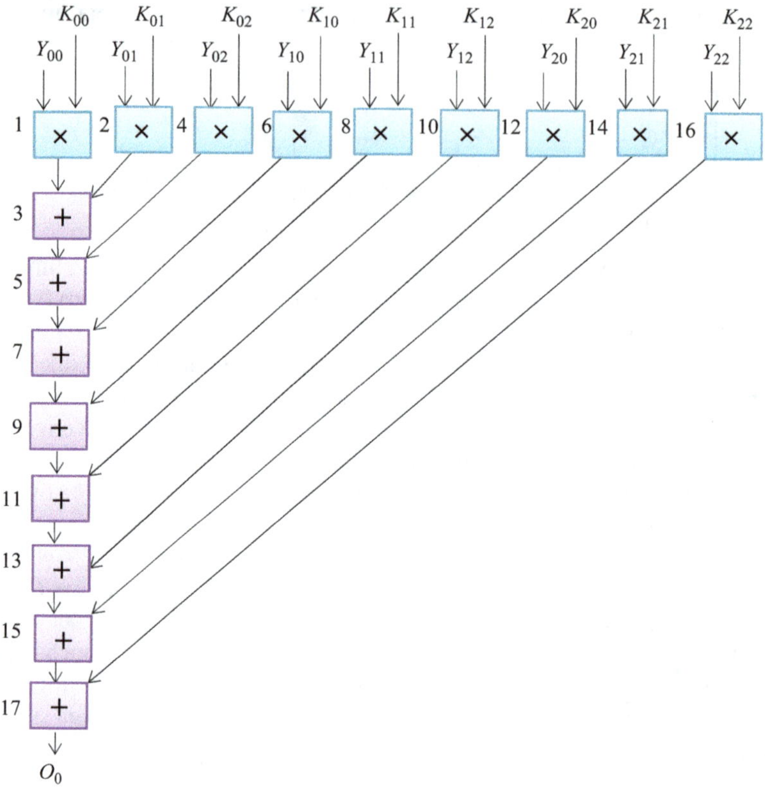

Figure 5.2 DFG of 3 × 3 filter hardware accelerator for image processing applications (Sengupta and Rathor, 2020)

kernel matrix $[K]_{3\times3}$ (given in (5.2)). As discussed earlier, variables p and q vary from 0 to $(N-1)$ and 0 to $(M-1)$, respectively, and variables r and s both vary from 0 to 2. For computing each output pixel value, the minimum and maximum values of both variables r and s are 0 and 2, respectively. However, the minimum and maximum values of variables p and q are not fixed during computation of each output pixel. Instead, the minimum value of variables p and q varies from to 0 to $(N-3)$ and $(M-3)$, respectively, and maximum values of p and q vary from to 2 to $(N-1)$ and $(M-1)$, respectively. While computing output pixels, the values of p and q vary in the window of 3 within their minimum and maximum values. This is because the size of kernel matrix is 3 × 3 and it slides over the modified input matrix by one position right while computing pixels in the same row and slides by one position below to compute pixels in the next column. For example, the variables p and q vary from 0 to 2 (i.e. minimum value is 0 and maximum value is 2 for both p and q) while computing first output pixel O_0, while computing second output pixel O_1, the variable p varies from 0 to 2 (i.e. minimum value is 0 and maximum value is 2) and q varies from 1 to 3 (i.e. minimum value is 1 and maximum value is 3). Similarly, while computing last output pixel $O_{[(N-n+1) \times (M-m+1)-1]}$, the variable p varies from $N-3$ to $N-1$ (i.e. minimum value of p is $N-3$ and maximum value is $N-1$) and q varies from $M-3$ to $M-1$ (i.e. minimum value of q is $M-3$ and maximum value is $M-1$). This is how the minimum and maximum values of variables p and q vary between 0 to $N-1$ and $M-1$, respectively, during the computation of all output pixels of filtered image (Sengupta and Rathor, 2020).

The generic equation of computing the first output pixel is given in the following equation (Sengupta and Rathor, 2020):

$$O_0 = \sum_{\substack{r=0 \\ p=0}}^{\substack{r=2 \\ p=2}} \left(\sum_{\substack{s=0 \\ q=0}}^{\substack{s=2 \\ q=2}} Y_{pq} \times K_{rs} \right) \tag{5.8}$$

The expansion of (5.8) is shown in (5.9) as follows (Sengupta and Rathor, 2020):

$$\begin{aligned} O_0 = {} & [(Y_{00} \times K_{00}) + (Y_{01} \times K_{01}) + (Y_{02} \times K_{02})] \\ & + [(Y_{10} \times K_{10}) + (Y_{11} \times K_{11}) + (Y_{12} \times K_{12})] \\ & + [(Y_{20} \times K_{20}) + (Y_{21} \times K_{21}) + (Y_{22} \times K_{22})] \end{aligned} \tag{5.9}$$

This equation is the basic computation function for computing pixels of a filtered image using 3 × 3 filter kernel. This equation shows the computation of first output pixel. However, the same computation function can be used to compute all output pixels just by changing the corresponding input pixel values Y_{pq}. The corresponding data-flow graph (DFG) representation of the basic computation function (Equation (5.9)) is shown in Figure 5.2. In order to generate hardware accelerator design for the image processing filter of kernel size 3 × 3, the DFG is fed into an HLS process which results into its equivalent RTL data path design. During the HLS process, the DFG is subjected to scheduling based on resource constraints of one multiplier and one adder. The scheduled DFG is shown in Figure 5.3. Post-HLS, the RTL design of 3 × 3 filter hardware accelerator is shown in Figure 5.4.

5.6 Designing functionally reconfigurable obfuscated (secured) 3 × 3 filter hardware accelerator

This section discusses the methodology of securing 3 × 3 filter hardware accelerators for image processing applications using structural obfuscation mechanism. Further, this section shows the process of designing an obfuscated 3 × 3 filter hardware accelerator in a functionally reconfigurable mode. Let us start discussion with the methodology of designing structurally obfuscated 3 × 3 filter hardware accelerators (Sengupta and Rathor, 2020).

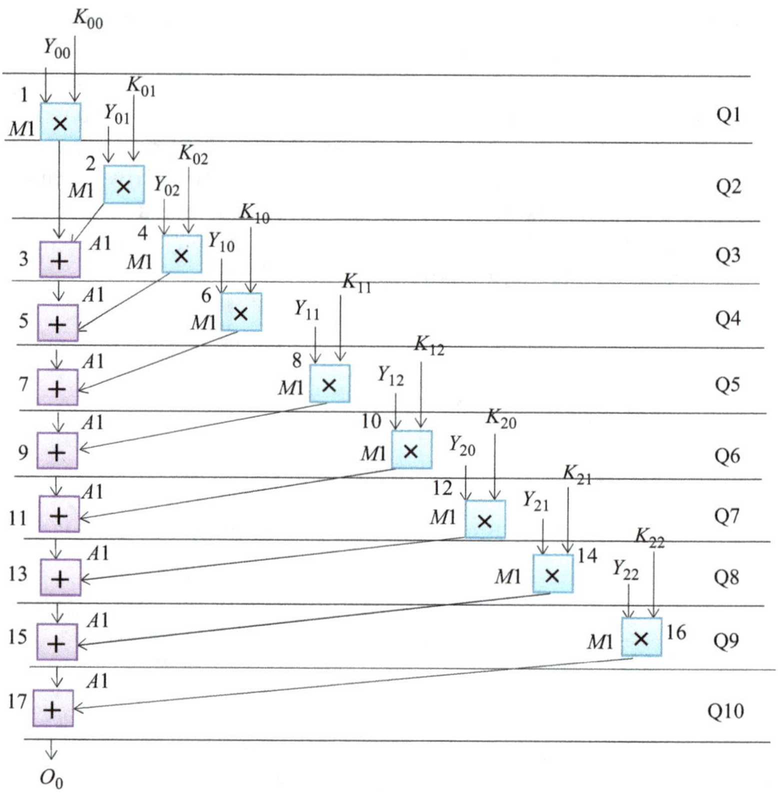

Figure 5.3 Scheduled DFG of 3 × 3 filter hardware accelerator based on resource constraints of 1M, 1A. Note: Q indicates control step

5.6.1 Structural obfuscation methodology for securing 3 × 3 filter hardware accelerators

A 3 × 3 filter hardware accelerator is structurally obfuscated by applying the following structural transformation techniques (Sengupta and Rathor, 2020):

1. loop-unrolling-based transformation
2. THT

These structural transformation techniques are applied during the high-level transformation phase of HLS. The security aware HLS design flow of generating secured 3 × 3 filter hardware accelerators is shown in Figure 5.5. As shown in the figure, the computation function of image processing filter (3 × 3 kernel size) is first converted into corresponding DFG representation. The 'for loop' in (5.7) with $n = m = 3$ represents the computation function of image processing filter for 3 × 3 kernel size. Further, loop-unrolling-based structural transformation is applied based on unrolling factor (UF = 2). It results into unrolled DFG that

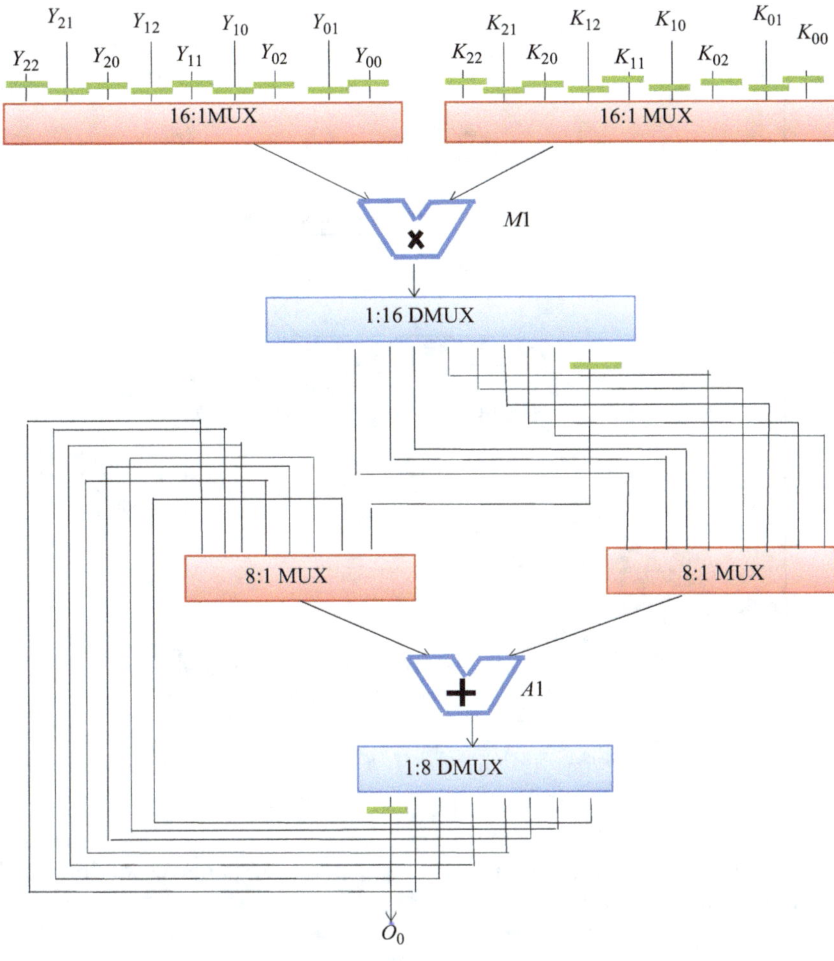

Figure 5.4 RTL data path of 3 × 3 filter hardware accelerators for image processing applications (based on resource constraints of 1M and 1A). Note: Green bars show the requirement of registers to store primary and intermediate inputs and outputs. Intermediate registers are only required if the intermediate outputs are not used in the next control step and they are required to be stored for future computation

computes two output pixel values (e.g. the first two output pixels are O_0 and O_1) concurrently. The DFG post applying loop-unrolling-based transformation is shown in Figure 5.6. For instance, the loop-unrolled DFG shows the computations of output pixels O_0 and O_1. In the loop-unrolled DFG, the sub-DFG computing output pixel O_0 is corresponding to the computation function shown in (5.8) or

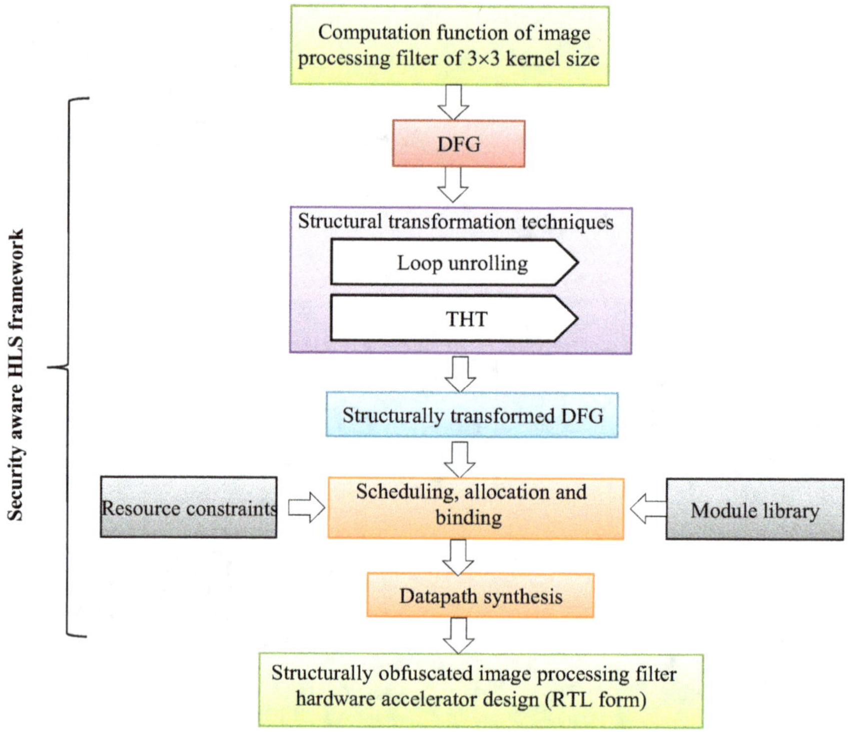

Figure 5.5 Structural obfuscation aware HLS methodology of designing secured 3 × 3 filter hardware accelerators for image processing applications

(5.9). Further, the sub-DFG computing output pixel O_1 is corresponding to the computation function given as follows:

$$O_1 = \sum_{\substack{r=0 \\ p=0}}^{\substack{r=2 \\ p=2}} \left(\sum_{\substack{s=0 \\ q=1}}^{\substack{s=2 \\ q=3}} Y_{pq} \times K_{rs} \right) \tag{5.10}$$

The expansion of (5.10) is shown in (5.11) as follows:

$$\begin{aligned} O_1 = {} & [(Y_{01} \times K_{00}) + (Y_{02} \times K_{01}) + (Y_{03} \times K_{02})] \\ & + [(Y_{11} \times K_{10}) + (Y_{12} \times K_{11}) + (Y_{13} \times K_{12})] \\ & + [(Y_{21} \times K_{20}) + (Y_{22} \times K_{21}) + (Y_{23} \times K_{22})] \end{aligned} \tag{5.11}$$

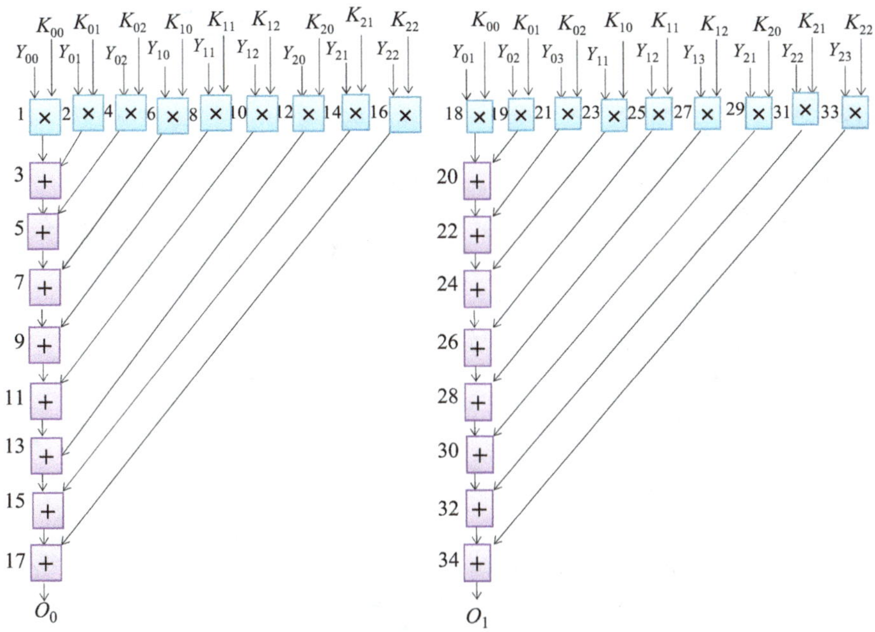

Figure 5.6 Loop-unrolled DFG of 3 × 3 filter hardware accelerator for image processing applications (Sengupta and Rathor, 2020)

The impact of the loop-unrolling-based transformation manifests change in the circuit in terms of parallel architecture causing structural obfuscation, during the computations of output pixel values (Sengupta and Rathor, 2020).

Further, the loop-unrolled DFG shown in Figure 5.6 is subjected to THT-based structural transformation. THT is a high-level transformation technique where the height of the DFG is subjected to alteration by breaking the sequential data dependency in the graph and forcing parallel dependency of some operations without affecting the design functionality. Figure 5.7 shows the DFG of the image processing filter of 3 × 3 kernel size, post applying both structural obfuscation techniques. The alteration in the height of the graph by modifying the data dependency of operation nodes incurs structural transformation–based obfuscation in the RTL structure post HLS. The THT-based high-level transformation technique contributes in structural obfuscation by incurring the following modifications in the RTL data path: modification in the interconnectivity of resources and size of multiplexers and demultiplexers. Hence, the resultant structurally obfuscated architecture of 3 × 3 filter hardware accelerators for image processing applications becomes highly unobvious, hindering RE by an attacker (Sengupta and Rathor, 2020). In order to generate the structurally obfuscated RTL structure of 3 × 3 filter hardware accelerator, first the obfuscated DFG shown in Figure 5.7

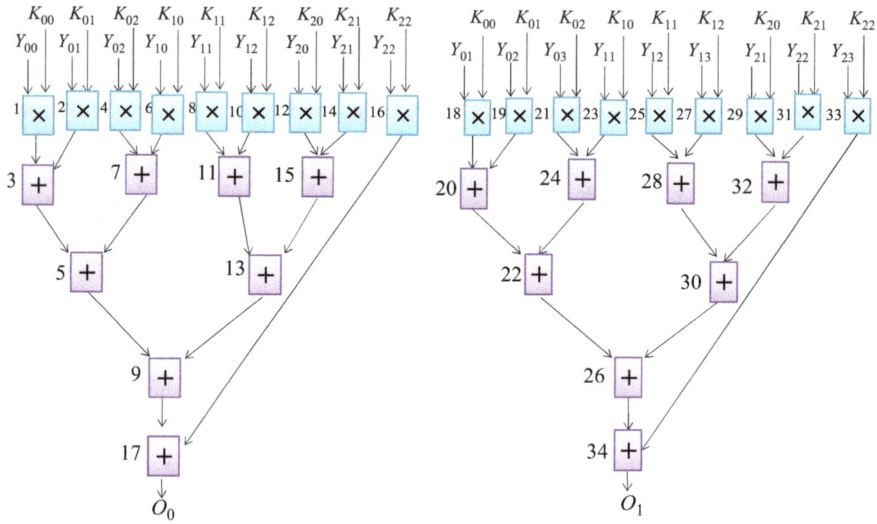

Figure 5.7 Structurally obfuscated loop-unrolled DFG of 3 × 3 filter hardware accelerator for image processing applications (Sengupta and Rathor, 2020)

is subjected to scheduling, allocation and binding based on module library and resource constraints of three multipliers and one adder. The scheduled and resource allocated DFG is shown in Figure 5.8. Further, the scheduled and resource-allocated DFG is subjected to data path synthesis phase of HLS, which results into structurally obfuscated RTL data path of 3 × 3 filter hardware accelerator.

5.6.2 Functionally reconfigurable processor mode of 3 × 3 filter hardware accelerators

Sengupta and Rathor (2020) presented a functionally reconfigurable processor of the structurally obfuscated 3 × 3 filter hardware accelerator. The structurally obfuscated 3 × 3 filter hardware accelerator architecture in functionally reconfigurable mode combines the functionality of five different image processing filters, viz. blur filter, VE filter, HE filter, sharpening filter and ED filter. More explicitly, the same functionally reconfigurable 3 × 3 filter hardware accelerator architecture is exploited to work as aforementioned five different filters, one at a time. The functionality of a specific filter is activated using a selection vector T[2 down to 0]. The reconfigurable functionality is enabled by multiplexing the kernel matrixes of all five different filters. The selection vector T functions as select lines for multiplexers to activate a specific 3 × 3 filter. The circuit of functionally reconfigurable structurally obfuscated 3 × 3 filter hardware accelerator is shown in Figure 5.9. As shown in the figure, the red colour boxes highlight the reconfigurable functionality

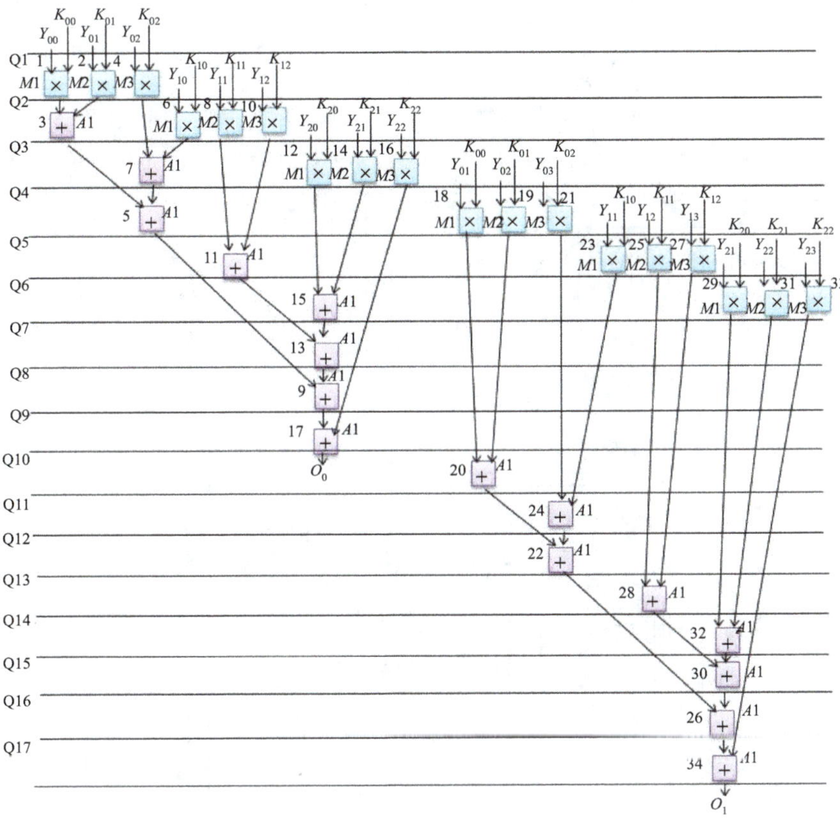

Figure 5.8 Scheduled and allocated obfuscated DFG of convolution filter based on 3M, 1A

of the 3 × 3 filter hardware accelerator. A specific value of 3-bit selection vector T enables the functionality of corresponding image processing filter. Table 5.1 shows the mapping of selection vector T to the respective type of 3 × 3 filter hardware accelerator for image processing.

5.6.3 How does structurally obfuscated 3 × 3 filter hardware accelerator thwarts Trojan insertion?

The structural transformation techniques applied to structurally obfuscate 3 × 3 filter hardware accelerator designs render the design structure unobvious and non-interpretable for an attacker. The impact of structural obfuscation can be observed by comparing the un-obfuscated 3 × 3 filter hardware accelerator data path shown in Figure 5.4 with the structurally obfuscated data path shown in Figure 5.9. As evident, the applied structural obfuscation techniques incur substantial (massive) circuit level alterations in the structure. The alterations due to obfuscation are in the form of changes in the (i) number of functional unit resources, (ii) number of

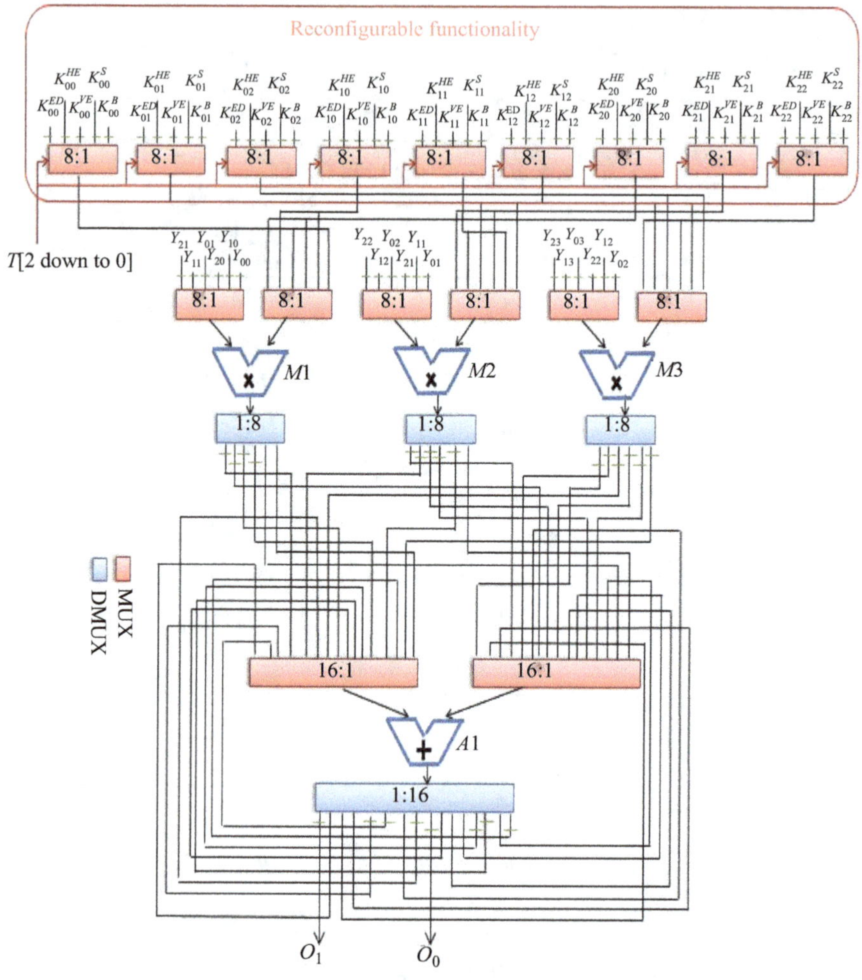

Figure 5.9 Reconfigurable structurally obfuscated RTL data path of 3 × 3 filter hardware accelerator based on 3M, 1A (Sengupta and Rathor, 2020). Note: Green bars indicate the registers

storage resources such as registers, (iii) number of interconnect-resources such as multiplexers and demultiplexers and (iv) size of multiplexers and demultiplexers, etc. Therefore, the structurally obfuscated data path is hugely different than the original (un-obfuscated) counterpart (while preserving functionality). Thereby the structural obfuscation confounds the attacker in deducing the original structure and functionality of the 3 × 3 filter hardware accelerator design through RE. This renders the attacker's attempt of secretly inserting Trojan into the 3 × 3 filter hardware accelerator design harder. This is because for a successful attack, the attacker needs to determine the correct locations for hiding the Trojan into the design. Otherwise, it fails to achieve the stealthy intention. The stealthy nature of Trojan helps in evading the detection during pre-silicon or post-silicon testing. The safer insertion of Trojan is only possible if the attacker becomes successful in fully understanding the original functionality and structure of the design through RE. However, the structural obfuscation techniques employed in the design makes the RE arduous and thwarts Trojan insertion.

5.7 Theory of 5 × 5 filter hardware accelerator

This section presents discussion on the theory of 5 × 5 filter hardware accelerators for image processing applications (Sengupta and Rathor, 2020). More explicitly, we discuss how a computation function for generating output pixels of filtered image using 5 × 5 filter kernels is derived. There are a number of image processing applications that use 5 × 5 filter kernels, such as blurring and deep embossment. Let us start discussing by defining the generic kernel matrix of size 5 × 5. The kernel matrix of 5 × 5 filter is defined using $[K]_{5\times5}$ which has been shown in the following equation (Sengupta and Rathor, 2020):

$$K = \begin{bmatrix} K_{00} & K_{01} & K_{02} & K_{03} & K_{04} \\ K_{10} & K_{11} & K_{12} & K_{13} & K_{14} \\ K_{20} & K_{21} & K_{22} & K_{23} & K_{24} \\ K_{30} & K_{31} & K_{32} & K_{33} & K_{34} \\ K_{40} & K_{41} & K_{42} & K_{43} & K_{44} \end{bmatrix}_{5\times5} \tag{5.12}$$

where K_{rs} indicates kernel value at the location of rth row and sth column in the kernel matrix. Here, the variables r and s both vary from 0 to 4.

The filtering of input images using a 5 × 5 filter is performed using 2D-convolution between input pixel matrix and the kernel matrix. Sengupta and Rathor (2020) applied 'same convolution' to perform filtering using 5 × 5 filter. Therefore, the size of a filtered (output) image remains same as input image. The generic input pixel matrix has been shown in (5.1). In order to apply the same convolution, first padding of additional rows and columns is performed in the input pixel matrix. The number of rows and columns to be padded additionally is determined by the padding rule shown in (5.3). Since filter kernel size is 5 × 5 (i.e. $w = 5$), L is computed to be 2. Hence, two rows are padded above and below in the input matrix, and two columns are padded to left and right in the input matrix. Hence, both the number of rows and columns of input matrix are increased by 4. Since the dimension of original matrix is $A \times B$, the modified dimension becomes $(A+4) \times (B+4)$. In general, the dimension of the modified input matrix post padding rows and columns is represented by $N \times M$. The modified matrix $[I]_{N\times M}$ of the

Table 5.1 Mapping of selection vector T *with the filter type*

3 × 3 filter type	Filter coefficient	Selection vector T[2 down to 0]
Blur	K_{rs}^{B}	000
Sharpening	K_{rs}^{S}	001
Vertical embossment	K_{rs}^{VE}	010
Horizontal embossment	K_{rs}^{HE}	011
Laplace edge detection	K_{rs}^{ED}	100

input image is given in the following equation (Sengupta and Rathor, 2020):

$$I = \begin{bmatrix} 0 & 0 & 0 & 0 & \cdots & 0 & 0 & 0 \\ 0 & 0 & 0 & 0 & \cdots & 0 & 0 & 0 \\ 0 & 0 & X_{00} & X_{01} & \cdots & X_{0(B-1)} & 0 & 0 \\ 0 & 0 & X_{10} & X_{11} & \cdots & X_{1(B-1)} & 0 & 0 \\ \vdots & \vdots & \vdots & \vdots & \ddots & \vdots & \vdots & \vdots \\ 0 & 0 & X_{(A-1)0} & X_{(A-1)1} & \cdots & X_{(A-1)(B-1)} & 0 & 0 \\ 0 & 0 & 0 & 0 & \cdots & 0 & 0 & 0 \\ 0 & 0 & 0 & 0 & \cdots & 0 & 0 & 0 \end{bmatrix}_{NB-} \tag{5.13}$$

The generic representation of modified matrix $[I]_{N\times M}$ is same as shown in (5.5). Once the size of input matrix $[I]_{A\times B}$ is extended based on the padding rule shown in (5.3), the modified input matrix $[I]_{N\times M}$ is subjected to 2D convolution with the filter kernel matrix shown in (5.12). As discussed earlier in Section 5.5, the size of output matrix of same convolution is denoted by $(N-n+1) \times (M-m+1)$. Thus the obtained output matrix represents the pixels of the filtered/processed image using 5 × 5 filter. The computation function used for computing output pixels O_V varying from O_0 to $O_{[(N-n+1)\times(M-m+1)-1]}$ is the same as given in (5.7); however, the value of n and m is 5, in case of 5 × 5 filter. As discussed earlier in Section 5.5, variables p and q in (5.7) vary from 0 to $(N-1)$ and 0 to $(M-1)$, respectively. However, in this case, variables r and s both vary from 0 to 4, as the filter kernel size is 5 × 5. During computation of each output pixel value using (5.7), the minimum and maximum values of both variables r and s are 0 and 4, respectively. Further, the minimum values of variables p and q vary from to 0 to $(N-5)$ and $(M-5)$, respectively, and maximum values of p and q vary from to 4 to $(N-1)$ and $(M-1)$, respectively. While computing output pixels, the values of p and q vary in the window of 5 within their minimum and maximum values. For example, the variables p and q vary from 0 to 4 (i.e. minimum value is 0 and maximum value is 4 for both p and q) while computing first output pixel O_0 using 5 × 5 filter, and while computing second output pixel O_1, p varies from 0 to 4 (i.e. minimum value is 0 and maximum value is 4) and q varies from 1 to 5 (i.e. minimum value is 1 and maximum value is 5). Similarly while computing last output pixel $O_{[(N-n+1)\times(M-m+1)-1]}$, the variable p varies from $N-5$ to $N-1$ (i.e. minimum value of p is $N-5$ and maximum value is $N-1$) and q varies from $M-5$ to $M-1$ (i.e. minimum value of q is $M-5$ and maximum value is $M-1$). This is how the minimum and maximum values of variables p and q vary between 0 to $N-1$ and $M-1$, respectively, during the computation of all output pixels of filtered image using 5 × 5 filter (Sengupta and Rathor, 2020).

The generic equation of computing the first output pixel is given in the following equation (Sengupta and Rathor, 2020):

$$O_0 = \sum_{\substack{r=0 \\ p=0}}^{\substack{r=4 \\ p=4}} \left(\sum_{\substack{s=0 \\ q=0}}^{\substack{s=4 \\ q=4}} Y_{pq} \times K_{rs} \right) \tag{5.14}$$

The expansion of (5.14) is shown in (5.15) as follows (Sengupta and Rathor, 2020):

$$\begin{aligned} O_0 = & [(Y_{00} \times K_{00}) + (Y_{01} \times K_{01}) + (Y_{02} \times K_{02}) + (Y_{03} \times K_{03}) + (Y_{04} \times K_{04})] \\ & +[(Y_{10} \times K_{10}) + (Y_{11} \times K_{11}) + (Y_{12} \times K_{12}) + (Y_{13} \times K_{13}) + (Y_{14} \times K_{14})] \\ & + [(Y_{20} \times K_{20}) + (Y_{21} \times K_{21}) + (Y_{22} \times K_{22}) + (Y_{23} \times K_{23}) + (Y_{24} \times K_{24})] \\ & +[(Y_{30} \times K_{30}) + (Y_{31} \times K_{31}) + (Y_{32} \times K_{32}) + (Y_{33} \times K_{33}) + (Y_{34} \times K_{34})] \\ & +[(Y_{40} \times K_{40}) + (Y_{41} \times K_{41}) + (Y_{42} \times K_{42}) + (Y_{43} \times K_{43}) + (Y_{44} \times K_{44})] \end{aligned} \tag{5.15}$$

This equation shows the computation of first output pixel using 5 × 5 filter. However, the same computation function can be used to compute all output pixels just by changing the corresponding input pixel values Y_{pq}. The corresponding DFG representation of the basic computation function (Equation (5.15)) is shown in Figure 5.10. Further, (5.15) can be expressed in a more simplified form as follows:

$$O_0 = [E1] + [E2] + [E3] + [E4] + [E5] \tag{5.16}$$

where all five sub-terms in (5.15) are represented using $E1$–$E5$ in (5.16).

5.8 Designing obfuscated (secured) 5 × 5 filter hardware accelerator

A 5 × 5 filter hardware accelerator is structurally obfuscated by applying following structural transformation techniques (Sengupta and Rathor, 2020):

1. Partitioning based transformation
2. THT

These structural transformation techniques are applied during the high-level transformation phase of HLS. The 'for loop' in (5.7) with $n = m = 5$ represents the computation function of image processing filter for 5 × 5 kernel size. Equation (5.7) can be reduced to a function that computes 1 pixel at a time. For example, the computation function in (5.15) shows the computation of first output pixel O_0.

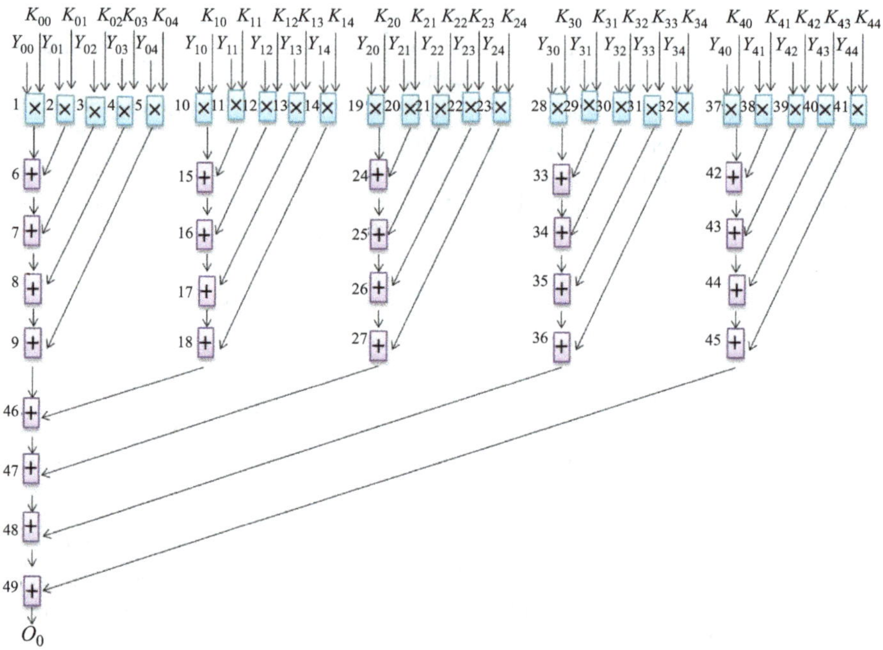

Figure 5.10 DFG of image processing convolution filter for 5 × 5 kernel size

The same function can be used to compute all the output pixels by changing the inputs. The corresponding DFG representation of the simplified computation function (Equation (5.15)) is shown in Figure 5.10. For instance, inputs and outputs of the DFG are corresponding to the first output pixel O_0. This DFG is subjected to partitioning and THT-based structure obfuscation techniques in order to generate structurally obfuscated 5 × 5 filter hardware accelerator.

First, partitioning-based structural obfuscation is applied on the DFG shown in Figure 5.10. Sengupta and Rathor (2020) performed partitioning by applying five cuts on the DFG. Applying five cuts produces six partitions (P1–P6) of the original DFG. Post partitioning, the resultant partitioned DFG is shown in Figure 5.11. The impact of DFG partitioning for structural obfuscation manifests obscurity in the RTL data path, in the form of modified interconnectivity of functional unit resources and modified size and change in number of multiplexers and demultiplexers. However, these modifications in the design do not incur any change in the functionality. Each partition of DFG is further subjected to THT-based structural transformation technique in order to enhance obscurity. Figure 5.12 shows the final obfuscated partitioned DFG of image processing filter of 5 × 5 kernel size. Further, the partitioned obfuscated DFG is subjected to scheduling based on resource constraints of 5*M* and 2*A*. The scheduled obfuscated DFG is shown in Figure 5.13. Further, each scheduled partition of the obfuscated DFG is individually subjected to data path synthesis phase of HLS process. The synthesized data paths of

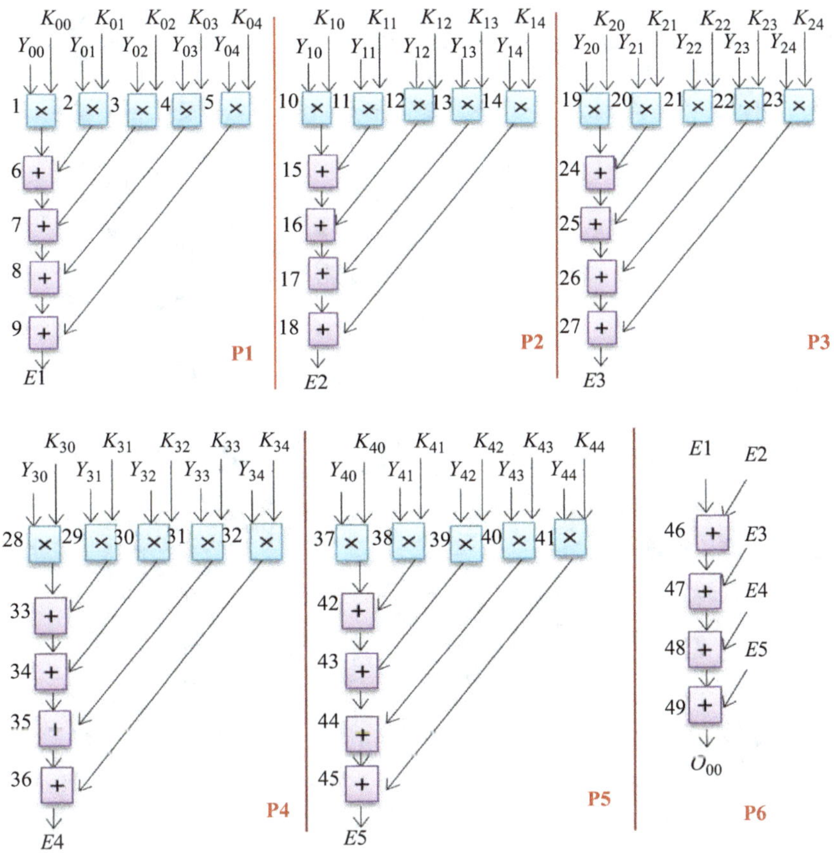

Figure 5.11 Partitioned DFG of image processing filter for 5 × 5 kernel size, post applying partitioning-based transformation

individual obfuscated partitions are shown in Figure 5.14. Subsequently, all individual data paths are combined into a single data path to generate a further structurally obfuscated RTL data path of 5 × 5 filter hardware accelerator for image processing applications. The structurally obfuscated RTL data path of 5 × 5 filter hardware accelerator is shown in Figure 5.15. Because of DFG partitioning and THT-based high-level transformation techniques, the count and size of interconnect hardware resources such as multiplexers and demultiplexers and count of storage hardware resource such as registers got modified in the RTL data path generated post-HLS. Moreover, the interconnection path among the different RTL modules also got altered due to structural obfuscation. However, the original functionality remains preserved in the structurally obfuscated design. Thus the obtained structurally obfuscated RTL data path of 5 × 5 filter hardware accelerator enhances security against RE (leading to secret Trojan insertion attack) (Sengupta and Rathor, 2020).

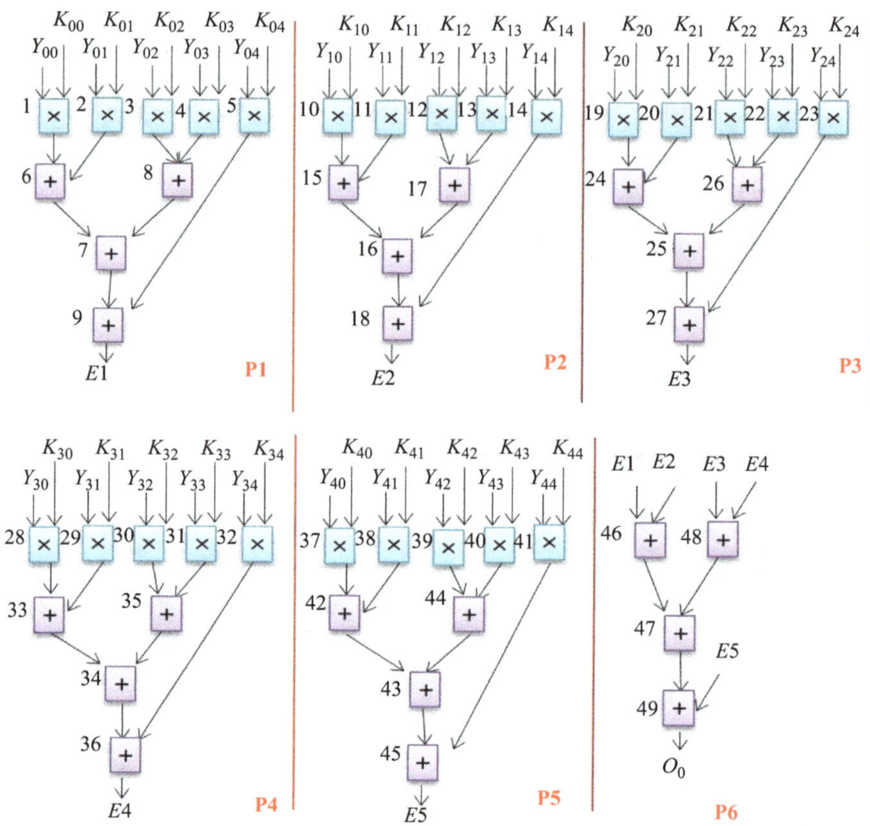

Figure 5.12 THT obfuscated DFG of image processing filter for 5 × 5 kernel size (Sengupta and Rathor, 2020)

5.9 Designing secured application specific filter hardware accelerators

Sengupta and Rathor (2020) proposed hardware accelerator design for each dedicated image processing filter, as application specific processor. More explicitly, dedicated individual 3 × 3 filter hardware accelerator for five different image processing applications such as blurring, sharpening, HE, VE and Laplace ED have been designed. Further, Sengupta and Rathor (2020) employed structural obfuscation (using loop-unrolling and THT techniques) to generate dedicated secured 3 × 3 filter hardware accelerator versions. Let us discuss the design process of dedicated structurally obfuscated 3 × 3 filter hardware accelerators, one by one, in the following subsections.

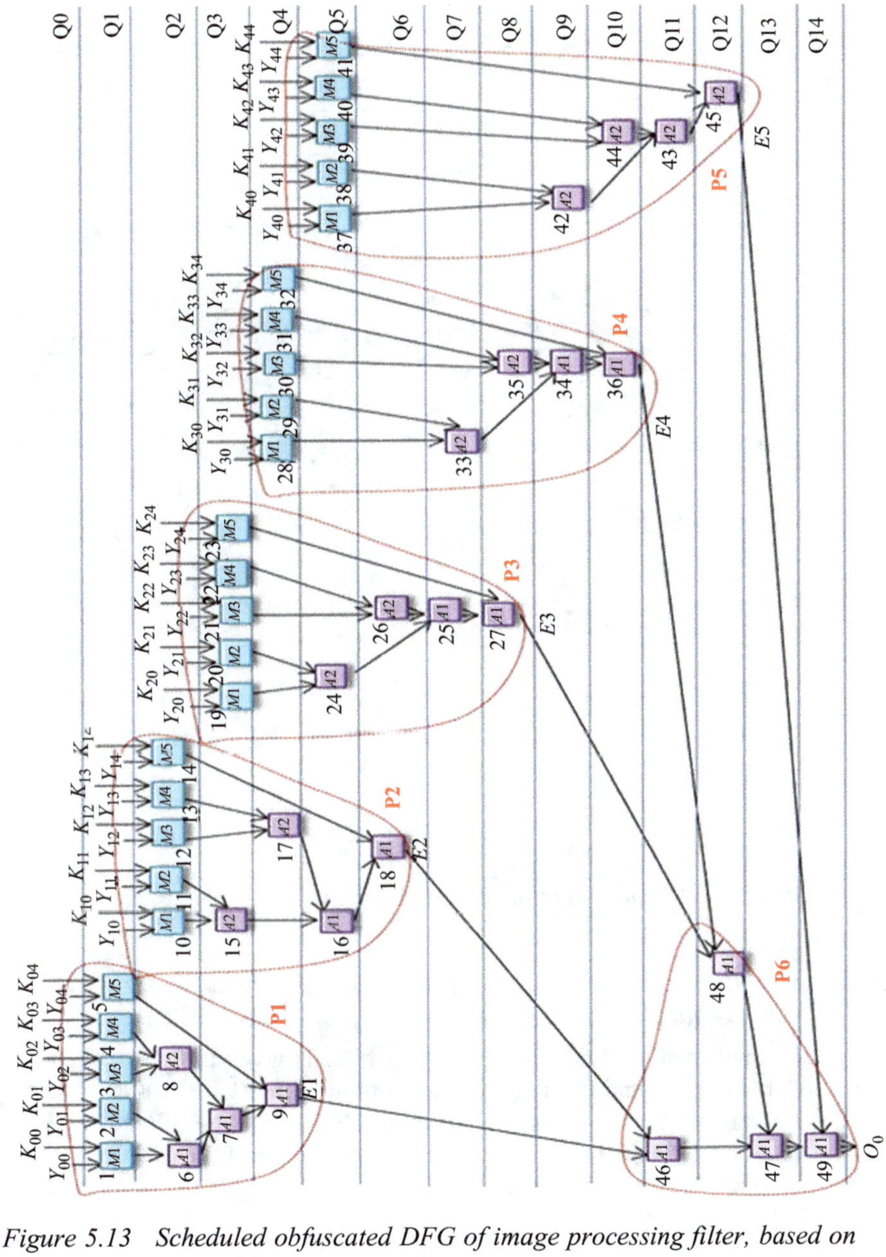

Figure 5.13 Scheduled obfuscated DFG of image processing filter, based on 5M and 2A

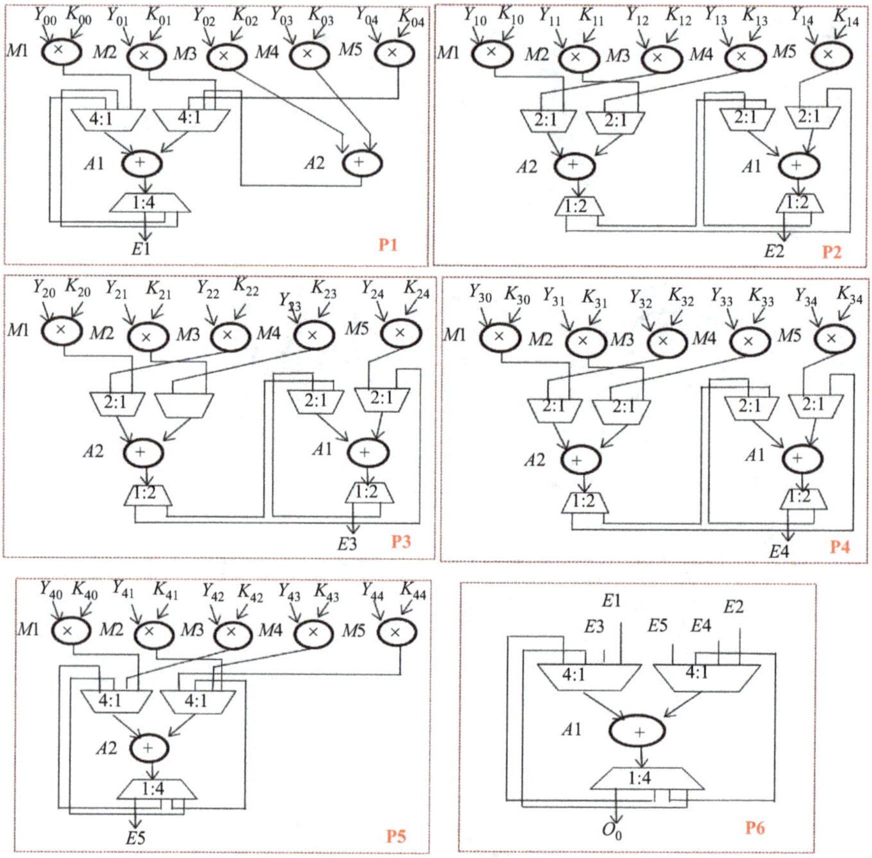

Figure 5.14 RTL data path of individual partitions of obfuscated DFG

5.9.1 Blur filter – mathematical function, RTL circuit and end-to-end demonstration

The generic mathematical function of a 3 × 3 filter is given in (5.7), which has been converted to (5.9) and (5.11) to enable loop-unrolling (UF = 2)-based structural transformation. The mathematical function of a 3 × 3 **blur filter (low pass filter – cuts of high frequency components)** can be derived from (5.9) and (5.11) based on its kernel matrix. The kernel matrix of a 3 × 3 blur filter (mean kernel) is given as follows:

$$K^B = 1/9\begin{bmatrix} 1 & 1 & 1 \\ 1 & 1 & 1 \\ 1 & 1 & 1 \end{bmatrix}_{3\times 3} \tag{5.17}$$

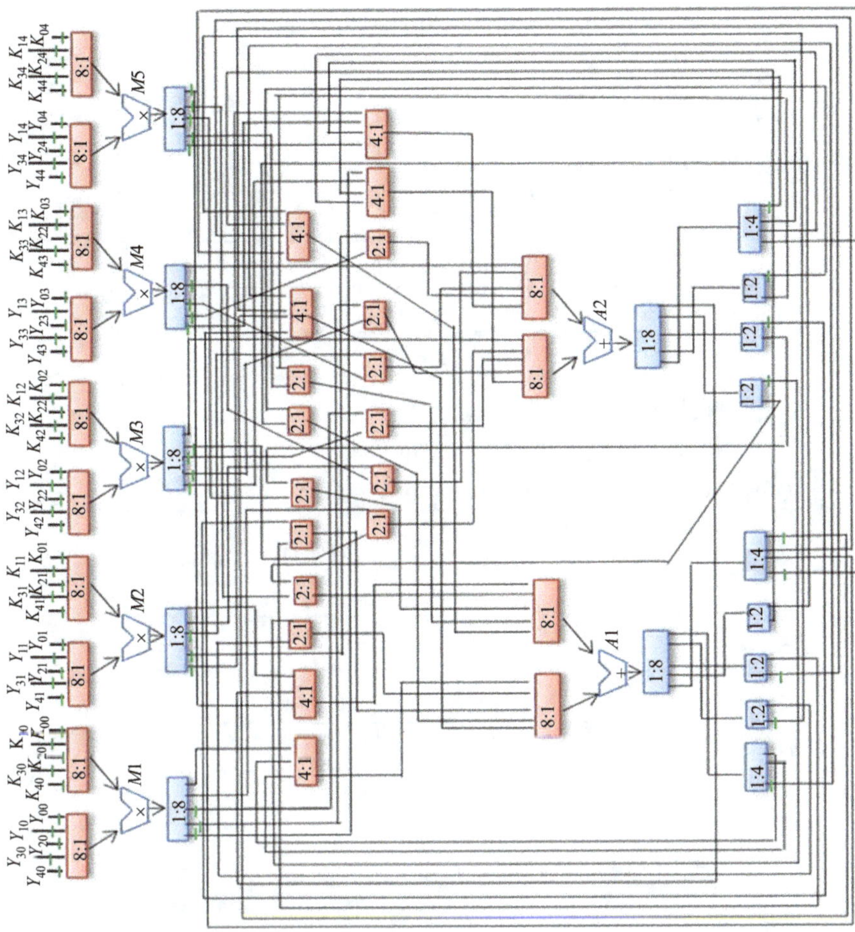

Figure 5.15 Structurally obfuscated RTL data path of 5 × 5 filter hardware accelerator (Sengupta and Rathor, 2020)

Based on this kernel matrix, (5.9) and (5.11) can be written as following equations:

$$O_0 = \left[\left(Y_{00} \times \left(\frac{1}{9}\right)\right) + \left(Y_{01} \times \left(\frac{1}{9}\right)\right) + \left(Y_{02} \times \left(\frac{1}{9}\right)\right)\right]$$
$$+ \left[\left(Y_{10} \times \left(\frac{1}{9}\right)\right) + \left(Y_{11} \times \left(\frac{1}{9}\right)\right) + \left(Y_{12} \times \left(\frac{1}{9}\right)\right)\right]$$
$$+ \left[\left(Y_{20} \times \left(\frac{1}{9}\right)\right) + \left(Y_{21} \times \left(\frac{1}{9}\right)\right) + \left(Y_{22} \times \left(\frac{1}{9}\right)\right)\right] \quad (5.18)$$

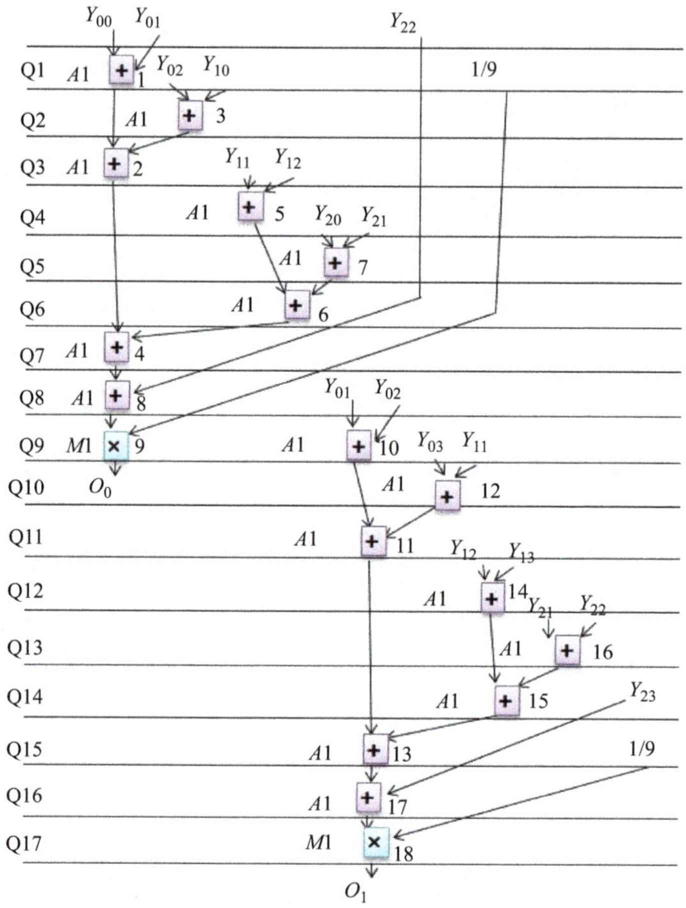

Figure 5.16 Scheduled DFG of 3 × 3 blur filter DFG based on 1M, 1A

$$O_1 = \left[\left(Y_{01} \times \left(\frac{1}{9}\right)\right) + \left(Y_{02} \times \left(\frac{1}{9}\right)\right) + \left(Y_{03} \times \left(\frac{1}{9}\right)\right)\right] + \left[\left(Y_{11} \times \left(\frac{1}{9}\right)\right) + \left(Y_{12} \times \left(\frac{1}{9}\right)\right) + \left(Y_{13} \times \left(\frac{1}{9}\right)\right)\right] + \left[\left(Y_{21} \times \left(\frac{1}{9}\right)\right) + \left(Y_{22} \times \left(\frac{1}{9}\right)\right) + \left(Y_{23} \times \left(\frac{1}{9}\right)\right)\right] \quad (5.19)$$

Equations (5.18) and (5.19) converge into the following equations which represent the mathematical function of a 3 × 3 blur filter with the loop-unrolling (UF = 2) transformation (Sengupta and Rathor, 2020):

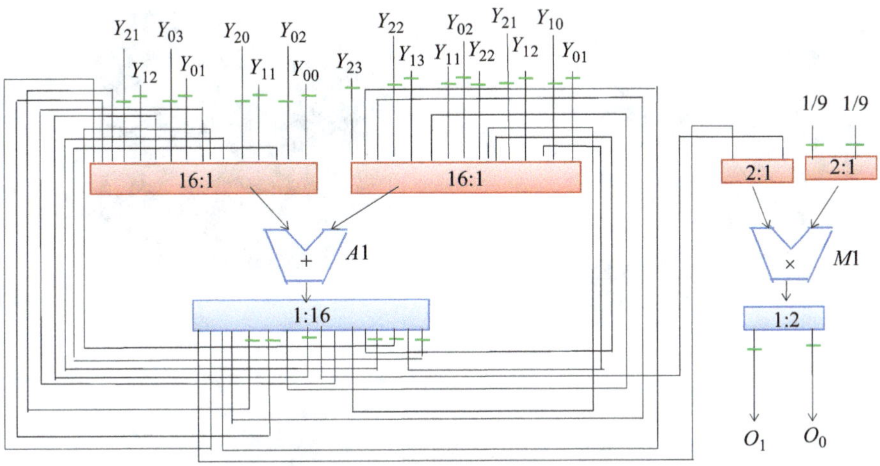

Figure 5.17 RTL circuit of a 3 × 3 blur filter hardware accelerator (Sengupta and Rathor, 2020)

$$O_0 = \left[(Y_{00} + Y_{01} + Y_{02} + Y_{10} + Y_{11} + Y_{12} + Y_{20} + Y_{21} + Y_{22}) \times \left(\frac{1}{9} \right) \right] \tag{5.20}$$

$$O_1 = \left[(Y_{01} + Y_{02} + Y_{03} + Y_{11} + Y_{12} + Y_{13} + Y_{21} + Y_{22} + Y_{23}) \times \left(\frac{1}{9} \right) \right] \tag{5.21}$$

In order to generate a hardware accelerator design of 3 × 3 blur filter, the above-mentioned mathematical function is converted into equivalent loop-unrolled DFG (the loop has been unrolled twice to generate two output pixels concurrently). The loop-unrolled DFG is further subjected to THT-based obfuscation. Thus, obtained obfuscated DFG of 3 × 3 blur filter is scheduled based on resource constraints of 1 multiplier and 1 adder. The scheduled DFG is shown in Figure 5.16. Further, the scheduled DFG of structurally obfuscated 3 × 3 blur filter is subjected to data path synthesis phase which results into a structurally obfuscated RTL data path of 3 × 3 blur filter hardware accelerator. The respective RTL circuit is shown in Figure 5.17. The end-to-end demonstration of image blurring using 3 × 3 blur filter hardware accelerator is shown in Figure 5.18.

5.9.2 Sharpening filter – mathematical function, RTL circuit and end-to-end demonstration

The mathematical function of a 3 × 3 sharpening filter **(high pass filter – cuts of low frequency components)** can be derived from (5.9) and (5.11) based on

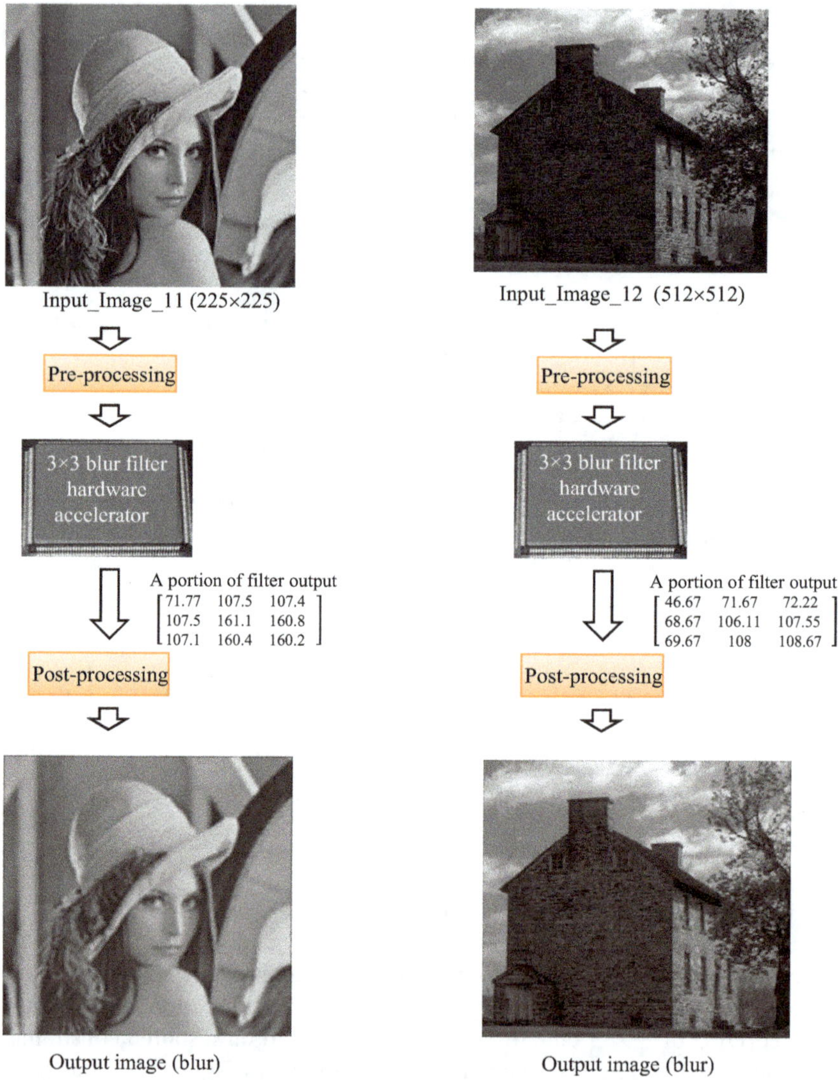

Figure 5.18 End-to-end demonstration of image blurring application using 3 × 3 blur filter hardware accelerator

its kernel matrix. The kernel matrix of a 3 × 3 sharpening filter is given as follows:

$$K^S = \begin{bmatrix} -1 & -1 & -1 \\ -1 & 9 & -1 \\ -1 & -1 & -1 \end{bmatrix}_{3\times 3} \tag{5.22}$$

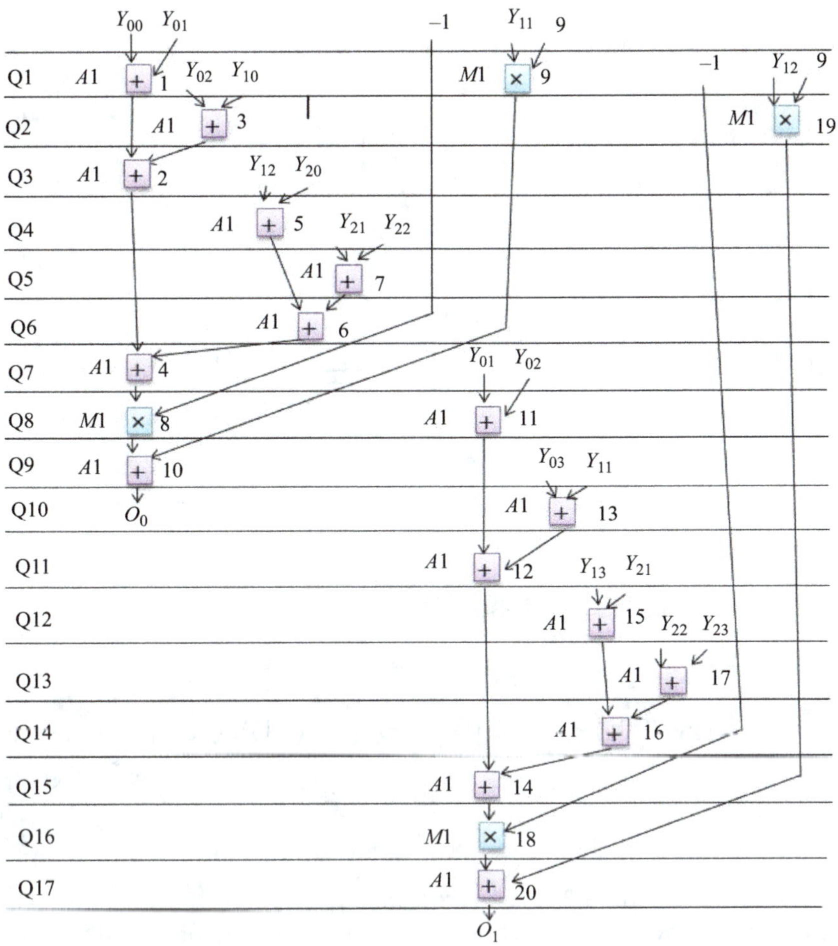

Figure 5.19 Scheduled DFG of 3 × 3 sharpening filter based on 1M, 1A

Based on this kernel matrix, (5.9) and (5.11) can be written as following equations:

$$O_0 = [(Y_{00} \times (-1)) + (Y_{01} \times (-1)) + (Y_{02} \times (-1))] + [(Y_{10} \times (-1)) + (Y_{11} \times 9) + (Y_{12} \times (-1))] + [(Y_{20} \times (-1)) + (Y_{21} \times (-1)) + (Y_{22} \times (-1))] \tag{5.23}$$

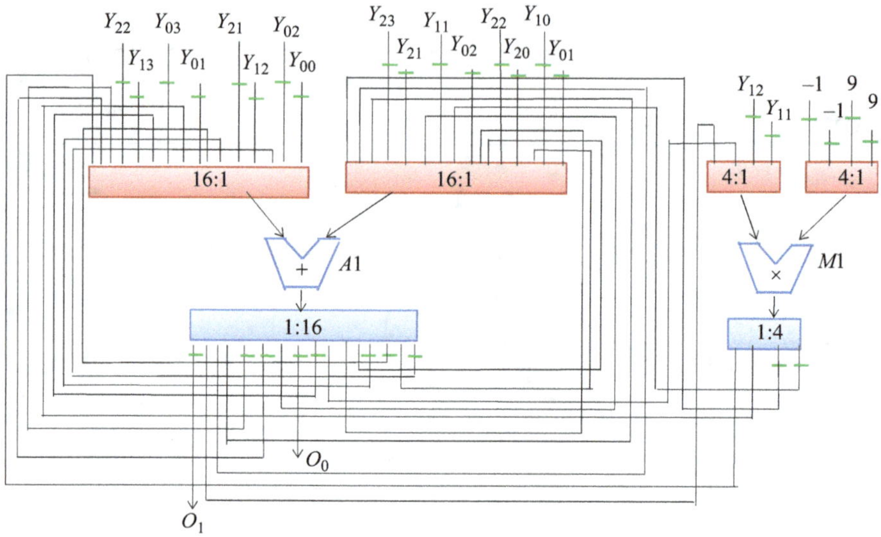

Figure 5.20 RTL circuit of a 3 × 3 sharpening filter hardware accelerator (Sengupta and Rathor, 2020)

$$\begin{aligned}O_1 = {} & [(Y_{01} \times (-1)) + (Y_{02} \times (-1)) + (Y_{03} \times (-1))] \\ & + [(Y_{11} \times (-1)) + (Y_{12} \times 9) + (Y_{13} \times (-1))] \\ & + [(Y_{21} \times (-1)) + (Y_{22} \times (-1)) + (Y_{23} \times (-1))] \end{aligned} \tag{5.24}$$

Equations (5.23) and (5.24) converge into the following equations which represent the mathematical function of a 3 × 3 sharpening filter, with the loop-unrolling (UF = 2)-based structural transformation (Sengupta and Rathor, 2020):

$$O_0 = [(Y_{00} + Y_{01} + Y_{02} + Y_{10} + Y_{12} + Y_{20} + Y_{21} + Y_{22}) \times (-1)] + (Y_{11} \times 9) \tag{5.25}$$

$$O_1 = [(Y_{01} + Y_{02} + Y_{03} + Y_{11} + Y_{13} + Y_{21} + Y_{22} + Y_{23}) \times (-1)] + (Y_{12} \times 9) \tag{5.26}$$

In order to generate an obfuscated hardware accelerator design of 3 × 3 sharpening filter, the abovementioned mathematical function is converted into equivalent loop-unrolled DFG which is further subjected to THT-based obfuscation. Thus, obtained obfuscated DFG of 3 × 3 sharpening filter is scheduled based on resource constraints of 1 multiplier and 1 adder as shown in Figure 5.19. Further, the scheduled obfuscated DFG of 3 × 3 sharpening filter is subjected to data path synthesis phase which results

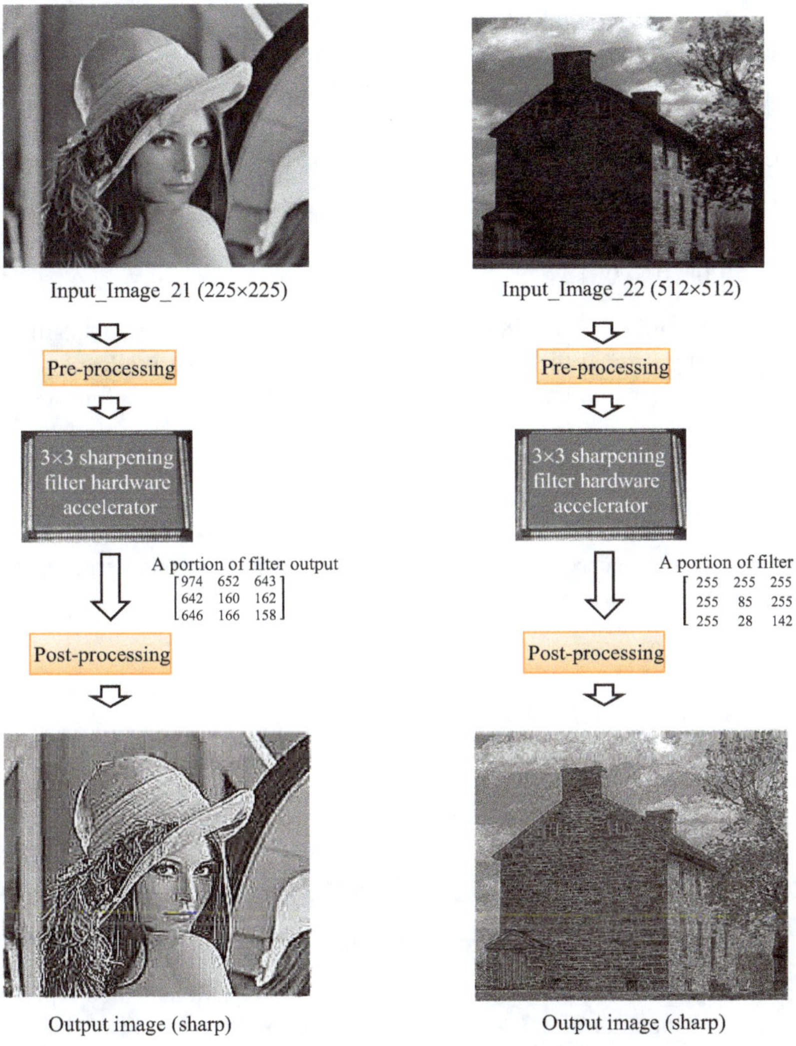

Figure 5.21 End-to-end demonstration of image sharpening application using a 3 × 3 filter hardware accelerator

into a structurally obfuscated RTL data path of 3 × 3 sharpening filter hardware accelerator as shown in Figure 5.20. The end-to-end demonstration of image sharpening using 3 × 3 filter hardware accelerator is shown in Figure 5.21.

5.9.3 Vertical embossment filter – mathematical function, RTL circuit and end-to-end demonstration

The mathematical function of a 3 × 3 VE filter can be derived from (5.9) and (5.11) based on its kernel matrix. The kernel matrix of a 3 × 3 VE filter is given as follows:

$$K^{VE} = \begin{bmatrix} 0 & 0 & 0 \\ 1 & 0 & -1 \\ 0 & 0 & 0 \end{bmatrix}_{3\times 3} \quad (5.27)$$

Since the kernel matrix of VE filter is not symmetric, therefore horizontal flip (or X-flip) followed by vertical flip (or Y-flip) are applied on the matrix. Post applying horizontal and vertical flip, the kernel matrix of VE filter is given as follows:

$$K^{VE} = \begin{bmatrix} 0 & 0 & 0 \\ -1 & 0 & 1 \\ 0 & 0 & 0 \end{bmatrix}_{3\times 3} \quad (5.28)$$

Thus obtained kernel matrix is used for convolution. Based on this kernel matrix, (5.9) and (5.11) can be written as following equations:

$$\begin{aligned} O_0 &= [(Y_{00} \times 0) + (Y_{01} \times 0) + (Y_{02} \times 0)] \\ &\quad + [(Y_{10} \times (-1)) + (Y_{11} \times 0) + (Y_{12} \times 1)] \\ &\quad + [(Y_{20} \times 0) + (Y_{21} \times 0) + (Y_{22} \times 0)] \end{aligned} \quad (5.29)$$

$$\begin{aligned} O_1 &= [(Y_{01} \times 0) + (Y_{02} \times 0) + (Y_{03} \times 0)] \\ &\quad + [(Y_{11} \times (-1)) + (Y_{12} \times 0) + (Y_{13} \times 1)] \\ &\quad + [(Y_{21} \times 0) + (Y_{22} \times 0) + (Y_{23} \times 0)] \end{aligned} \quad (5.30)$$

Equations (5.29) and (5.30) converge into the following equations which represent the mathematical function of a 3 × 3 VE filter, with the loop-unrolling (UF = 2) structural transformation (Sengupta and Rathor, 2020):

$$O_0 = [(Y_{10} \times (-1))] + [(Y_{12})] \quad (5.31)$$

$$O_1 = [(Y_{11} \times (-1))] + [(Y_{13})] \quad (5.32)$$

In order to generate a dedicated obfuscated hardware accelerator design of 3 × 3 VE filter, the previous equations show mathematical function is converted into equivalent loop-unrolled DFG. However, THT-based obfuscation is not applicable on the loop-unrolled DFG of 3 × 3 VE filter. Further, the obfuscated DFG of 3 × 3 VE filter is scheduled based on resource constraints of 1 multiplier and 1 adder as shown in Figure 5.22. Further, the scheduled DFG of 3 × 3 VE filter is subjected to data path synthesis phase which results into a structurally obfuscated RTL data path

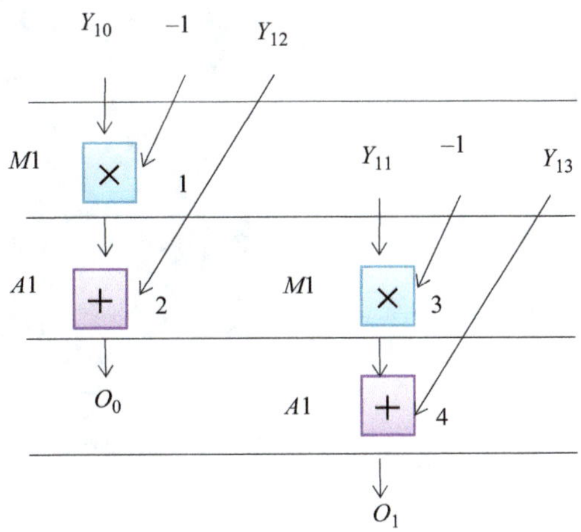

Figure 5.22 Scheduled DFG of 3 × 3 VE filter based on 1M, 1A

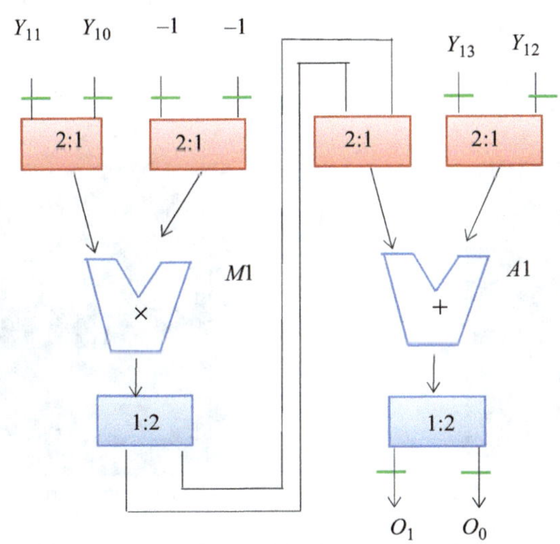

Figure 5.23 RTL circuit of a 3 × 3 VE filter hardware accelerator (Sengupta and Rathor, 2020)

of 3 × 3 VE filter hardware accelerator as shown in Figure 5.23. The end-to-end demonstration of VE of images using 3 × 3 filter hardware accelerator is shown in Figure 5.24.

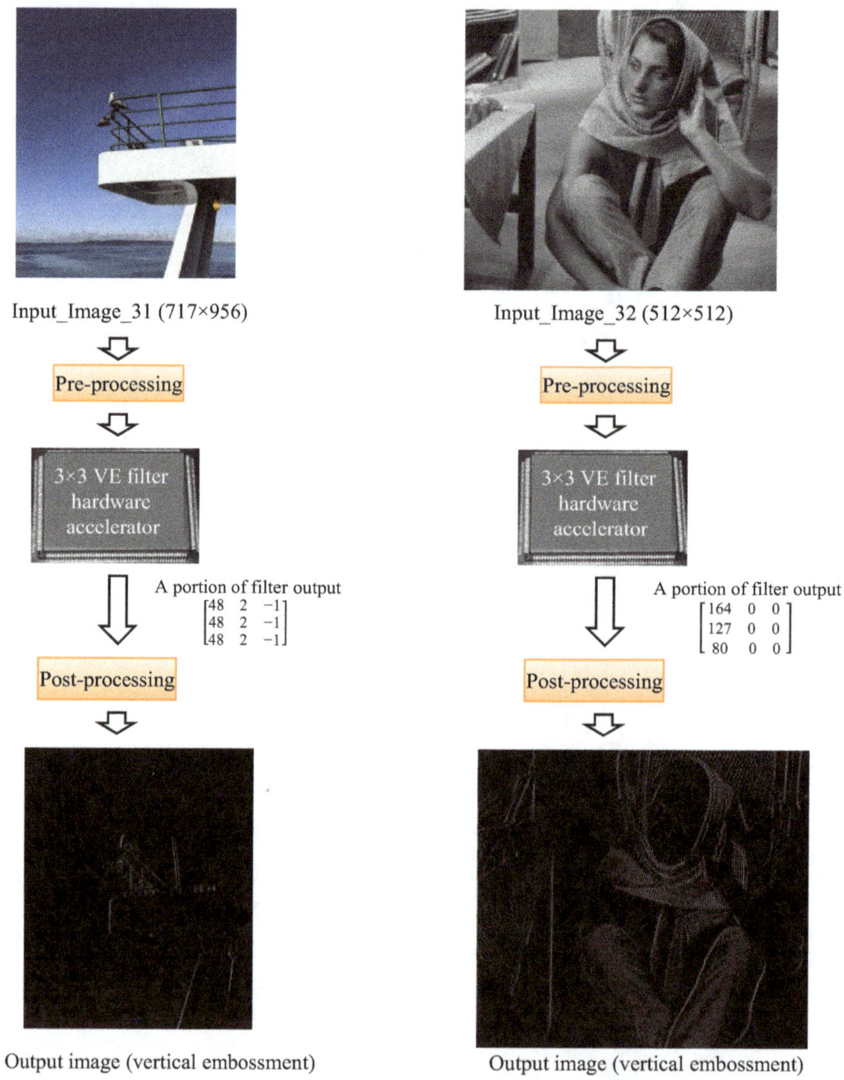

Figure 5.24 End-to-end demonstration of vertical embossment application using 3 × 3 filter hardware accelerator

5.9.4 Horizontal embossment filter – mathematical function, RTL circuit and end-to-end demonstration

The mathematical function of a 3 × 3 HE filter can be derived from (5.9) and (5.11) based on its kernel matrix. The kernel matrix of a 3 × 3 HE filter is given as follows:

$$K^{HE} = \begin{bmatrix} 0 & 1 & 0 \\ 0 & 0 & 0 \\ 0 & -1 & 0 \end{bmatrix}_{3\times 3} \tag{5.33}$$

Since the kernel matrix of HE filter is not symmetric, horizontal flip (or *X*-flip) followed by vertical flip (or *Y*-flip) are applied on the matrix. Post applying horizontal and vertical flip, the kernel matrix of HE filter is given as follows:

$$K^{HE} = \begin{bmatrix} 0 & -1 & 0 \\ 0 & 0 & 0 \\ 0 & 1 & 0 \end{bmatrix}_{3\times 3} \tag{5.34}$$

Based on this kernel matrix, (5.9) and (5.11) can be written as following equations:

$$\begin{aligned} O_0 = {} & [(Y_{00} \times 0) + (Y_{01} \times (-1)) + (Y_{02} \times 0)] \\ & + [(Y_{10} \times 0) + (Y_{11} \times 0) + (Y_{12} \times 0)] \\ & + [(Y_{20} \times 0) + (Y_{21} \times 1) + (Y_{22} \times 0)] \end{aligned} \tag{5.35}$$

$$\begin{aligned} O_1 = {} & [(Y_{01} \times 0) + (Y_{02} \times (-1)) + (Y_{03} \times 0)] \\ & + [(Y_{11} \times 0) + (Y_{12} \times 0) + (Y_{13} \times 0)] \\ & + [(Y_{21} \times 0) + (Y_{22} \times 1) + (Y_{23} \times 0)] \end{aligned} \tag{5.36}$$

Equations (5.35) and (5.36) converge into the following equations which represent the mathematical function of a 3 × 3 HE filter, with the loop-unrolling (UF = 2) structural transformation (Sengupta and Rathor, 2020):

$$O_0 = [(Y_{21})] + [(Y_{01} \times (-1))] \tag{5.37}$$

$$O_1 = [(Y_{22})] + [(Y_{02} \times (-1))] \tag{5.38}$$

In order to generate a dedicated obfuscated hardware accelerator design of 3 × 3 HE filter, the previous mathematical function is converted into its equivalent loop-unrolled DFG. However, THT-based obfuscation is not applicable on the loop-unrolled DFG of 3 × 3 HE filter. Thus, the obfuscated DFG of 3 × 3 HE filter is next scheduled based on resource constraints of 1 multiplier and 1 adder as shown in Figure 5.25. Further, the scheduled DFG of 3 × 3 HE filter is subjected to data path synthesis phase which results into a structurally obfuscated RTL data path of 3 × 3 HE filter hardware accelerator as shown in Figure 5.26. The end-to-end demonstration of HE of images using 3 × 3 filter hardware accelerator is shown in Figure 5.27.

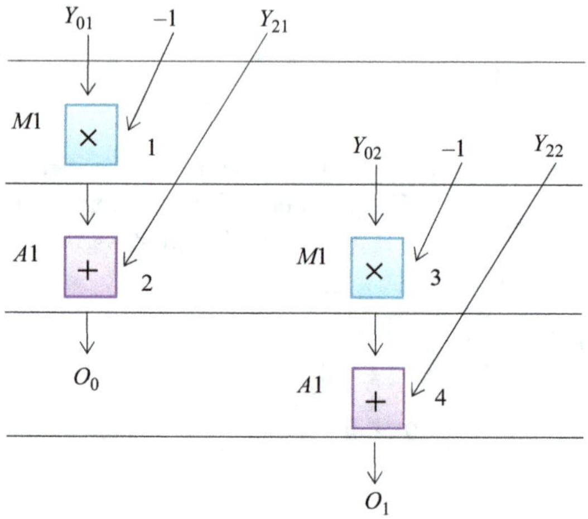

Figure 5.25 Scheduled DFG of 3 × 3 HE filter based on 1M, 1A

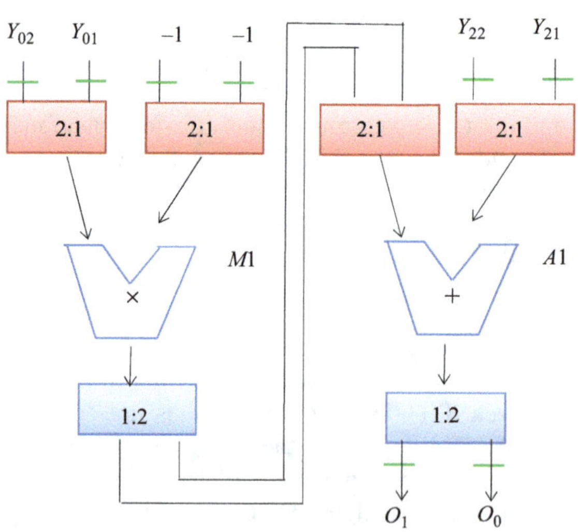

Figure 5.26 RTL circuit of 3 × 3 HE filter hardware accelerator (Sengupta and Rathor, 2020)

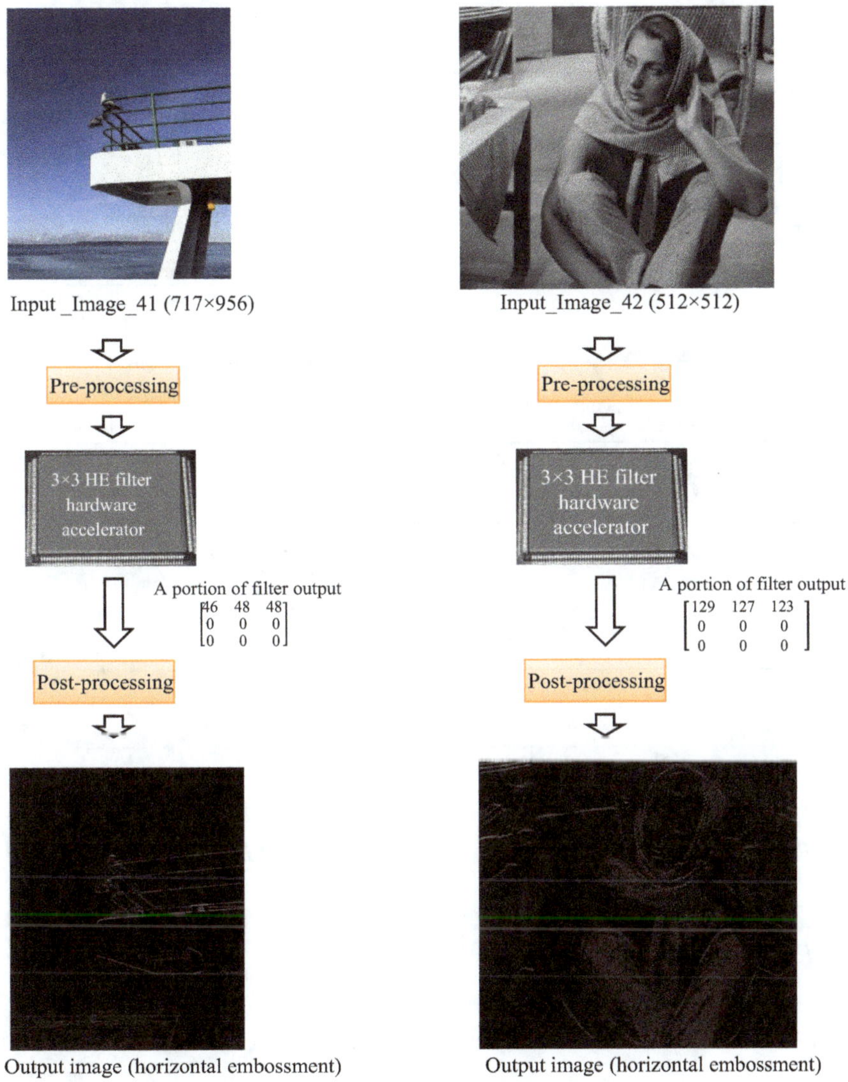

Figure 5.27 End-to-end demonstration of horizontal embossment application using 3 × 3 filter hardware accelerator

5.9.5 Laplace edge-detection filter – mathematical function, RTL circuit and end-to-end demonstration

The mathematical function of a 3 × 3 Laplace ED filter can be derived from (5.9) and (5.11) based on its kernel matrix. The kernel matrix of a 3 × 3 Laplace ED filter is given as follows:

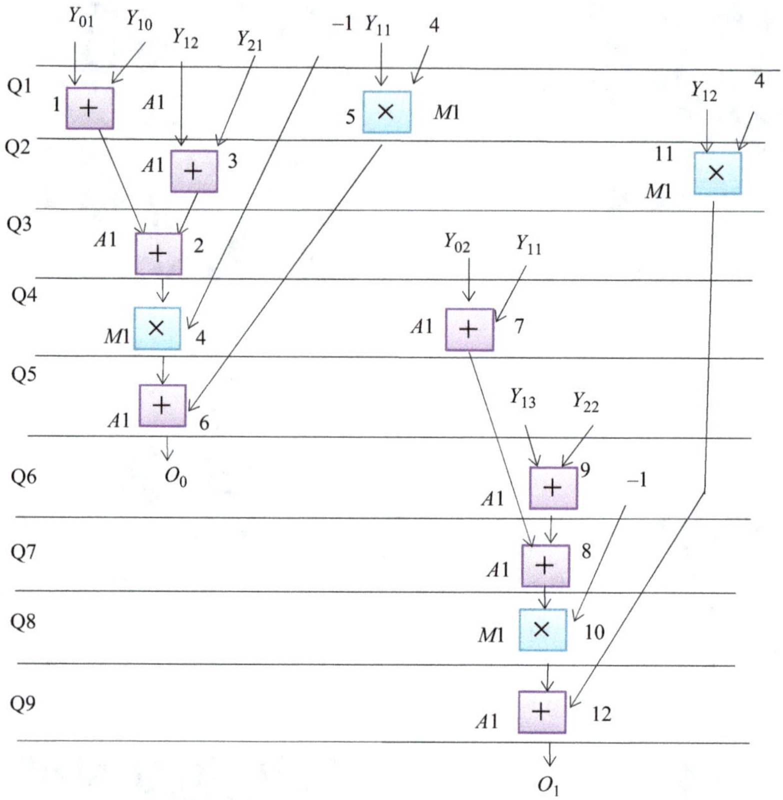

Figure 5.28 Scheduled DFG of 3 × 3 edge-detection filter based on 1M, 1A

$$K^{ED} = \begin{bmatrix} 0 & -1 & 0 \\ -1 & 4 & -1 \\ 0 & -1 & 0 \end{bmatrix}_{3\times 3} \tag{5.39}$$

Based on this kernel matrix, (5.9) and (5.11) can be written as following equations:

$$\begin{aligned} O_0 = {} & [(Y_{00} \times 0) + (Y_{01} \times (-1)) + (Y_{02} \times 0)] \\ & + [(Y_{10} \times (-1)) + (Y_{11} \times 4) + (Y_{12} \times (-1))] \\ & + [(Y_{20} \times 0) + (Y_{21} \times (-1)) + (Y_{22} \times 0)] \end{aligned} \tag{5.40}$$

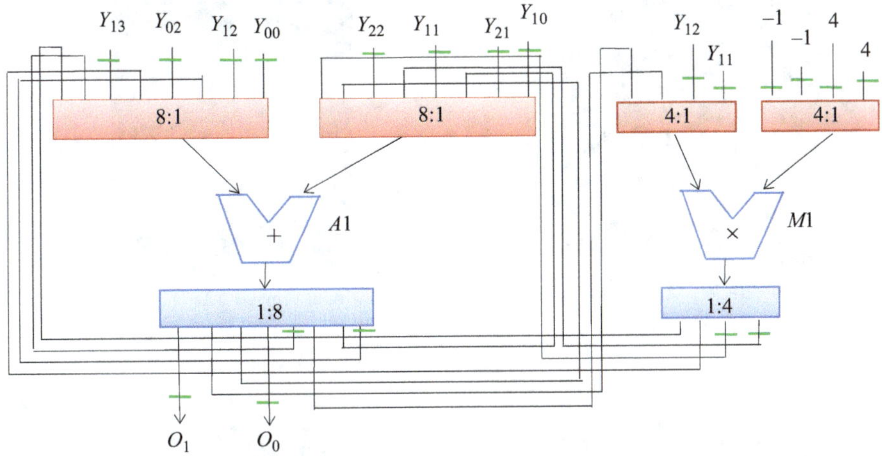

Figure 5.29 RTL circuit of 3 × 3 edge-detection filter hardware accelerator (Sengupta and Rathor, 2020)

$$O_1 = [(Y_{01} \times 0) + (Y_{02} \times (-1)) + (Y_{03} \times 0)] + [(Y_{11} \times (-1)) + (Y_{12} \times 4) + (Y_{13} \times (-1))] + [(Y_{21} \times 0) + (Y_{22} \times (-1)) + (Y_{23} \times 0)] \quad (5.41)$$

Equations (5.40) and (5.41) converge in to following equations which represent the mathematical function of a 3 × 3 ED filter, with the loop-unrolling (UF = 2) structural transformation (Sengupta and Rathor, 2020):

$$O_0 = [(Y_{01} + Y_{10} + Y_{12} + Y_{21}) \times (-1)] + (Y_{11} \times 4) \quad (5.42)$$

$$O_1 = [(Y_{02} + Y_{11} + Y_{13} + Y_{22}) \times (-1)] + [(Y_{12} \times 4)] \quad (5.43)$$

In order to generate a dedicated obfuscated hardware accelerator design of 3 × 3 ED filter, the previous mathematical function is converted into its equivalent loop-unrolled DFG, which is further subjected to THT-based obfuscation. Thus, the obtained obfuscated DFG of 3 × 3 ED filter is scheduled based on resource constraints of 1 multiplier and 1 adder as shown in Figure 5.28. Further, the scheduled DFG of 3 × 3 ED filter is subjected to data path synthesis phase which results into a structurally obfuscated RTL data path of 3 × 3 ED filter hardware accelerator as shown in Figure 5.29. The end-to-end demonstration of ED of images using a dedicated obfuscated 3 × 3 filter hardware accelerator is shown in Figure 5.30.

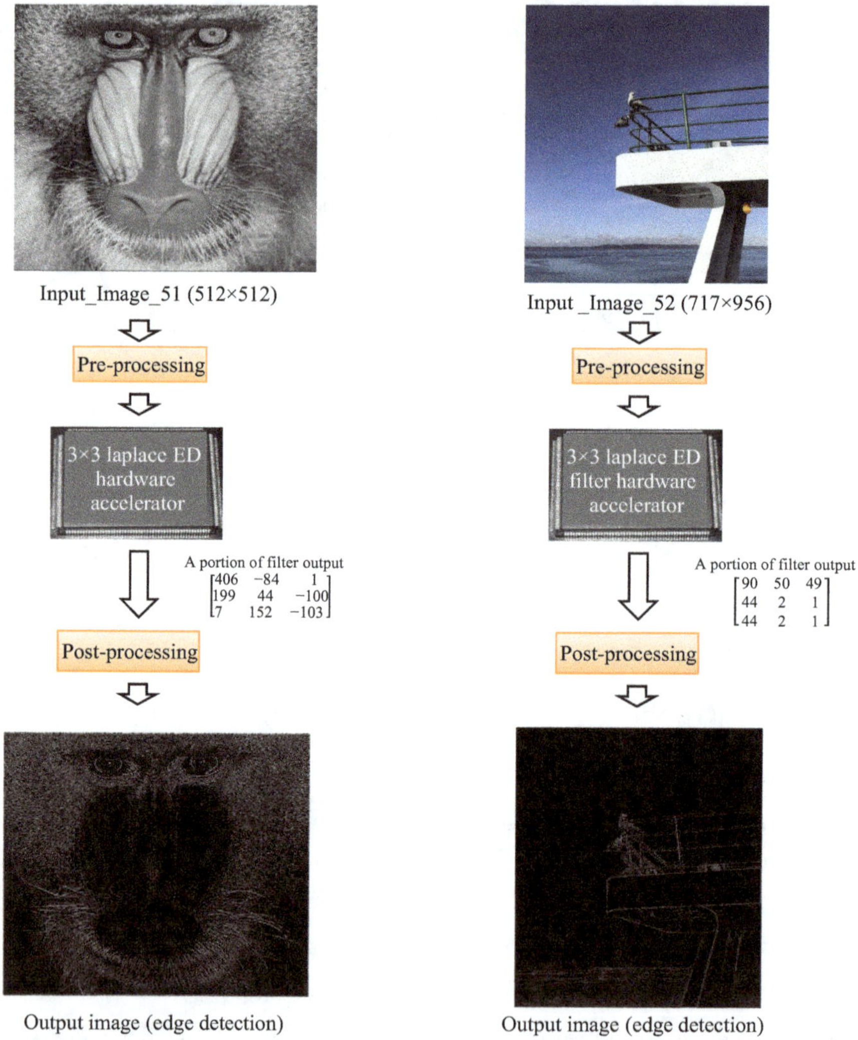

Figure 5.30 End-to-end demonstration of edge detection using 3 × 3 Laplace edge-detection filter hardware accelerator

5.10 Equivalent MATLAB codes for image processing filters

This section provides the equivalent MATLAB codes for various image processing applications.

5.10.1 Blur filter

```
line1: clc;
line2: I=imread('Input_Image_11.jpg') % Read input image in a matrix I
line3: S = im2uint8(I) % converting pixel values into unsigned 8-bit integer
line4: W = rgb2gray(S) % converting 3D (RGB) to grayscale
line5: filename= 'Pixel_matrix1.xlsx'
line6: xlswrite(filename,W); % To write pixel matrix MxN of input image in an
excel sheet
line7: A=readmatrix(' Pixel_matrix1.xlsx');% To read pixel matrix MxN of input
image from the excel sheet to matrix A
line8: B = ones(3)/9; %3x3 matrix: mean filter/kernel for image blurring
line9: C = conv2(A,B,'same') %performing 2D convolution (same) between mean
filter/kernel and input pixel matrix
line10: D =uint8(C) % converts to unsigned 8-bit integer values from 'double
datatype' before displaying
line11: display(D); % displaying pixel values of filtered image post converting
into unsigned 8-bit integer
line12: imwrite(D, 'sample_blur.jpg') % constructing filtered image from the
output pixel matrix D
```

Note: In the MATLAB code, lines 2–4 are the pre-processing steps and lines 10–12 are the post-processing steps (refer to Figure 5.18). Line numbers 5–7 are optional. They are just to show how an input pixel matrix can be recorded in an excel sheet and how the pixel values in the excel sheet can be read into a matrix variable. In the line number 9, the input matrix W can directly be applied in the convolution process. The matrix B is same as the kernel matrix K^B of blur filter, discussed in Section 5.9.

Demonstration of convolution function on an image (Input_Image_11)

After pre-processing, the input pixel matrix I is subjected to 2D convolution with the Blur filter kernel matrix K^B (given in (5.17)). The input image pixel matrix I, output matrix O and the blur filter kernel matrix K^B have been denoted by matrix A, matrix C and matrix B, respectively, in the MATLAB code. The demonstration of 'same' convolution function

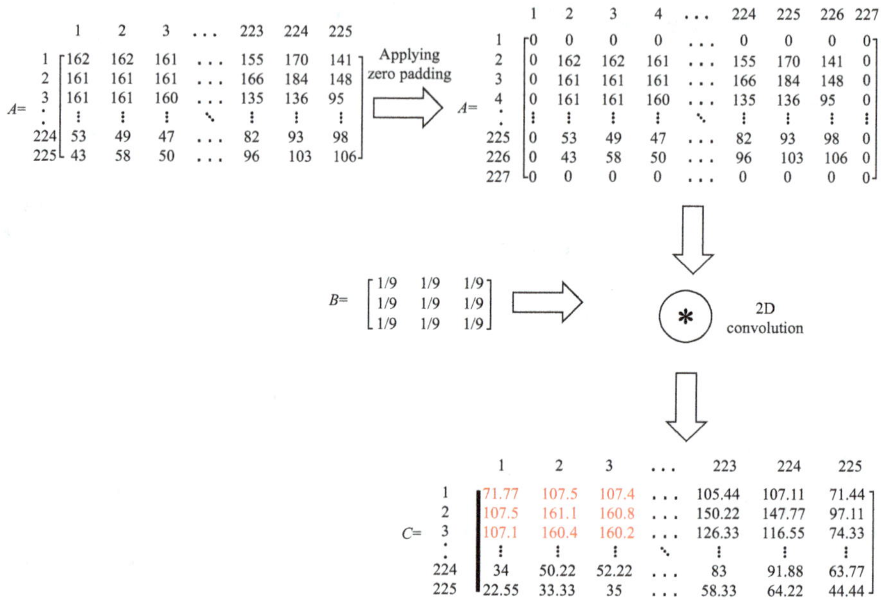

Figure 5.31 Demonstration of 'same' convolution using 3 × 3 blur filter on Input_Image_11 (225 × 225)

(C = conv2(A, B, 'same')) used in the MATLAB code, on an sample Input_Image_11, is shown in Figure 5.31. The output of same convolution using MATLAB function is same as that of blur filter hardware accelerator presented in Section 5.9. For demonstration, a portion (first 3 × 3 values highlighted in red in matrix C) of the output of MATLAB convolution function is shown in Figure 5.31. It is same as the corresponding portion (first 3 × 3 values) of convolution output (shown in Figure 5.18) generated by blur filter hardware accelerator.

5.10.2 Sharpening filter

```
line1: clc;

line2: I=imread('Input_Image_21.jpg') % Read input image in a matrix I

line3: S = im2uint8(I) % converting pixel values into unsigned 8-bit integer

line4: W = rgb2gray(S) % converting 3D (RGB) to grayscale

line5: filename= 'Pixel_matrix2.xlsx'

line6: xlswrite(filename,W); % To write pixel matrix MxN of input image in an
excel sheet

line7: A=readmatrix('Pixel_matrix2.xlsx');% To read pixel matrix MxN of input
image from the excel sheet to matrix A

line8: B=readmatrix('B_sharp_kernel.xlsx'); % Reading 3x3 filter/kernel of
image sharpening, from excel sheet to matrix B

line9: C = conv2(A,B,'same') %performing 2D convolution (same) between
filter/kernel and input pixel matrix

line10: D =uint8(C) % converts to unsigned 8-bit integer values from 'double
datatype' before displaying

line11: display(D); % displaying pixel values of filtered image post converting
into unsigned 8-bit integer

 line12: imwrite(D, 'sample_sharp.jpg') % constructing filtered image from the
                                   output pixel matrix D
```

Note: In the MATLAB code, lines 2–4 are the pre-processing steps and lines 10–12 are the post-processing steps (refer to Figure 5.21). Line numbers 5–7 are optional. Matrix B is same as the kernel matrix K^S of 3 × 3 sharpening filter, discussed in Section 5.9.

Demonstration of convolution function on an image (Input_Image_21)

After pre-processing, the input pixel matrix I is subjected to 2D convolution with the sharpening filter kernel matrix K^S (given in (5.22)). The input image pixel matrix I, output matrix O and the sharpening filter kernel matrix K^S have been denoted by matrix A, matrix C and matrix B respectively in the MATLAB code. The demonstration of 'same' convolution function (C = conv2(A, B, 'same')) used in the MATLAB code, on a sample Input_Image_21, is given in Figure 5.32. The output of 'same convolution' using MATLAB function is same as that of sharpening filter hardware accelerator, discussed in Section 5.9. For demonstration, a portion (first 3 × 3 values highlighted in red in matrix C) of the output of MATLAB convolution function is shown in Figure 5.32. It is same as the corresponding portion (first 3 × 3 values) of convolution output (shown in Figure 5.21) generated by sharpening filter hardware accelerator.

5.10.3 Vertical embossment filter

```
line1: clc;
line2: I=imread('Input_Image_31.jpg') % Read input image in a matrix I
line3: S = im2uint8(I) % converting pixel values into unsigned 8-bit integer
line4: W = rgb2gray(S) % converting 3D (RGB) to grayscale
line5: filename= 'Pixel_matrix3.xlsx'
line6: xlswrite(filename,W); % To write pixel matrix MxN of input image in an excel sheet
line7: A=readmatrix('Pixel_matrix3.xlsx');% To read pixel matrix MxN of input image from the excel sheet to matrix A
line8: B=readmatrix('B_VE_kernel.xlsx'); % Reading 3x3 filter/kernel of vertical embossment, from excel sheet to matrix B
line9: C = conv2(A,B,'same') %performing 2D convolution (same) between filter/kernel and input pixel matrix
line10: D =uint8(C) % converts to unsigned 8-bit integer values from 'double datatype' before displaying
line11: display(D); % displaying pixel values of filtered image post converting into unsigned 8-bit integer
line12: imwrite(D, 'sample_VE.jpg') % constructing filtered image from the output pixel matrix D
```

Note: In the MATLAB code, lines 2–4 are the pre-processing steps and lines 10–12 are the post-processing steps (refer to Figure 5.24). Line numbers 5–7 are optional. The matrix B is same as the kernel matrix K^{VE} of 3 × 3 VE filter, discussed in Section 5.9.

Demonstration of convolution function on an image (Input_Image_31)

After pre-processing, the input image pixel matrix I is subjected to 2D convolution with the VE filter kernel matrix K^{VE} (given in (5.28)). The input image pixel matrix I, output matrix O and the filter kernel matrix K^{VE} have been denoted by matrix A, matrix C and matrix B, respectively, in the MATLAB code. The demonstration of the 'same' convolution function (C = conv2(A, B, 'same')) used in the MATLAB code, on an sample Input_Image_31 is given in Figure 5.33. The output of same convolution using

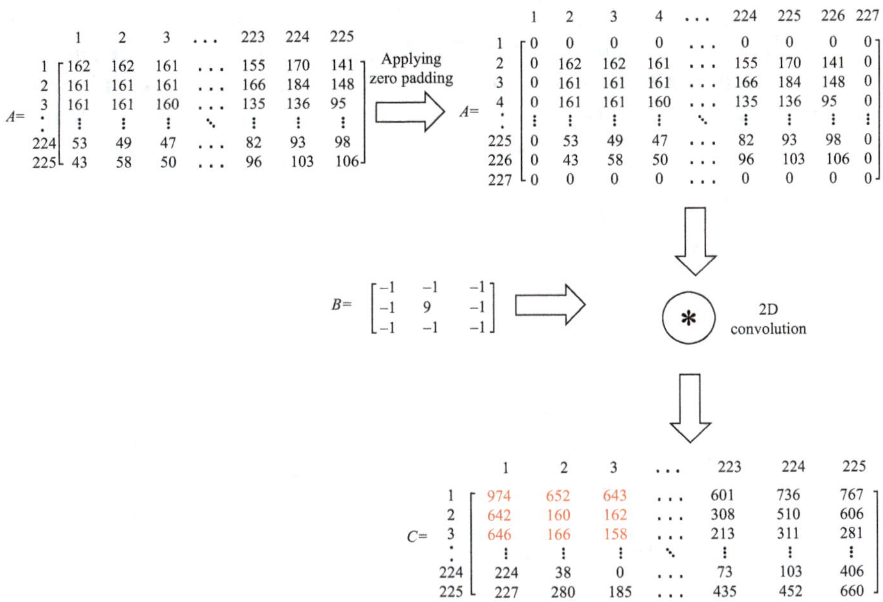

Figure 5.32 Demonstration of 'same' convolution using 3 × 3 sharpening filter on Input_Image_21 (225 × 225)

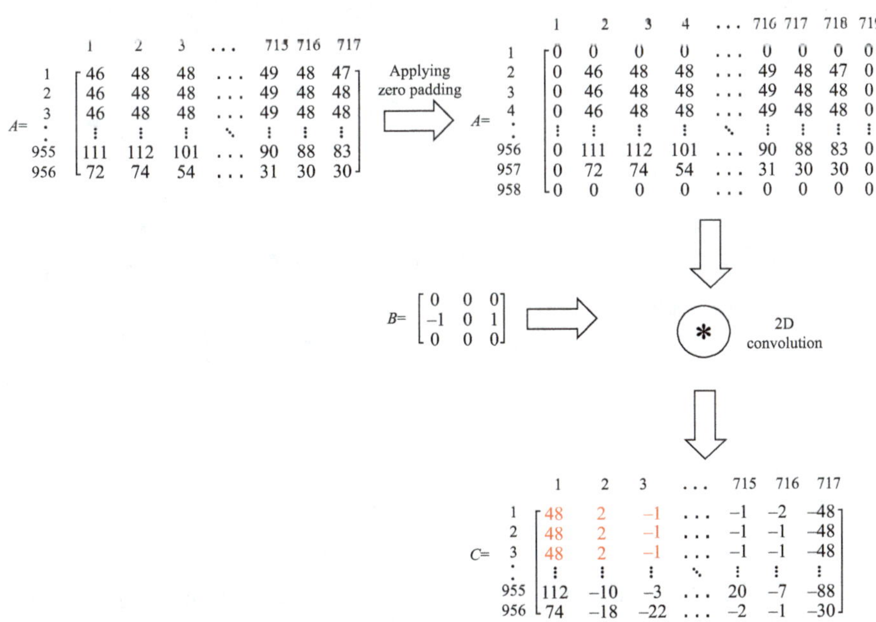

Figure 5.33 Demonstration of 'same' convolution using 3 × 3 VE filter on Input_Image_31 (717 × 956)

MATLAB function is same as that of VE filter hardware accelerator presented in Section 5.9. For demonstration, a portion (first 3 × 3 values highlighted in Red in matrix *C*) of the output of MATLAB convolution function is shown in Figure 5.33. It is same as the corresponding portion (first 3 × 3 values) of convolution output (shown in Figure 5.24) generated by VE filter hardware accelerator.

5.10.4 Horizontal embossment filter

```
line1: clc;
line2: I=imread('Input_Image_41.jpg') % Read input image in a matrix I
line3: S = im2uint8(I) % converting pixel values into unsigned 8-bit integer
line4: W = rgb2gray(S) % converting 3D (RGB) to grayscale
line5: filename= 'Pixel_matrix4.xlsx'
line6: xlswrite(filename,W); % To write pixel matrix MxN of input image in an
excel sheet
line7: A=readmatrix('Pixel_matrix4.xlsx');% To read pixel matrix MxN of input
image from the excel sheet to matrix A
line8: B=readmatrix('B_HE_kernel.xlsx'); % Reading 3x3 filter/kernel of
horizontal embossment, from excel sheet to matrix B
line9: C = conv2(A,B,'same') %performing 2D convolution (same) between
filter/kernel and input pixel matrix
line10: D =uint8(C) % converts to unsigned 8-bit integer values from 'double
datatype' before displaying
line11: display(D); % displaying pixel values of filtered image post converting
into unsigned 8-bit integer
line12: imwrite(D, 'sample_HE.jpg') % constructing filtered image from the
output pixel matrix D
```

Note: In the MATLAB code, lines 2–4 are the pre-processing steps and lines 10–12 are the post-processing steps (refer to Figure 5.27). Line numbers 5–7 are optional. The matrix *B* is same as the kernel matrix K^{HE} of 3 × 3 HE filter, discussed in Section 5.9.

Demonstration of convolution function on an image (Input_Image_41)

After pre-processing, the input image pixel matrix I is subjected to 2D convolution with the HE filter kernel matrix K^{HE} (given in (5.34)). The input image pixel matrix I, output matrix O and the filter kernel matrix K^{HE} have been denoted by matrix A, matrix C and matrix B, respectively, in the MATLAB code. The demonstration of the 'same' convolution function (C = conv2(A, B, 'same')) used in the MATLAB code, on an sample Input_Image_41, is given in Figure 5.34. The output of the same convolution using MATLAB function is same as that of HE filter hardware accelerator presented in Section 5.9. For demonstration, a portion (first 3×3 values in matrix C) of the output of MATLAB convolution function is shown in Figure 5.34. It is same as the corresponding portion (first 3×3 values) of convolution output (shown in Figure 5.27) generated by HE filter hardware accelerator.

5.10.5 Laplace edge-detection filter

```
line1: clc;
line2: I=imread('Input_Image_51.jpg') % Read input image in a matrix I
line3: S = im2uint8(I) % converting pixel values into unsigned 8-bit integer
line4: W = rgb2gray(S) % converting 3D (RGB) to grayscale
line5: filename= 'Pixel_matrix5.xlsx'
line6: xlswrite(filename,W); % To write pixel matrix MxN of input image in an
excel sheet
line7: A=readmatrix('Pixel_matrix5.xlsx');% To read pixel matrix MxN of input
image from the excel sheet to matrix A
line8: B=readmatrix('B_Laplcae_ED_kernel.xlsx'); % Reading 3x3 filter/kernel
of Laplace edge detection, from excel sheet to matrix B
line9: C = conv2(A,B,'same') %performing 2D convolution (same) between
filter/kernel and input pixel matrix
line10: D =uint8(C) % converts to unsigned 8-bit integer values from 'double
datatype' before displaying
line11: display(D); % displaying pixel values of filtered image post converting
into unsigned 8-bit integer
line12: imwrite(D, 'sample_ED.jpg') % constructing filtered image from the
output pixel matrix D
```

Note: In the MATLAB code, lines 2–4 are the pre-processing steps and lines 10–12 are the post-processing steps (refer to Figure 5.30). Line numbers 5–7 are optional. The matrix B is same as the kernel matrix K^{ED} of 3 × 3 Laplace ED filter, discussed in Section 5.9.

Demonstration of convolution function on an image (Input_Image_51):

After pre-processing, the input image pixel matrix I is subjected to 2D convolution with the Laplace ED filter kernel matrix K^{ED} (given in (5.39)). The input image pixel matrix I, output matrix O and the filter kernel matrix K^{ED} have been denoted by matrix A, matrix C and matrix B, respectively, in the MATLAB code. The demonstration of 'same' convolution function (C = conv2(A, B, 'same')) used in the MATLAB code, on an sample Input_Image_51, is given in Figure 5.35. The output of same convolution using MATLAB function is same as that of Laplace ED filter hardware accelerator presented in Section 5.9. For demonstration, a portion (first 3 × 3 values highlighted in red in matrix C) of the output of MATLAB convolution function is shown in Figure 5.35. It is same as the corresponding portion (first 3 × 3 values) of convolution output (shown in Figure 5.30) generated by Laplace ED filter hardware accelerator.

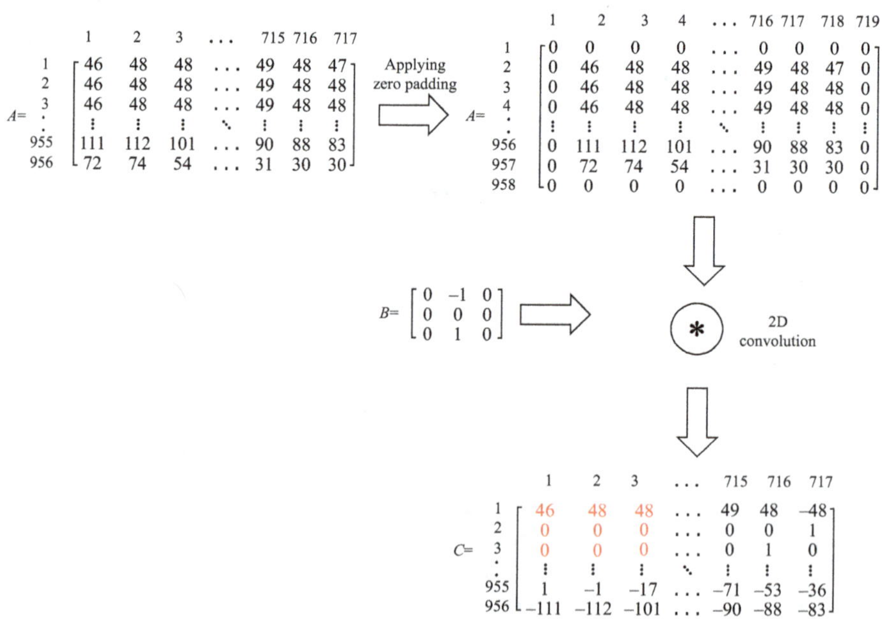

Figure 5.34 Demonstration of 'same' convolution using 3 × 3 HE filter on Input_Image_41 (717 × 956)

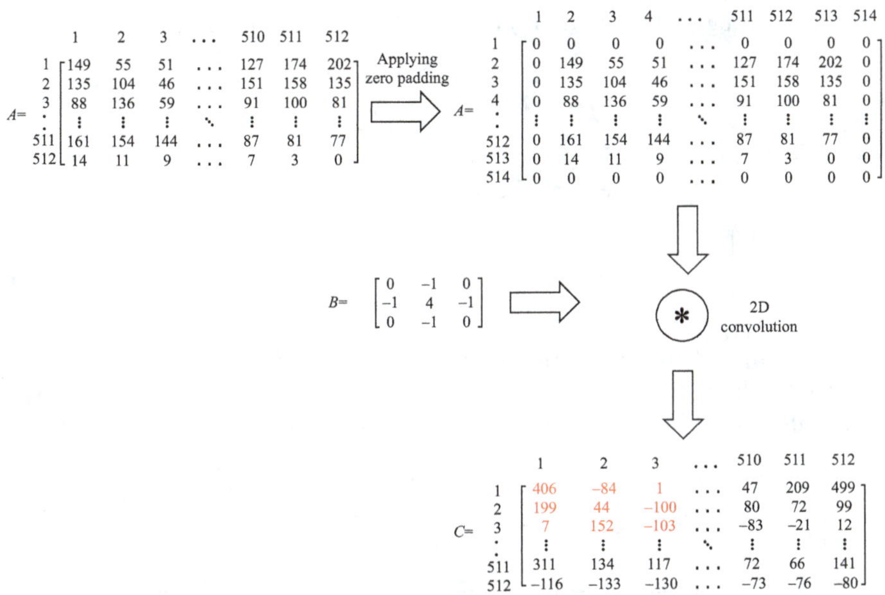

Figure 5.35 Demonstration of 'same' convolution using 3 × 3 Laplace ED filter on Input_Image_51 (512 × 512)

5.11 Additional information on image processing convolution filters

5.11.1 Deriving Laplace filter kernel matrix

A derivative is defined as a slope as given in the following equation:

$$f'(a) = \lim_{h \rightarrow i} \frac{f(a+h) - f(a)}{h} \tag{5.44}$$

In a discrete grid, the smallest h is 1. Therefore, the first derivative is

$$f'(x) = f(x+1) - f(x) \tag{5.45}$$

where x represents a position of pixel value.

Further,

$$f'(x+1) = f(x+2) - f(x+1) \tag{5.46}$$

The second derivative is the first derivative of the first derivative given in (5.45).

$$f''(x) = f'(x+1) - f'(x) \tag{5.47}$$

Upon substituting (5.45) and (5.46) into (5.47), we obtain the following equation:

$$\begin{aligned} f''(x) &= f(x+2) - f(x+1) - f(x+1) + f(x) \\ f''(x) &= f(x+2) - 2f(x+1) + f(x) \end{aligned} \tag{5.48}$$

In an array form, $f''(x) = [1, -2, 1]$

In the matrix form, the second derivative is

$$f''(x) = \begin{bmatrix} 0 & 0 & 0 \\ 1 & -2 & 1 \\ 0 & 0 & 0 \end{bmatrix} \tag{5.49}$$

The matrix of x derivative is row based. The transpose of matrix in (5.49) is y derivative (column based).

Adding matrix in (5.49) and its transpose results into Laplace operator matrix.

$$\begin{bmatrix} 0 & 0 & 0 \\ 1 & -9 & 1 \\ 0 & 0 & 0 \end{bmatrix} + \begin{bmatrix} 0 & 1 & 0 \\ 0 & -9 & 0 \\ 0 & 1 & 0 \end{bmatrix} = \begin{bmatrix} 0 & 1 & 0 \\ 1 & -9 & 1 \\ 0 & 1 & 0 \end{bmatrix} \tag{5.50}$$

Note: When a blurred image is applied on a Laplace filter, it results into a sharpened image.

5.11.2 Difference between convolution and correlation

Suppose, a generic kernel matrix is as follows:

$$K = \begin{bmatrix} k1 & k2 & k3 \\ k4 & k5 & k6 \\ k7 & k8 & k9 \end{bmatrix} \tag{5.51}$$

And a generic input matrix is as follows:

$$I = \begin{bmatrix} y1 & y2 & y3 \\ y4 & y5 & y6 \\ y7 & y8 & y9 \end{bmatrix} \tag{5.52}$$

Applying correlation between matrix [I] and matrix [K]:

$$I \circledast K = \begin{bmatrix} y1 & y2 & y3 \\ y4 & y5 & y6 \\ y7 & y8 & y9 \end{bmatrix} \circledast \begin{bmatrix} k1 & k2 & k3 \\ k4 & k5 & k6 \\ k7 & k8 & k9 \end{bmatrix}$$

$$I \circledast K = y1k1 + y2k2 + y3k3 + y4k4 + y5k5 + y6k6 + y7k7 + y8k8 + y9k9 \tag{5.53}$$

Applying convolution between matrix [I] and matrix [K]:

To apply convolution, first the kernel matrix is horizontally flipped (x-flip) followed by vertical flip (y-flip).

The kernel matrix post applying horizontal flip:

$$K = \begin{bmatrix} k7 & k8 & k9 \\ k4 & k5 & k6 \\ k1 & k2 & k3 \end{bmatrix}$$

Further, post applying vertical flip:

$$K = \begin{bmatrix} k9 & k8 & k7 \\ k6 & k5 & k4 \\ k3 & k2 & k1 \end{bmatrix}$$

The horizontal flip followed by vertical flip rotates the matrix by 180°.

Now applying convolution:

$$I \circledast K = \begin{bmatrix} y1 & y2 & y3 \\ y4 & y5 & y6 \\ y7 & y8 & y9 \end{bmatrix} \circledast \begin{bmatrix} k9 & k8 & k7 \\ k6 & k5 & k4 \\ k3 & k2 & k1 \end{bmatrix}$$

$$I \circledast K = y1k9 + y2k8 + y3k7 + y4k6 + y5k5 + y6k4 + y7k3 + y8k2 + y9k1 \tag{5.54}$$

The difference between correlation and convolution between I and K is evident from (5.53) and (5.54).

Note: The convolution and correlation for symmetric filter remains same, since the symmetrical kernel matrix does not change post applying horizontal and vertical flip.

5.12 Analysis of case studies

This section discusses security and design cost analysis of image processing filter hardware accelerators. The security analysis has been discussed for both 3 × 3 and 5 × 5 filter hardware accelerators. Further, the security analysis has been discussed for dedicated 3 × 3 filter hardware accelerators designed for individual image processing applications such as blurring, sharpening, VE, HE and Laplace ED. In addition, power consumption and design cost analysis have been discussed for each dedicated 3 × 3 filter hardware accelerators.

5.12.1 Security analysis

Sengupta and Rathor (2020) secured image processing filter hardware accelerators using structural obfuscation mechanism based on the following high-level transformation techniques applied during HLS design process: loop unrolling, DFG partitioning and THT. The applied structural obfuscation techniques lead to significant structural transformations in the RTL data path in terms of change in (i) count of function unit resource such as multipliers and adders, (ii) size and count of interconnect hardware resources such as multiplexers and demultiplexers, (iii) count of storage hardware resources such as registers and latches, (iv) interconnection path among the various RTL components, etc. without disturbing the original functionality. Further, Sengupta and Rathor (2020) presented the strength of structural obfuscation in terms of modification in gate count post-obfuscation. The gate count is modified in terms of change in the number and size of RTL components; hence, the change in gate count does not follow any particular pattern. Thereby, the design netlist post structural obfuscation becomes significantly different than the original version. Therefore, an attempt to analyse/identify any pattern in the change of gate count does not help an attacker in deducing the original design. This hinders the process of RE by an attacker, hence preventing against Trojan insertion.

In the following subsections, security metric has been analysed for the generic 3 × 3 and 5 × 5 filter hardware accelerators and for dedicated 3 × 3 filter hardware accelerators.

5.12.1.1 Analysing security metric of generic 3 × 3 and 5 × 5 filter hardware accelerators

Sengupta and Rathor (2020) analysed security metric of generic 3 × 3 and 5 × 5 filter hardware accelerators in terms of (i) change in RTL components and (ii) gate count modified post-obfuscation. Figure 5.36 shows the change in RTL components and gate count of generic 3 × 3 filter hardware accelerators post applying structural obfuscation. The changes in functional unit resources, interconnect hardware resources and storage hardware resources have been shown in Figure 5.36(a)–(c), respectively. Further, Figure 5.36(d) shows the change in gate count post applying structural obfuscation on generic 3 × 3 filter hardware accelerator. Similarly, Figure 5.37 shows the change in RTL components and gate count of generic 5 × 5 filter hardware accelerators post applying structural obfuscation. The changes in functional unit resources, interconnect hardware resources and storage hardware resources have been shown in Figure 5.37(a)–(c), respectively. Further, Figure 5.37(d) shows the change in gate count post applying structural obfuscation techniques on generic 5 × 5 filter hardware accelerator. As shown in Figures 5.36 and 5.37, significant changes in the RTL structure and gate count incur (without affecting functionality) due to employed structural obfuscation. This indicates higher strength of obfuscation for thwarting against RE (and secret Trojan insertion) by an adversary.

5.12.1.2 Analysing security metric of dedicated 3 × 3 filter hardware accelerator

Sengupta and Rathor (2020) analysed security metric of dedicated 3 × 3 filter hardware accelerators in term of gate count modified post-obfuscation. Figure 5.38 compares the gate count of dedicated 3 × 3 filter hardware accelerators, viz. blur, sharpening, VE, HE and Laplace ED filter, before and after applying structural obfuscation. Figure 5.39 reports the change in the gate count post-structural obfuscation. As shown in the figure, structural obfuscation techniques applied on dedicated 3 × 3 filter hardware accelerators affect gate count hugely (without

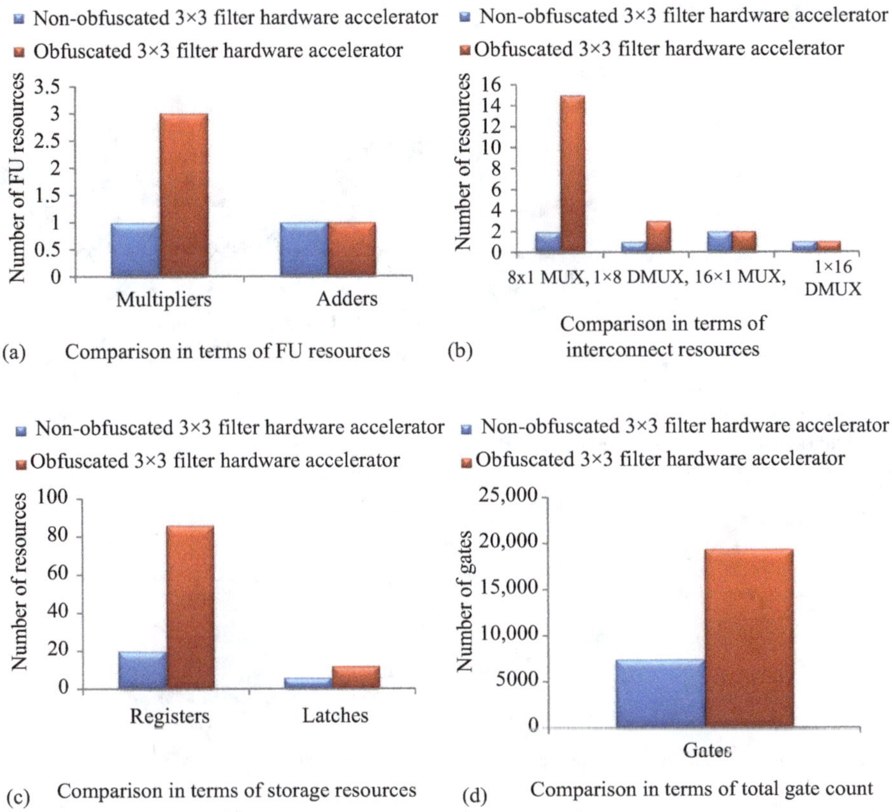

Figure 5.36 Comparison of un-obfuscated and structurally obfuscated 3 × 3 filter hardware accelerators in terms of RTL components and gate count (Sengupta and Rathor, 2020)

affecting functionality) of the design. This shows that significant obscurity is obtained in the design netlist, hence thwarting against RE and Trojan insertion.

5.12.2 Design cost analysis

The design cost needs to be evaluated in order to analyse the impact of the structural obfuscation technique on the design overhead of image processing filter hardware accelerators. Following equation is used to evaluate the design cost (Sengupta and Rathor, 2020):

$$C_d(U_i) = \rho_1 \frac{L_d}{L_m} + \rho_2 \frac{A_d}{A_m} \tag{5.55}$$

where $C_d(U_i)$ is the design cost calculated for resource constraints U_i, further L_d and L_m are the design latency at specified resource constraints and maximum design

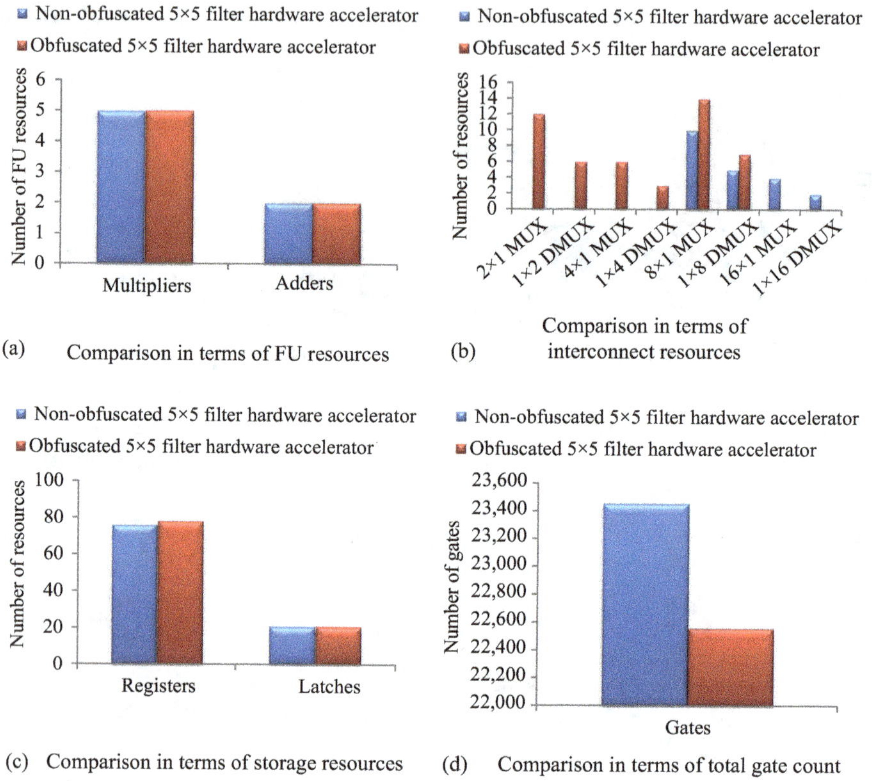

Figure 5.37 Comparison of un-obfuscated and structurally obfuscated 5 × 5 filter hardware accelerators in terms of RTL components and gate count (Sengupta and Rathor, 2020)

latency, respectively, A_d and A_m are the design area at specified resource constraints and maximum area, respectively, and ρ_1, ρ_2 are the weights which are fixed at 0.5. Figure 5.40 shows the design cost of dedicated 3 × 3 filter hardware accelerators. Further, Figure 5.41 shows the power consumption of dedicated 3 × 3 filter hardware accelerators. As shown in Figures 5.40 and 5.41, design cost and power consumption of structurally obfuscated dedicated 3 × 3 filter hardware accelerators is quite low.

5.13 Conclusion

Image processing filters have important applications in this era of modern technology. The image processing filters in the form of hardware accelerators help in achieving high performance at low power consumption. This chapter discussed multimodal image processing filter hardware accelerators for two different kernel

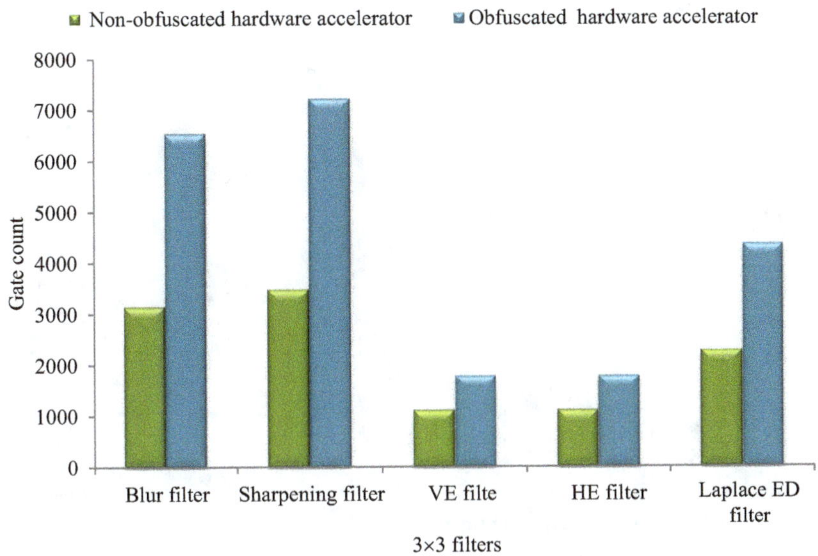

Figure 5.38 Comparison of un-obfuscated and structurally obfuscated dedicated 3 × 3 filter hardware accelerators in term of gate count (Sengupta and Rathor, 2020)

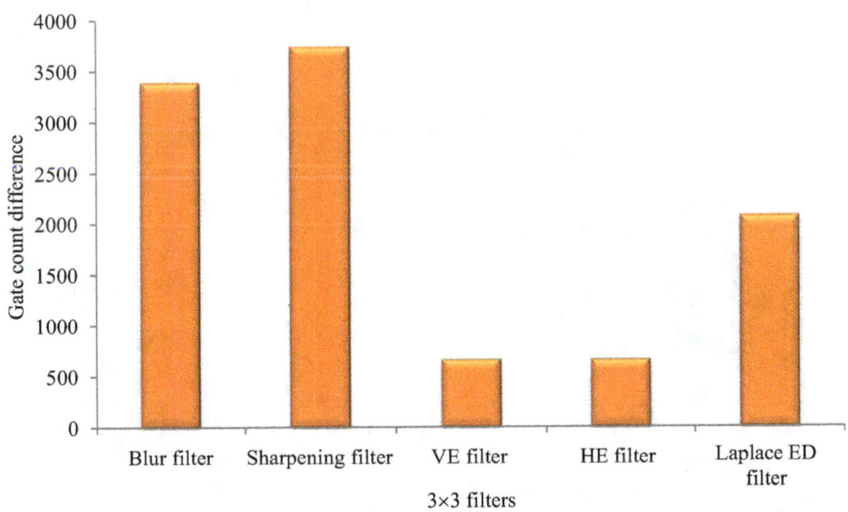

Figure 5.39 Gate count difference of un-obfuscated and structurally obfuscated dedicated 3 × 3 filter hardware accelerators (Sengupta and Rathor, 2020)

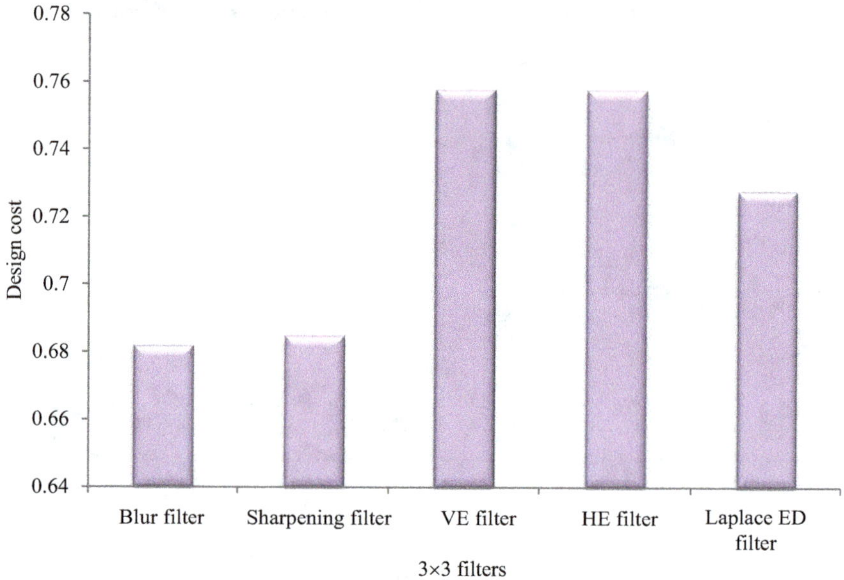

Figure 5.40 Design cost of dedicated 3 × 3 filter hardware accelerators (Sengupta and Rathor, 2020)

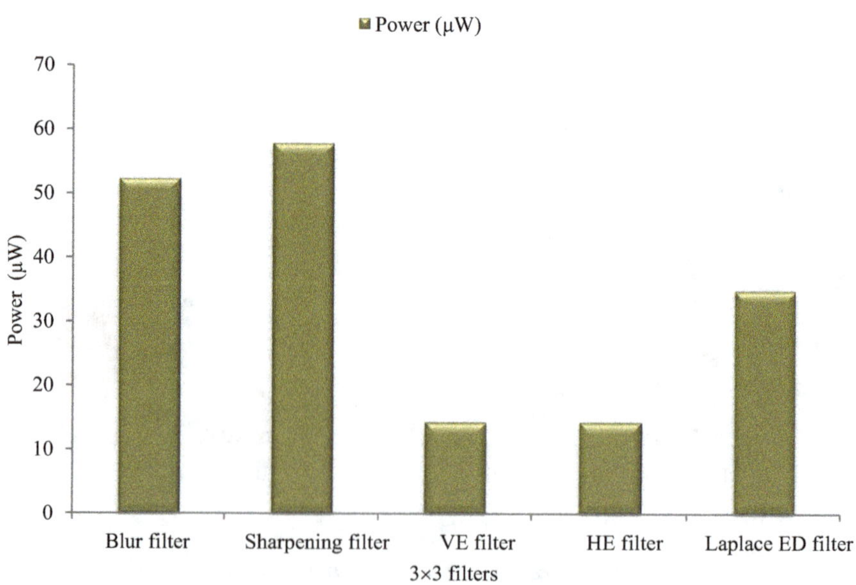

Figure 5.41 Power consumption of dedicated 3 × 3 filter hardware accelerators (Sengupta and Rathor, 2020)

sizes, viz. 3 × 3 and 5 × 5. A functionally reconfigurable processor mode of 3 × 3 filter hardware accelerators has been discussed, where five different filters are combined into a single hardware architecture and a specific filter architecture is enabled at a single instant of time using a selection vector. This chapter also discussed the structurally obfuscated versions of 3 × 3 and 5 × 5 filter hardware accelerators. The motive of employing structural obfuscation into the design of 3 × 3 and 5 × 5 filter hardware accelerators is to secure them against RE (and hardware Trojan threat). Further, dedicated 3 × 3 filter hardware accelerators as application specific architecture have been presented for image processing applications, viz. blurring, sharpening, VE, HE and ED. The structurally obfuscated versions of dedicated 3 × 3 filter hardware accelerators have also been discussed in this chapter. Further a discussion on the analysis of structurally obfuscated image processing filter hardware accelerators has been presented, in terms of security and design cost. The case study analysis showed that high obfuscation (security) at low design cost is achieved for structurally obfuscated image processing filter hardware accelerator architectures.

At the end of this chapter, a reader becomes competent to develop an understanding on the following concepts:

- need of image processing filter hardware accelerators;
- need of security for image processing filter hardware designs;
- theory of 3 × 3 filter including derivation of its generic computation function;
- designing a generic 3 × 3 filter hardware accelerator and its structurally obfuscated version using loop-unrolling and THT-based high-level transformations;
- designing a 3 × 3 filter hardware accelerator as a functionally reconfigurable processor architecture;
- theory of 5 × 5 filter including derivation of its generic computation function;
- designing a generic 5 × 5 filter hardware accelerator and its structurally obfuscated version using DFG partitioning and THT-based high-level transformations;
- deriving computation functions of dedicated 3 × 3 filters for image blurring, Laplace ED, sharpening, VE and HE filter;
- designing dedicated 3 × 3 filter hardware accelerators for image blurring, Laplace ED, sharpening, VE and HE filter with end-to-end demonstration on benchmark images;
- MATLAB codes for image blurring, Laplace ED, sharpening, VE and HE filter as well as demonstration of convolution filtering;
- analysis on case studies in term of security achieved using structural obfuscation and
- analysis on case studies in terms of the design cost.

5.14 Questions and exercise

1. What are the different types of image processing filters used in camera?
2. How does security affect image processing filter hardware?
3. What is a 2D convolution filter?

4. What is a filter kernel?
5. Represent an image matrix with zero padding to perform 'same convolution'.
6. What is the zero padding rule?
7. What steps are performed in the pre-processing steps?
8. What steps are performed in the post-processing steps?
9. Represent a filter kernel matrix generically.
10. What is the computation function for first pixel of an image processing filter hardware?
11. What type of structural transformation techniques are applied on image processing filter hardware? Explain each one of them.
12. Design an RTL data path of 3 × 3 filter hardware accelerators for image processing applications for two adder and one multiplier resources constraints.
13. Design an obfuscated RTL data path of 3 × 3 filter hardware accelerators for image processing applications for two adder and one multiplier resources constraints.
14. How does loop-unrolling affect the hardware architecture of filters?
15. How does functionally reconfigurable processor of the structurally obfuscated 3 × 3 filter hardware accelerator work?
16. What is the role of selection vector in functionally reconfigurable processor of the structurally obfuscated 3 × 3 filter hardware accelerator?
17. How is partitioning based transformation useful in designing RTL data path of 5 × 5 filter hardware accelerators?
18. Derive the mathematical function of a 3 × 3 blur filter with the loop-unrolling (UF = 2) transformation.
19. What is the function of blur filter?
20. Derive the mathematical function of a 3 × 3 sharpening filter, with the loop-unrolling (UF = 2) transformation.
21. What is the function of sharpening filter?
22. What is the difference between correlation and convolution? How does the kernel matrix change?
23. Derive the mathematical function of a 3 × 3 HE filter, with the loop-unrolling (UF = 2) transformation.
24. What is 3 × 3 HE filter?
25. Derive the Laplace ED filter kernel matrix.
26. What is 3 × 3 VE filter?
27. Write the equivalent MATLAB codes for image processing filter functions.

References

M. Azizabadi and A. Behrad (2013), 'Design and VLSI implementation of new hardware architectures for image filtering,' *Proc. MVIP*, Zanjan, pp. 110–115.

L. Benda, P. Mudry and A. Ijspeert (2008), 'Hardware acceleration for image processing,' *Technical Report EPFL*, [online] Available: http://biorob2.epfl.ch/pages/studproj90/birg67936/rapport.pdf.

C. Cappetta, G. D. Licciardo and L. Di Benedetto (2017), 'Hardware accelerator using Gabor filters for image recognition applications,' *Proc. ISSCS*, Iasi, pp. 1–4.

R. S. Chakraborty and S. Bhunia (2009), 'Security against hardware Trojan through a novel application of design obfuscation,' *Proc. of the International Conference on Computer-Aided Design*, ACM, pp. 113–116.

R. S. Chakraborty and S. Bhunia (2011), 'Security against hardware Trojan attacks using key-based design obfuscation,' *J. Electron. Test.*, vol. 27(6), pp. 767–785.

H. Dutta, F. Hannig, J. Teich, B. Heigl and H. Hornegger (2006), 'A design methodology for hardware acceleration of adaptive filter algorithms in image processing,' *Proc. ASAP'06*, Steamboat Springs, CO, pp. 331–340.

Y. Lao and K. K. Parhi (2015), 'Obfuscating DSP circuits via high-level transformations' *IEEE Trans. Very Large Scale Integr. (VLSI) Syst.*, vol. 23(5), pp. 819–830.

H. R. Mahdiany, A. Hormati and S. M. Fakhraie (2001), 'A hardware accelerator for DSP system design,' *Proc. ICM*, pp. 141–144.

V. K. Mishra and A. Sengupta (2014), 'MO-PSE: adaptive multi-objective particle swarm optimization based design space exploration in architectural synthesis for application specific processor design,' *J. Adv. Eng. Soft.*, 67, 111–124.

S. Ortega-Cisneros, M. A. Carrazco-Díaz, A. Pedroza de-la-Crúz, J. J. Raygoza-Panduro, F. Sandoval-Ibarra, and J. Rivera-Domínguez (2014), 'Real time hardware accelerator for image filtering,' *Progress in Pattern Recognition, Image Analysis, Computer Vision, and Applications*, CIARP, pp. 80–87.

A. Sengupta (2016), 'Intellectual property cores: protection designs for CE products,' *IEEE Consum. Electron. Mag.*, vol. 5(1), pp. 83–88.

A. Sengupta (2017), 'Hardware security of CE devices [hardware matters],' *IEEE Consum. Electron. Mag.*, vol. 6(1), pp. 130–133.

A. Sengupta and S. P. Mohanty (2019), 'Trojan security aware DSP IP core and integrated circuits', *IP Core Protection and Hardware-Assisted Security for Consumer Electronics*, e-ISBN: 9781785618000, pp. 79–121, doi: 10.1049/PBCS060E_ch3.

A. Sengupta (2020), 'Frontiers in securing IP cores – forensic detective control and obfuscation techniques', The Institute of Engineering and Technology (IET), ISBN-10: 1-83953-031-6, ISBN-13: 978-1-83953-031-9.

A. Sengupta and M. Rathor (2019b), 'Protecting DSP kernels using robust hologram-based obfuscation,' *IEEE Trans. Consum. Electron.*, vol. 65(1), pp. 99–108.

A. Sengupta and M. Rathor (2020), 'Obfuscated hardware accelerators for image processing filters – application specific and functionally reconfigurable processors,' *IEEE Trans. Consum. Electron.*, Accepted, 2020, doi: 10.1109/TCE.2020.3027760.

A. Sengupta and D. Roy (2017), 'Protecting an intellectual property core during architectural synthesis using high-level transformation based obfuscation,' *Electron. Lett.*, vol. 53(13), pp. 849–851.

A. Sengupta, D. Roy, S. P. Mohanty and P. Corcoran (2017), 'DSP design security in CE through algorithmic transformation based structural obfuscation,' *IEEE Trans. Consum. Electron.*, vol. 63(4), pp. 467–476.

A. Sengupta, D. Roy, S. P. Mohanty and P. Corcoran (2018), 'Low-cost obfuscated JPEG CODEC IP core for secure CE hardware,' *IEEE Trans. Consum. Electron.*, vol. 64(3), pp. 365–374.

D. Tsiktsiris, D. Ziouzios and M. Dasygenis (2018), 'A portable image processing accelerator using FPGA,' *Proc. MOCAST*, Thessaloniki, pp. 1–4.

J. V. Vourvoulakis, J. Lygouras and J. A. Kalomiros (2012), 'acceleration of image processing algorithms using minimal resources of custom reconfigurable hardware,' *Proc. PCI*, Piraeus, pp. 68–73.

L. Wu, J. Liu and Y. Luo (2009), 'The design of co-processor for the image processing single chip system,' *Proc. ICCIT*, Seoul, pp. 943–946.

X. Zhang and M. Tehranipoor (2011), 'Case study: detecting hardware Trojans in third-party digital IP cores,' *IEEE International Symposium on Hardware-Oriented Security and Trust*, San Diego, CA, pp. 67–70.

Chapter 6

Fingerprint biometric for securing hardware accelerators

Anirban Sengupta[1]

The chapter describes a promising approach for securing hardware accelerators against intellectual property (IP) piracy and fraud IP ownership threat using biometric fingerprinting. Securing a hardware IP using biometric fingerprint is a paradigm shift from the contemporary approaches. This chapter lucidly explains the biometric-fingerprint-based process using real-life digital signal processing (DSP)/image processing application.

The chapter is organized as follows: Section 6.1 discusses about the introduction of the chapter; Section 6.2 highlights the salient features of this chapter; Section 6.3 presents discussion on some contemporary approaches; Section 6.4 explains the threat model where biometric-fingerprint-based security approach is useful; Section 6.5 provides a high-level perspective of biometric fingerprint approach for securing hardware accelerators; Section 6.6 presents the details of biometric-fingerprinting-based IP protection approach; Section 6.7 presents analysis on case studies; Section 6.8 discusses the advantages of this approach and Section 6.9 concludes the chapter.

6.1 Introduction

A very large-scale integration (VLSI) design process is accomplished through various phases such as high-level synthesis (HLS), logic synthesis, physical/layout synthesis, mask generation and integrated circuit (IC) or system-on-chip (SoC) fabrication. An SoC is a heterogeneous architecture which is the integration of various kinds of designs procured from different third-party IP vendors. The heterogeneous SoC architecture includes (i) computational intensive designs such as DSP or multimedia cores and (ii) non-computational intensive designs such as general purpose processors. The hardware of computational intensive designs (IP cores) are also known as hardware accelerators (Mahdiany *et al.,* 2001; Sengupta and Rathor, 2020b).

[1]Computer Science and Engineering, Indian Institute of Technology Indore, Indore, India

DSP and multimedia cores such as joint photographic experts group (JPEG) compression–decompression (codec)-processor-based designs are extensively employed as hardware accelerators in consumer electronics systems such as digital camera, laptop and tablet. The motive of employing DSP and multimedia cores in the form of hardware accelerators is to achieve high performance and throughput, as hardware accelerators are dedicated to execute computational part of applications such as image codec and audio–video encoding–decoding (Schneiderman, 2010; Sengupta *et al.,* 2018).

Growing hardware threats such as false claim of ownership and piracy are sabotaging the investment and R&D involved in the IP core and SoC designs. This pushes the IP/IC designers towards ensuring protection against aforementioned hardware threats (Plaza and Markov, 2015; Sengupta, 2016, 2017; Sengupta and Rathor, 2019). The false claim of ownership and piracy threats may have manifold impact on VLSI and electronics industry such as (i) false claim of ownership threat challenges the ownership rights of an original IP/IC/SoC supplier or vendor and promotes adversaries towards deceiving genuine owners, (ii) pirated/counterfeited designs sabotage the income and status of an authentic IC/IP suppliers in the industry and (iii) counterfeited designs are not reliable hence may lead to some critical failure and consumer safety issues. The underlying reason is that the malicious logic (Trojan) may be hidden in the counterfeited IC/IP by an adversary (who is associated with the untrustworthy offshore design house or foundry). Additionally, counterfeited designs are not passed through stringent testing and reliability checks unlike authentic designs, to ensure the quality as well as safety of critical systems and end users. The previous discussion stresses as to why employing security mechanisms during the design process of IPs/ICs is so crucial. The integration of a security algorithm within the design during HLS process is vital from the perspective of achieving a higher security at minimal design overhead (Pilato *et al.,* 2018; Sengupta, 2020).

This chapter discusses a new paradigm of security for DSP and multimedia hardware accelerators based on IP vendor's biometric fingerprint (Sengupta and Rathor, 2020a). This is a robust security technique which leads to the ending of false claim of ownership threat and offers a promising detective control of counterfeited and cloned designs. The biometric-fingerprinting-based security mechanism opens a new branch (field) of hardware security where a natural biometric fingerprint of an IP/IC vendor or designer is used for securing a design. This new branch of security requires knowledge of multiple domains such as biometric fingerprint, image processing and VLSI design (specifically HLS).

6.2 Salient features of the chapter

This chapter focuses on the discussion of converting a biometric fingerprint into its equivalent hardware security constraints and the process of embedding them into hardware accelerator design through HLS framework. Further, the impact of embedding biometric fingerprint into a hardware accelerator design is also discussed in the chapter. This section highlights the salient features of the chapter which are as follows (Sengupta and Rathor, 2020a):

- The chapter includes details of the range of fingerprinting constraints obtained from several different biometric fingerprints. The analysis revealed that the range varies between the minimum 350 and the maximum 895 bits.
- The chapter includes discussion on security because of embedding biometric fingerprint in terms of probability of coincidence (*Pc*) metric. The impact on *Pc* has been discussed for several different fingerprints. Analysis on case studies shows that a very low *Pc* value in the range of 2.22E−3 to 4.35E−6 is achieved.
- This chapter includes the impact of various resource constraints on the design cost overhead of biometric-fingerprint-embedded JPEG compression processor. Case studies show that the design cost overhead of biometric-fingerprinting-based security approach is 0% irrespective of the applied resource constraints.
- This chapter also compares the biometric-fingerprint-based security with a contemporary security approach. The comparison has been made in terms of *Pc* for several different constraint sizes. Comparison reveals that the minimum 3.9E+2 times and the maximum 6.9E+4 times lower *Pc* has been achieved compared with the contemporary security approach.

6.3 Discussion on contemporary approaches

Author's (vendor's)-signature-based hardware watermarking is one of the most predominant approaches for the security of hardware accelerator designs against false claim of ownership and piracy. Apart from signature-based hardware watermarking, a crypto-digital-signature-based approach for securing DSP hardware accelerators is also proposed in the literature. This section discusses on how biometric-fingerprint-based security approach is distinct from contemporary signature-based watermarking and crypto-digital-signature-based security approaches.

6.3.1 Biometric-fingerprinting-based IP protection v/s hardware watermarking

For securing DSP hardware accelerators, some hardware watermarking approaches (Koushanfar *et al.,* 2005; Le Gal and Bossuet, 2012; Sengupta and Bhadauria, 2016; Sengupta *et al.,* 2016; Sengupta and Roy, 2017; Sengupta and Mohanty, 2019) based on author's signature have been proposed, which leverage HLS framework while embedding watermarking constraints. The security achieved in these approaches is based on secret author's signature which is embedded into the design in the form of watermarking constraints. The author's signature can comprise two or multiple variables, where each variable has a unique mapping to corresponding hardware security constraints. The IP designer/author selects a specific signature combination and length. The robustness of the hardware watermarking depends on the following factors: (i) covertness of the number of variables used in constructing an author's signature, (ii) covertness of the mapping rules used for converting the signature digits into respective watermarking security constraints and (iii) covertness of the signature strength (length), i.e. the number of corresponding watermarking constraints embedded. However, in the scenario, if the number of signature variables, their mapping rules and the signature length are compromised or leaked to an adversary, the objective of embedding watermark gets defeated. This is because an adversary can fraudulently claim the IP ownership or can embed watermark of a genuine author into his/her fake/counterfeited designs once he/she obtains required information of original watermark dishonestly. In this scenario, hardware watermarking approaches prove to be weak (Saha and Sur-Kolay, 2011; Sengupta *et al.*, 2016, 2019). Further, the author's signature contains only few digits and has a predetermined length; therefore, it is vulnerable to tampering. Moreover, in the case of hardware watermarking, one cannot easily make a prior estimation about the optimal signature combination and length that would lead to higher security (hidden digital evidence) at minimal design overhead (Sengupta and Rathor, 2019).

Compared with hardware watermarking approaches, the biometric-fingerprinting-based approach (Sengupta and Rathor, 2020a) offers robust security because of the following reasons: a biometric fingerprint is naturally unique; therefore, the fingerprinting constraints are not vulnerable. Leakage or extraction of fingerprinting constraints cannot help an adversary as she/he cannot claim or prove her/his right over the secret constraints because of natural uniqueness involved in a biometric fingerprint. Second, the number of biometric fingerprint constraints to be embedded is not prefixed by a designer, unlike hardware watermarking approaches where signature length is predetermined in advance. Rather, the number of constraints to be embedded in the biometric fingerprinting approach depends on the number of minutiae points in the fingerprint of an IP/IC vendor. Finally, the hardware security constraints corresponding to a biometric fingerprint are capable of identifying the genuine IP owner or vendor because of the following reasons (Sengupta and Rathor, 2020a):

1. *Uniqueness of the digital template*: The digital template obtained from a biometric fingerprint not only includes evidence of the total number of 1s and 0s but also the information of each bit position. Although two digital templates may contain equal number of 1s and 0s, nevertheless their respective positions in both digital templates can never match. This leads to the conclusion that a digital template corresponding to a biometric fingerprint is always unique.
2. *Robust correlation between a digital template and its equivalent hardware security constraints*: This is because of the following three reasons: (i) the number of hardware security constraints depends on quantity of 1s and 0s in the digital template; (ii) conversion of digital template into security constraints also depends on ordered list of storage variables in the coloured interval graph (CIG) of the target design. This ordering of storage variables is specified by the designer/vendor and is used during converting bits of a digital template into its equivalent security constraints. The vendor specifies the particular order in

which the storage variables of CIG are arranged in the list. This is because the arrangements of storage variables in a list can be made in many different ways such as sorting in increasing order, sorting in decreasing order, sorting as per control steps number in scheduling, sorting in alternate arrangements of storage variables and sorting based on functional units (FUs); (iii) the correlation also depends on the total number of storage variables present in the CIG of the target design, which, in turn, depends on the resource constraints used during scheduling. This resource constraint is either designer specified or derived from design architecture exploration process.

6.3.2 Biometric-fingerprinting-based IP protection v/s crypto digital signature (Sengupta and Rathor, 2020a)

Cryptographic digital-signatures-based approaches such as discussed in Sengupta *et al.* (2019) depend on encoding of target design into a bitstream (digital message) followed by secure hashing using secure hash algorithm (SHA)-512 and Rivest, Shamir, Adleman (RSA) encryption (using vendor's private key). The generated encrypted message/bitstream is embedded into the target design as a digital signature. The key differences between a 'cryptographic digital-signature-based approach' with a 'biometric-fingerprinting-based approach' are as follows:

- Security achieved due to cryptographic digital signature (Sengupta *et al.*, 2019) is highly dependent on diverse intermediate and auxiliary factors. These factors include the selection of scheduling algorithm, encoding rules employed to produce bitstream corresponding to the scheduled design, selection of secure hashing algorithm such as SHA-512 to produce hash-digest, and size by which hash-digest is divided into equal partitions as well as private key strength of RSA encryption chosen by the designer. The generated digital signature is sensitive to the variation in any of these intermediate factors. In contrast, security achieved due to the biometric fingerprint does not have much dependency on intermediate steps. The digital template corresponding to a biometric fingerprint is directly obtained from its minutiae points. This leads to a highly robust security mechanism which is less sensitive towards auxiliary intermediate factor.
- The complexity of cryptographic digital signature generation process (Sengupta *et al.*, 2019) is high because of the involvement of manifold intermediate steps. In contrast, generation of digital template corresponding to a biometric fingerprint is a simple process with the maximum security achieved. The generation of digital template is directly based on the natural biometric features (minutiae points) which are unique for each fingerprint.
- Biometric fingerprinting approach (Sengupta and Rathor, 2020a) generates unique hardware security constraints corresponding to a biometric fingerprint, without involving any complex algorithm of generating unique digest such as SHA. On the contrary, cryptographic digital signature generation process exploits complex SHA-512 algorithm to generate the unique hash-digest corresponding to each different input message (encoded bitstream). The

complexity of SHA-512 algorithm is high due to the involvement of several modules such as hash buffers, word computation function and round function computation (which is further based on summation function, rotation function, majority function and condition function) (Sengupta *et al.,* 2019).

- The detection process of cryptographic digital signature embedded into the design is complex. This is because the detection algorithm depends on the following factors: (i) public key, (ii) knowledge of points of concatenation of bitstreams and (iii) strength of the signature (Sengupta *et al.,* 2019). On the contrary, the biometric fingerprint constraints can be smoothly detected. This is because only comparison of digital templates is sufficient to prove true authorship (Sengupta and Rathor, 2020a).

6.4 Threat model

The biometric-fingerprint-based security approach (Sengupta and Rathor, 2020a) discussed in this chapter targets the following threats to an IP/IC or a hardware accelerator.

1. *IP piracy (counterfeiting or cloning)*: Today's VLSI design chain involves offshore design houses and foundries in order to increase productivity, reduce design complexity and fulfil time to market requirement. These offshore entities cannot be blindly trusted as an adversary present there may pirate the design to generate fake IPs/ICs or hardware accelerators. The biometric-fingerprinting-based approach disables the piracy threat by embedding fingerprint constraints into the design. The natural biometric fingerprint embedded into the designs can be detected to identify the fake designs.
2. *False claim of IP ownership*: A dishonest user may cheat the true IP owner by claiming IP ownership fraudulently. Embedding natural biometric fingerprint is a robust solution against nullifying false claim of IP ownership threat.

6.5 High-level perspective of biometric fingerprinting approach for securing hardware accelerators

An abstract view of securing hardware accelerator designs using biometric fingerprinting approach (Sengupta and Rathor, 2020a) is shown in Figure 6.1. As shown in the figure, this security approach generates a digital template corresponding to the IP vendor's biometric fingerprint. Then the obtained fingerprint's digital template is converted into hardware security constraints and implanted into the hardware accelerator design. Thus, a hardware accelerator is secured using biometric fingerprint. The embedded biometric fingerprint is used to nullify false claim of IP ownership because of its natural uniqueness. Additionally, the biometric fingerprint embedded in the form of authentic secret mark also enables the detection of IP counterfeiting and cloning.

Figure 6.2 depicts the design flow of obtaining a secured hardware accelerator design (in register transfer level (RTL)) using vendor's biometric fingerprint

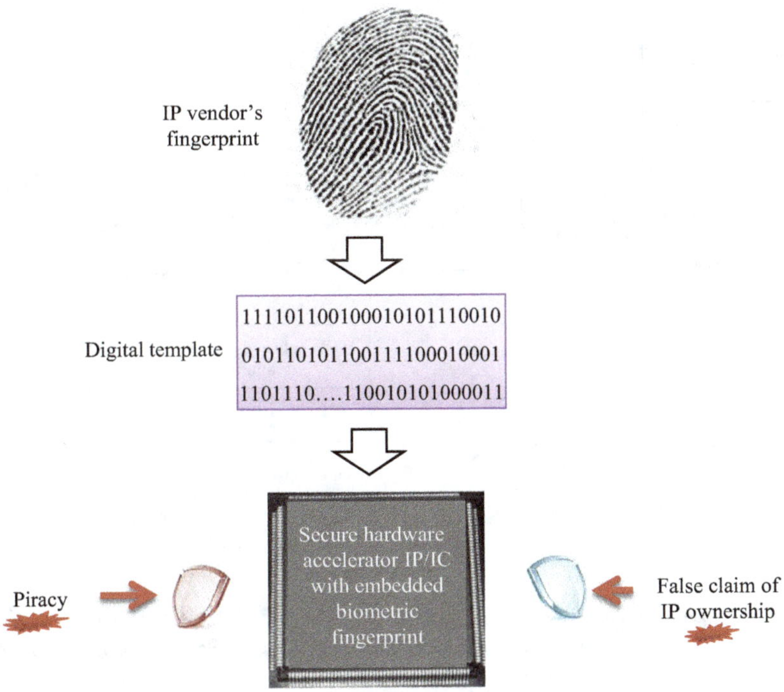

Figure 6.1 Abstract view of securing hardware accelerators using biometric fingerprinting (Sengupta and Rathor, 2020a)

(Sengupta and Rathor, 2020a). As shown in the figure, the vendor's biometric fingerprint is implanted through HLS process. The following inputs are required to generate a fingerprint-embedded hardware accelerator design in RTL:

1. high-level or algorithmic description (such as C/C++/computation function) of target DSP or image processing application;
2. module library;
3. resource constraints and
4. IP vendor's/designer's biometric fingerprint.

The first step in the secure design flow of biometric fingerprinting approach is to generate a control data-flow graph (DFG), an intermediate representation, from the algorithmic description of the intended hardware accelerator application. Next, an HLS framework is used to implant the digital template of biometric fingerprint during register allocation phase. The generic steps of obtaining the digital template of a biometric fingerprint followed by implanting into a design are summarized as follows (Sengupta and Rathor, 2020a):

1. The first step towards the generation of digital template is to capture biometric fingerprint of the IP vendor using an optical scanner device.

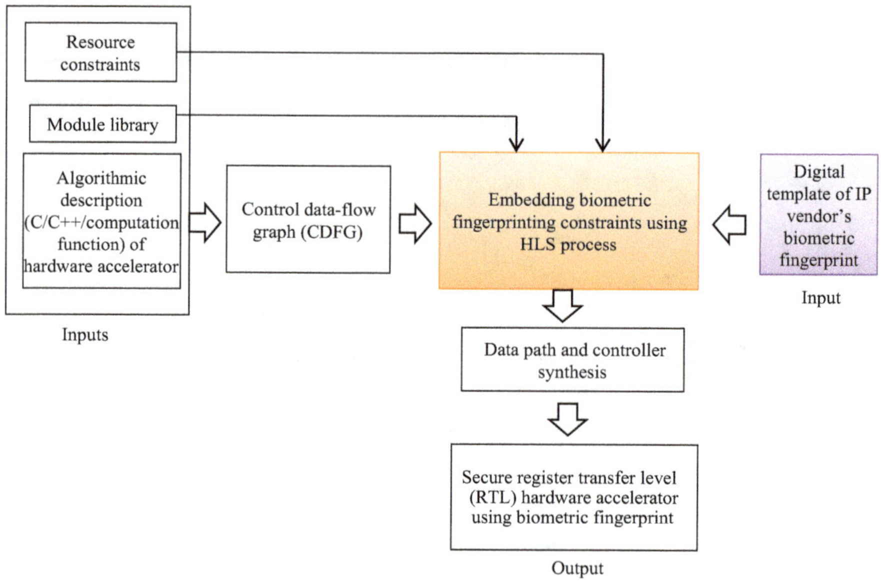

Figure 6.2 Design flow of obtaining secured hardware accelerator design (in RTL form) using vendor's biometric fingerprint (Sengupta and Rathor, 2020a)

2. Second, the obtained biometric fingerprint image is subjected to some pre-processing steps in order to remove unwanted noises and improve the image quality. Some of the major pre-processing steps include image enhancement using fast Fourier transform (FFT), binarization and thinning.
3. Third, the minutiae points (which are natural/unique biometric features) are extracted from the enhanced, binarized and thinned fingerprint image. In a biometric fingerprint, the minutiae points are the unique features which are leveraged to uniquely identify a person during the detection process. The biometric-fingerprinting-based security approach (Sengupta and Rathor, 2020a) uses minutiae points of the vendor's fingerprint to obtain corresponding digital template and implant into a hardware accelerator design in the form of secret hardware security constraints.
4. Next, minutiae points of biometric fingerprint are converted into corresponding digital template using conversion algorithm proposed by Sengupta and Rathor (2020a).
5. Thereafter, mapping of bits in the digital template of the vendor's biometric fingerprint is performed in order to obtain equivalent secret hardware constraints. The mapping of digital template into hardware security constraints is based on mapping rules proposed by Sengupta and Rathor (2020a).
6. Subsequently, the register allocation phase of HLS framework is exploited to implant hardware security constraints into the design of hardware accelerator.

Post-implanting constraints, biometric-fingerprint-implanted scheduled and resource-allocated design is generated.

7. Finally, biometric-fingerprint-implanted scheduled and resource-allocated design is subjected to data path and controller synthesis of HLS process that results into a secure RTL data path of the hardware accelerator design.

6.6 Details of biometric fingerprinting approach for securing hardware accelerators

The details of biometric fingerprinting approach for securing hardware accelerators are discussed under the following subsections (Sengupta and Rathor, 2020a).

6.6.1 Background on biometric fingerprint

Biometric fingerprint: A biometric fingerprint is defined as a distinct pattern of ridges (single curved segment) and valleys (the area between two adjacent ridges) on the finger surface of an individual. Biometric fingerprint of each individual is always unique. In a fingerprint identification system, minutiae features of a biometric fingerprint are extracted and matched with the incoming (test) fingerprint. A biometric fingerprint representation based on minutia points results into a small template size. Most of the fingerprint identification and verification systems use minutiae-based representation.

Minutiae points: The points in a biometric fingerprint where the ridge lines 'terminate' or 'fork' are referred to as minutia points. These are the important and distinct features of a fingerprint image. Some important types of minutiae points are as follows:

- *Ridge bifurcation*: The point where a single ridge bifurcates into two or more branch ridges.
- *Ridge termination*: The point where a ridge terminates abruptly.
- *Ridge dot*: A ridge of minor size.
- *Ridge island*: A ridge slightly longer than dots and exists in a middle space between two bifurcating ridges.
- *Pond or lake*: An empty space between two bifurcating ridges.
- *Bridges*: Small joining segments of two longer adjacent ridges.
- *Crossovers*: Crossings of two ridges.

The most commonly exploited minutia types are **ridge terminations** and **ridge bifurcations**. This is because the combinations of ridge termination/ending and bifurcation result into all other types of minutiae features. In general, the ridge terminations and ridge bifurcations are called major minutiae features.

Attributes of each minutiae point: Each minutiae point in a biometric fingerprint is identified using the following four attributes:

1. x-Coordinate of the location of a minutia point, denoted as X_m.
2. y-Coordinate of the location of a minutia point, denoted as Y_m.

3. Minutiae type, denoted as M_t.
4. Ridge direction or angle, denoted as R_a. The definitions of termination and bifurcation angles are as follows:
   (i) *Termination angle*: It is measured (either clockwise or anticlockwise) between the horizontal and the direction of the ridge ending or termination.
   (ii) *Bifurcation angle*: It is measured (either clockwise or anticlockwise) between the horizontal and the direction of the valley ending between the ridge bifurcations.

Pre-processing of biometric fingerprint: A biometric fingerprint is subjected to some pre-processing before extraction of the minutiae points. The pre-processing steps are applied for the purpose of removal of unwanted noises and enhancement of the fingerprint image quality. This results into minimal chances of extracting false minutiae points. Some general pre-processing steps are highlighted here as follows:

1. *FFT enhancement*: In order to perform FFT enhancement, the fingerprint image is partitioned into little transforming blocks (e.g. 32 $\times$ 32 pixels or lower), and each block is subjected to two-dimensional (2D)-FFT. The application of 2D-FFT leads to the following enhancements in the fingerprint image: (i) filling up minor holes in ridges (i.e. reconnection of broken ridges), (ii) fine separation of parallel ridges and (iii) enhancing thickness of the ridges.
2. *Segmentation*: In each captured fingerprint image, there are generally two regions: (i) foreground (the fingerprint region) and (ii) background. Segmentation process is applied to distinguish between the fingerprint region and background area. This is useful to circumvent the extraction of minutiae points from the noisy background region. Traditionally, segmentation algorithm makes partitions of captured fingerprint image into a number of blocks. Further, mean and variance value for each block is evaluated. In order to distinguish between the fingerprint region and background area, the variance value is compared with the threshold value. Threshold value is the boundary intensity of fingerprint region and background area directly.
3. *Binarization*: Binarization is a process of transforming a greyscale image into binary image. An image bearing only two intensity values is referred to as the binary image. The binary image has only pure black and pure white pixels. The intensity value of a pure black pixel is denoted by 0 and of pure white pixel is represented by 255. The underlying concept of binarization is based on the comparison of the pixel intensities with the threshold value. Those pixels intensities of which are less than the threshold intensity are set to 0 (pure black) and others are set to 255 (pure white). This is how ridge lines can be separated from the background of fingerprint region.
4. *Thinning*: Thinning process performs deletion of pixels at the edge of ridge lines in order to reduce the thickness of each ridge line to 1 pixel width only.

6.6.2 *Detailed methodology of biometric-fingerprint-based hardware security*

This section illustrates the methodology of biometric-fingerprint-based hardware security in detail. The process of extraction of minutiae points and formation of digital template is illustrated using a sample biometric fingerprint. Further, demonstration of embedding fingerprint constraints is presented on JPEG codec hardware accelerator design. Figure 6.3 depicts the steps of performing biometric-fingerprinting-based hardware security (Sengupta and Rathor, 2020a). As shown in the figure, the captured fingerprint image is first pre-processed in order to enhance the image quality for mitigating the possibility of extraction of false minutiae points. This is because the captured vendor's fingerprint may be noisy and it is important to do some pre-processing in order to locate minutiae points correctly. Therefore, the captured biometric fingerprint is subjected to pre-processing steps such as FFT enhancement, binarization and thinning before locating/extracting the minutiae points, which serve as the unique features of an individual's fingerprint.

6.6.2.1 Minutiae extraction process

Once the fingerprint image is pre-processed, it is subjected to minutiae extraction process as shown in Figure 6.3. Sengupta and Rathor (2020a) applied crossing number (*CN*) algorithm (Zhao and Tang, 2007) for the extraction of minutiae points from the fingerprint. The value of *CN* decides whether a minutiae point is ridge termination or ridge bifurcation. For a pixel q, calculation of *CN* is performed using the following function (Zhao and Tang, 2007):

$$CN = 0.5 * \sum_{i=1}^{8} |q_i - q_{i+1}| \tag{6.1}$$

q_4	q_3	q_2
q_5	q	q_1
q_6	q_7	q_8

The previous 3 × 3 block shows the pixel values q_i in the neighbourhood of the pixel q. In (6.1), the value of q_i is either 0 or 1 and $q_1 = q_9$.

In order to extract minutiae points, scanning of fingerprint thinning image is performed throughout the area of interest, and *CN* value is calculated (using (6.1)) for all minutiae points. If the value of *CN* is calculated to be '1', then the minutia point is a ridge termination. If the value of *CN* is calculated to be '3', then the minutia point is a ridge bifurcation. This is how all the minutiae points with their respective types are extracted from the fingerprint image.

For demonstration, a fingerprint image (101_1.tif) has been chosen from a publicly available database (Alilou, 2019). The chosen fingerprint image is shown in Figure 6.4(a). First, the input fingerprint image is enhanced using pre-processing

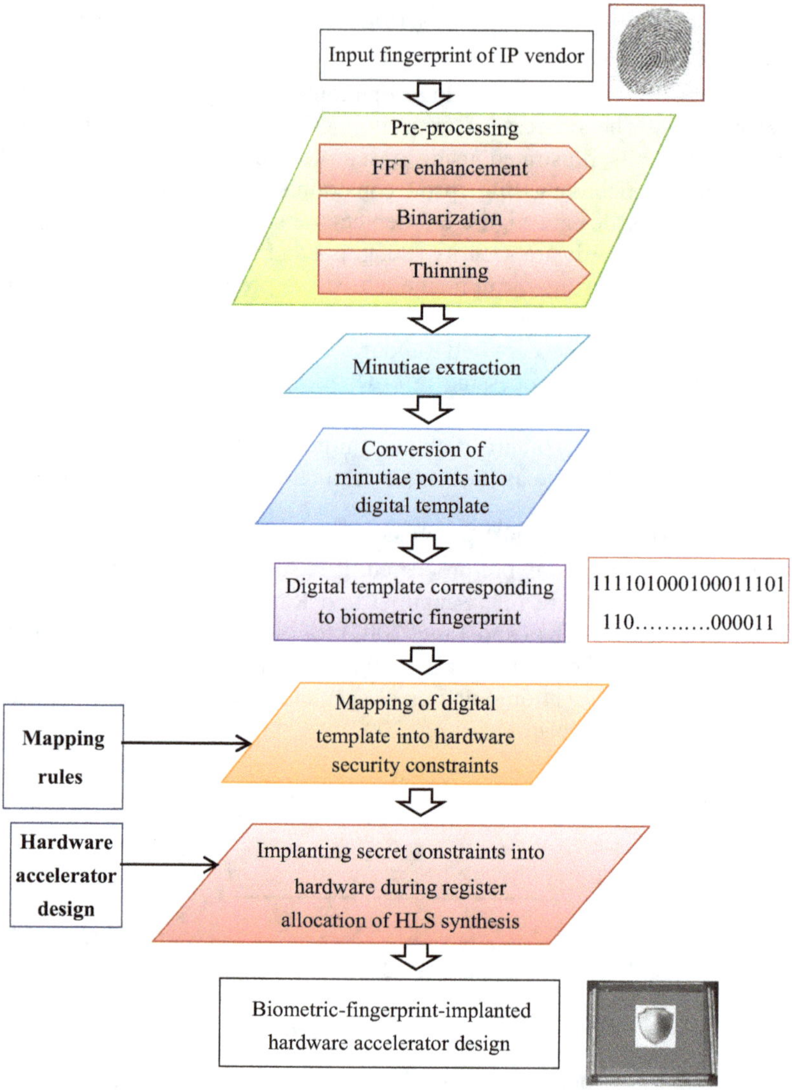

Figure 6.3 The major steps of performing biometric-fingerprinting-based security (Sengupta and Rathor, 2020a)

steps such as FFT enhancement. Next, binarization is performed, which results into a binary fingerprint image bearing only two intensity values, 0 and 255. The binary fingerprint image has been shown in Figure 6.4(b). Further, thinning process is performed on the binary fingerprint image. The thinned image has been shown in Figure 6.4(c). Subsequently, *CN* algorithm is applied to extract minutiae points from the processed thinned fingerprint image. The extracted minutiae points have been located in Figure 6.4(d). As shown in the figure, a total of 22 minutiae points

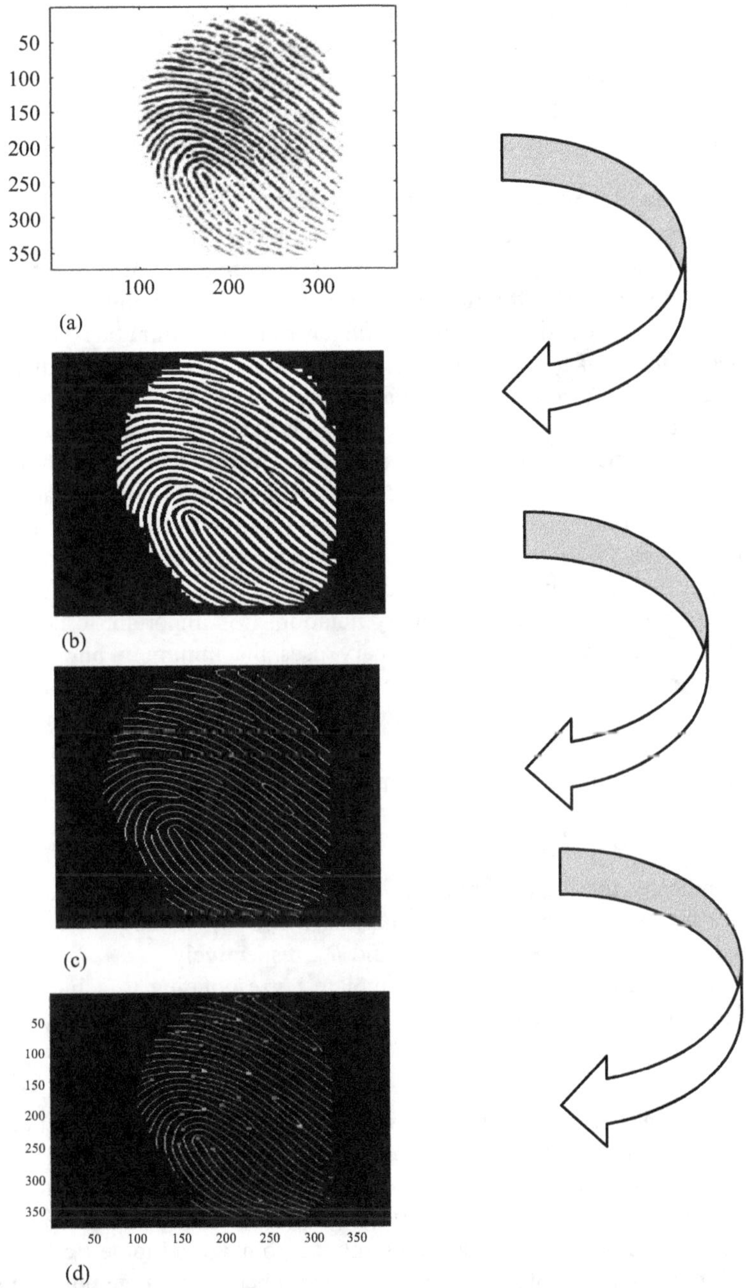

Figure 6.4 Minutiae points extraction flow: (a) captured (input; 101_1) fingerprint image, (b) binary fingerprint image post enhancement, (c) fingerprint mage post applying thinning and (d) fingerprint mage with minutiae points located

have been highlighted. The ridge terminations have been highlighted using red colour and ridge bifurcations have been highlighted using blue colour. A total of 16 minutiae points are extracted as ridge terminations, and 6 are extracted as ridge bifurcations as shown in Figure 6.4(d). *Note*: the boundary minutiae points have not been considered as they increase the chances of false minutiae point extraction.

6.6.2.2 Digital template generation process

This subsection discusses the process of generating digital template from a biometric fingerprint. Once all minutiae points are extracted from the fingerprint image, they are converted to the corresponding digital template. As discussed earlier in the background of biometric fingerprint, each minutiae point can be represented in terms of four attributes, (i) x-coordinate (X_m), (ii) y-coordinate (Y_m), (iii) minutiae type (M_t) and (iv) direction/angle of ridge (R_a), where x and y coordinates indicate the location of minutiae points in the fingerprint image, minutiae type indicates whether the minutiae point is ridge bifurcation or termination and the ridge angle indicates the orientation of minutiae points. Sengupta and Rathor (2020a) measured ridge angle in clockwise direction from the horizontal axis. Thus, obtained attributes of minutiae points are leveraged to form a digital template corresponding to the fingerprint. In order to do so, first decimal value of each attribute is represented in terms of binary notation. It is important to note that the conversion process from decimal to binary uses the minimum number of bits, required for representing a decimal value into corresponding binary notation. Once all attributes of a minutiae point are converted into binary representation, they are concatenated to form a digital template. The following function is used to form digital templates of individual minutiae points:

$$Di = \left(X_m^b\right)\left(Y_m^b\right)\left(M_t^b\right)\left(R_a^b\right) \tag{6.2}$$

where Di indicates the digital template of ith minutiae point and symbol '&' indicates the concatenation operator. Further, X_m^b, Y_m^b, M_t^b and R_a^b indicate binary representations of attributes X_m, Y_m, M_t and R_a, respectively.

In order to generate final digital template corresponding to a biometric fingerprint, the following function is used:

$$DT =_{i=1}^{n} \& Di \tag{6.3}$$

where DT indicates the final digital template, and n indicates the total number of minutiae points extracted from a fingerprint image. As evident from (6.3), the final digital template is produced by concatenating individual digital templates Di of n number of minutiae points of a biometric fingerprint. Thus, generated final digital template contains information of all the minutiae points of a biometric fingerprint. The size of the digital template depends on the number of minutiae points. Further, the positions of '0' and '1' bits in the digital template encapsulate the information of all four attributes of each minutiae point. Thereby, the digital template generation process proposed by Sengupta and Rathor (2020a) generates a highly robust digital representation of a biometric fingerprint.

Let us see the demonstration of digital template generation corresponding to the biometric fingerprint and its minutiae points shown in Figure 6.4. Table 6.1 reports all 22 minutiae points with their respective attributes, viz. X_m, Y_m, M_t and R_a. For example, the first minutiae point in the table is ridge bifurcation attributes of which are as follows:

1. x-coordinate $X_m = 216$
2. y-coordinate $Y_m = 46$
3. minutiae type $M_t = 3$
4. angle of ridge $R_a = 29°$

The attributes (in decimal representation) of each minutiae point are converted to corresponding binary representation and concatenated using (6.2) to form a digital template. For example, the digital template of the first minutiae point is '1101100010111011111101' which has been obtained (using (6.2)) as follows: $D1 = (11011000)\&(101110)\&(11)\&(11101)$. Similarly, individual digital template of all minutiae points is formed. Table 6.2 shows the individual digital templates $D1$–$D22$ of all 22 minutiae points. Further to generate the final digital template, individual digital templates $D1$–$D22$ are concatenated using (6.3), where $n = 22$. The generated final digital template corresponding to the biometric fingerprint

Table 6.1 Attributes of extracted minutiae points from fingerprint image (101_1)

No.	X_m	Y_m	Minutiae type name	Minutiae type colour	Minutiae type (M_t) number	Ridge angle in degree (R_a)
1	216	46	Ridge bifurcation	Blue	3	29
2	190	49	Ridge ending	Red	1	205
3	146	64	Ridge ending	Red	1	187
4	247	80	Ridge ending	Red	1	40
5	173	86	Ridge ending	Red	1	21
6	302	93	Ridge ending	Red	1	48
7	176	127	Ridge bifurcation	Blue	3	16
8	227	131	Ridge bifurcation	Blue	3	32
9	164	135	Ridge ending	Red	1	190
10	117	140	Ridge ending	Red	1	330
11	216	169	Ridge ending	Red	1	42
12	256	170	Ridge bifurcation	Blue	3	223
13	196	181	Ridge ending	Red	1	214
14	176	187	Ridge bifurcation	Blue	3	26
15	151	195	Ridge ending	Red	1	328
16	285	215	Ridge bifurcation	Blue	3	45
17	227	218	Ridge ending	Red	1	47
18	152	219	Ridge ending	Red	1	131
19	169	233	Ridge ending	Red	1	237
20	147	242	Ridge ending	Red	1	264
21	186	250	Ridge ending	Red	1	239
22	240	332	Ridge ending	Red	1	29

Table 6.2 Digital template corresponding to each minutiae point

No.	Digital templates of individual minutiae points		Number of bits
1	D1	11011000-101110-11-11101	21
2	D2	10111110-110001-1-11001101	23
3	D3	10010010-1000000-1-10111011	24
4	D4	11110111-1010000-1-101000	22
5	D5	10101101-1010110-1-10101	21
6	D6	100101110-1011101-1-110000	23
7	D7	10110000-1111111-11-10000	22
8	D8	11100011-10000011-11-100000	24
9	D9	10100100-10000111-1-10111110	25
10	D10	1110101-10001100-1-101001010	25
11	D11	11011000-10101001-1-101010	23
12	D12	100000000-10101010-11-11011111	27
13	D13	11000100-10110101-1-11010110	25
14	D14	10110000-10111011-11-11010	23
15	D15	10010111-11000011-1-101001000	26
16	D16	100011101-11010111-11-101101	25
17	D17	11100011-11011010-1-101111	23
18	D18	10011000-11011011-1-10000011	25
19	D19	10101001-11101001-1-11101101	25
20	D20	10010011-11110010-1-100001000	26
21	D21	10111010-11111010-1-11101111	25
22	D22	11110000-101001100-1-11101	23
Total bits in the digital template			526

(101_1) is shown in Figure 6.5. The total size of the digital template is 526 bits, where the number of 0s is 224 and the number of 1s is 302.

6.6.2.3 Conversion process of digital template into hardware security constraints

The conversion process of digital template into hardware security constraints depends on a designer-selected phase in the design process for implanting the secret constraints. In the biometric-fingerprinting-based security approach (Sengupta and Rathor, 2020a), digital template corresponding to an IP vendor's biometric fingerprint is converted into hardware security constraints for implantation in the

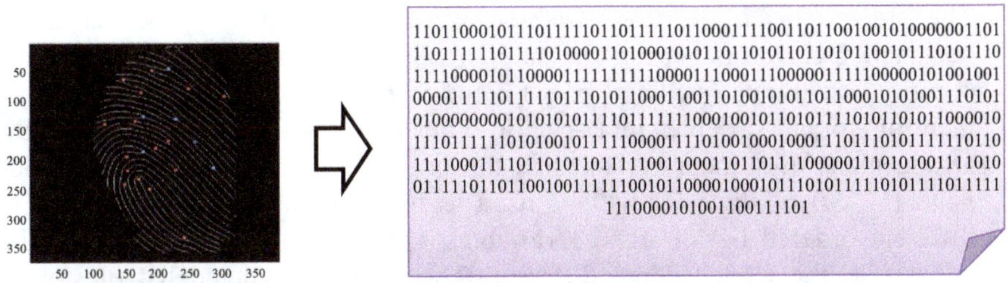

Fingerprint image with minutiae points located

Corresponding final digital template (526 bits)
Number of 0s=224 and Number of 1s=302

Figure 6.5 Digital template corresponding to the biometric fingerprint (101_1)

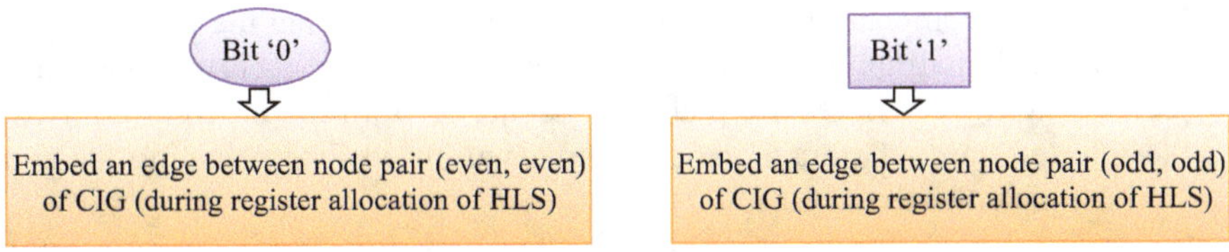

Figure 6.6 Mapping of bits of digital template into hardware security constraints (Sengupta and Rathor, 2020a)

register allocation phase of HLS process. The implantation of secret fingerprint constraints is accomplished using a CIG framework. Therefore, in order to understand the conversion/mapping of digital template into security constraints, knowledge of CIG framework is a prerequisite. A summary of CIG is given as follows: a CIG is created using the information of allocation of storage variables of the target design to the registers. Nodes in the CIG represent storage variables (*S*) of the design, and the colour of a node represents its assignment to a register. Therefore, the total number of distinct colours used in the CIG is equal to the minimum number of registers required to accommodate all storage variables of the design. Further, overlapping of lifetime of storage variables is indicated by drawing edges between nodes in the CIG.

In order to obtain hardware security constraints to be implanted into a CIG framework, the conversion or mapping rules are used for digital template (such as shown in Figure 6.6). Based on the mapping rules shown in Figure 6.6, each bit of the digital template is converted into an artificial edge (representing covert security constraints) for adding into the CIG. It is important to note that the conversion of each bit into the hardware security constraints not only depends on the number of 0s and 1s in the digital template but also depends on the following additional factors:

1. Storage variable count in the design, which, in turn, depends on the resource constraints used during scheduling.

2. The ordered list of storage variables, which is specified by the vendor or designer. Before converting the bits of digital template into the hardware security constraints (artificial edges between node pairs (*Si*, *Sj*)), an arrangement of storage variables in a specific order is necessary. This is because the arrangement of storage variables in the ordered list determines which node pairs (*Si*, *Sj*) of the CIG will be used to implant the constraint edges. For different ordered list of storage variables, the resulting implanted constraint edges into the CIG become different. The different ordered list of storage variables can be created in many ways, for example (i) sorting of storage variables in increasing order, (ii) sorting in decreasing order, (iii) sorting as per control step's number in scheduling, (iv) sorting in alternate arrangements of storage variables and (v) sorting based on functional units. An IP designer or vendor selects a specific ordered list of storage variables before converting the bits of digital template into the security constraints. During the conversion of bits of the digital template into corresponding hardware security constraints, pairs of storage variables are picked in sequence from the ordered list. Next, the selected pairs of storage variables/nodes are used for implanting the constraint edges into the CIG.

Although there might be a possibility that two digital templates containing the same number of 0s and 1s, the corresponding hardware security constraints are unique because of additional dependence of the conversion process on (i) the number of storage variables in the design and (ii) ordered list of storage variable. Henceforth, the mapping or correlation between the digital template (corresponding to a biometric fingerprint) and hardware security constraints is highly robust and unique.

6.6.2.4 Implanting process

The fingerprint constraints (or hardware security constraints) corresponding to bits '0' and '1' of the digital template are implanted into the target design during register allocation phase of HLS process. In order to do so, the hardware security constraints obtained from the mapping of bit '1' and '0' of the digital template are implanted into the CIG one-by-one. Each constraint is implanted by adding an artificial edge into the CIG. It is noteworthy that some constraint edges may not need to be additionally implanted due to the existence of those edges in the CIG by default. In other words, the CIG of the target design may already satisfy some hardware security constraints by default. Further, conflicts also may arise while implanting some constraint edges. For example, if a constraint edge has to be implanted between two nodes of the same colours in the CIG, then it leads to conflict. This is because an edge can only be drawn in a CIG between two nodes of different colours. When conflict of adding any constraint edge arises, the colour/register of one of the node/storage variables in the respective node pair is essentially swapped with another node in the same control step to enable the implantation of the constraint edge into the CIG. This swapping results into alterations in the register allocation of some storage variables. These alterations in the CIG or register allocation highlight the embedded fingerprint constraint into the design.

Further in some cases, the conflict of adding of a constraint edge between two nodes of the same colour may not be resolved by only performing swapping of the colours/registers within the respective control step. In such a scenario, implantation of the constraint edge is enabled by using extra colour or register in the CIG. This results into register overhead. However, in most of the designs, existing registers are sufficient to accommodate all constraint edges by performing local alterations. The probability of register overhead due to implanting security constraints greatly reduces with increase in the size of the target design (DSP application).

6.6.2.5 Implanting fingerprint constraints into JPEG codec

This subsection discusses the process of implanting hardware security constraints into a JPEG codec hardware accelerator. For demonstration, hardware security constraints corresponding to the digital template shown in Figure 6.5 are implanted into the design of JPEG codec hardware accelerator. Before discussing the implantation process, a summary of JPEG codec hardware accelerator and its DFG representation are presented in the following subsections.

Summary of JPEG codec

The JPEG compression process performs the compression by performing transformation of the input images from spatial representation to frequency representation followed by quantization. The image pixels transformation from spatial to frequency representation is accomplished using 2D-DCT (discrete cosine transform) algorithm (8-point). Further, the actual compression phase is performed by quantization process which discards less important frequency components and keeps only the most important frequency components. The step-by-step process of JPEG compression is summarized as follows (Sengupta and Rathor, 2020a).

First, an input image is transformed into an $N \times N$ matrix of pixels, where N indicates the size of square matrix. The intensity values of pixels range between 0 and 255, where 0 and 255 indicate pure dark (black) and pure bright (white) pixels, respectively (for a greyscale image). Next, the $N \times N$ pixels of input image are partitioned in to non-overlapping 8×8 blocks. This is because an 8-point DCT operates on an 8×8 matrix at a time. Further, each 8×8 block of pixels is subjected to 2D-DCT transformation using a DCT coefficient matrix B (Sengupta and Rathor, 2020a). The computation function of 2D-DCT transformation is as follows:

$$W = (B * P) * B' \tag{6.4}$$

where W indicates the DCT-transformed 8×8 block of image pixels, P indicates an 8×8 block of image pixels and B' represents the transpose of 2D-DCT coefficient matrix B. The function of computing the first pixel value W_{11} of the DCT-transformed 8×8-matrix W is given as follows:

$$\begin{aligned} W_{11} &= b4 * g11 + b4 * g12 + b4 * g13 + b4 * g14 + b4 * g15 \\ &\quad + b4 * g16 + b4 * g17 + b4 * g18 \end{aligned} \tag{6.5}$$

where $g11$–$g18$ indicate the elements in the first row of the matrix $[B*P]$ and $b4$ indicates the elements in the first column (repeating throughout the column) of matrix B'. The function of computing the element $g11$ of the matrix $[B*P]$ is as follows:

$$\begin{aligned} g11 &= b4 * p11 + b4 * p21 + b4 * p31 + b4 * p41 + b4 * p51 \\ &\quad + b4 * p61 + b4 * p71 + b4 * p81 \end{aligned} \tag{6.6}$$

where $p11$–$p81$ indicate the pixel values in the first column of the input matrix P, and $b4$ indicates the elements in the first row (repeating throughout the row) of matrix B.

Once the DCT-transformed matrix is computed for each 8×8 block of pixels, the next step in the JPEG compression process is to perform quantization. The quantization is performed using a quantization matrix T which is chosen on the basis of the required level of compression and quality of image. To perform quantization, each pixel value in the DCT-transformed matrix W is divided by the respective value (t) in the quantization matrix T. Post division, each value is rounded-off to the closest integer value. The computation function to perform quantization on each DCT-transformed pixel value is as follows:

$$W'_{ij} = W_{ij} * \frac{1}{t_{ij}} \tag{6.7}$$

where W'_{ij} indicates a pixel value of compressed image after quantization, W_{ij} indicates the corresponding pixel value after DCT transformation and t_{ij} indicates a coefficient in the quantization matrix at the respective position ij.

Equations (6.5)–(6.7), corresponding to the 2D-DCT transformation and quantization functions, are used to perform computational intensive part of the JPEG compression process. Hence, the DCT transformation and quantization part of the JPEG compression process is realized using hardware accelerator. In order to generate the JPEG compression hardware accelerator, the algorithmic description in the form of (6.5)–(6.7) is converted into a DFG representation as shown in Figure 6.7. The entire DFG form is referred to as macro-IP which generates computed compressed pixel values of an image. The macro-IP uses micro-IP underneath to perform a part of DCT transformation. Further as shown in the figure, operation 135 produces the first pixel value (W_{11}) computed post DCT transformation. To generate the first pixel of compressed JPEG image (W'_{11}), the operation 136 in the DFG performs quantization on the DCT-transformed value (output of operation 135) using respective element t_{11} of the quantization matrix T.

As evident from Figure 6.7, the DFG of JPEG compression processor comprises eight micro-IPs underneath, where each micro-IP executes 16 operations. Hence, total operations in the eight micro-IPs are 128, and the entire JPEG compression processor has $128+8=136$ operations as shown in Figure 6.7. Further, the DFG of JPEG compression processor is subjected to scheduling, hardware allocation and binding phases of HLS using resource constraints of three adders and three multipliers. The scheduling of 136 operations using resource constraints of three

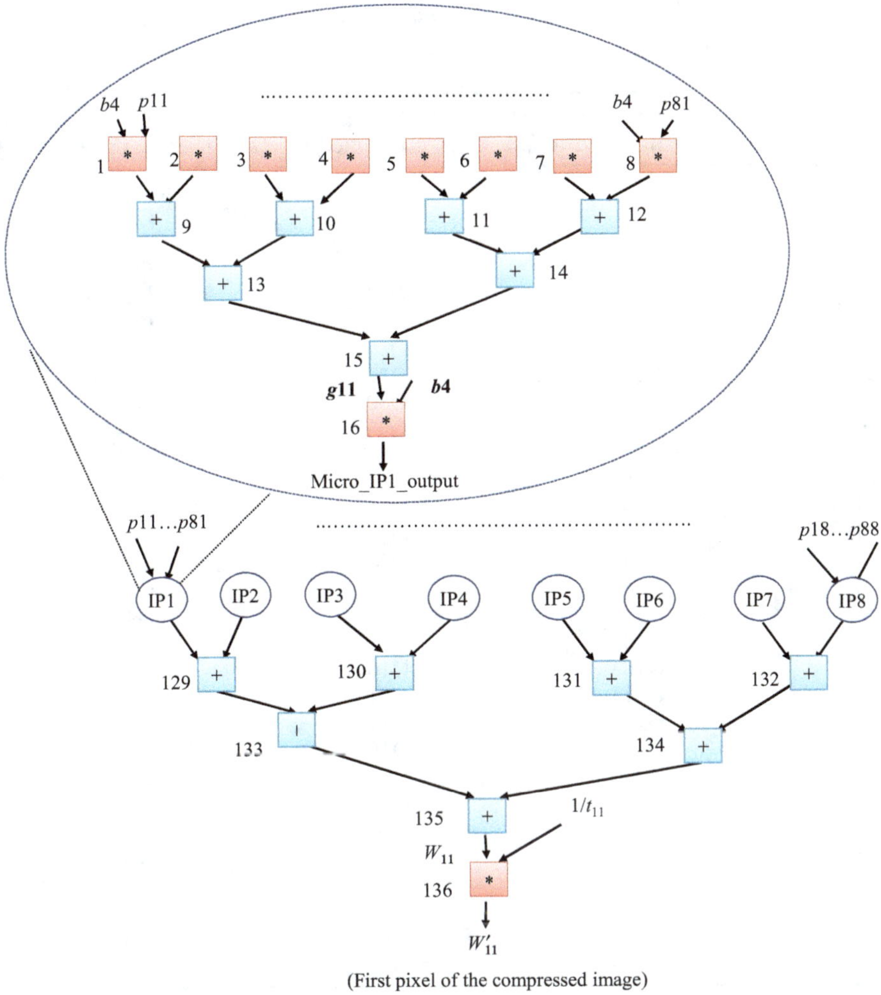

Figure 6.7 DFG form of JPEG compression algorithm (Sengupta and Rathor, 2020a)

adders ($A1$, $A2$ and $A3$) and three multipliers ($M1$, $M2$ and $M3$) takes 30 control steps (Q). Scheduling, allocation and binding of 136 operations of JPEG compression processor using three multipliers and three adders are shown in Tables 6.3 and 6.4, respectively. Further, there are a total of 209 primary and intermediate inputs and outputs in the design; therefore, 209 storage variables are required to store them. A CIG representation is used to graphically show the assignment of 209 storage variables of the design to 73 registers. The constructed CIG comprises 209 nodes ($S0$–$S208$) and 73 colours (registers). The reason as to why the JPEG

Table 6.3 Scheduling and allocation of multiplication operations of JPEG compression processor

Control steps	Operations (O) assigned to *M*1	Operations (O) assigned to *M*2	Operations (O) assigned to *M*3
Q1	O1	O2	O3
Q2	O4	O5	O6
Q3	O7	O8	O17
Q4	O18	O19	O20
Q5	O21	O22	O23
Q6	O24	O33	O34
Q7	O35	O36	O37
Q8	O38	O39	O40
Q9	O49	O50	O51
Q10	O52	O53	O54
Q11	O55	O56	O65
Q12	O66	O67	O68
Q13	O69	O70	O71
Q14	O72	O81	O82
Q15	O83	O84	O85
Q16	O86	O87	O88
Q17	O97	O98	O99
Q18	O100	O101	O102
Q19	O103	O104	O113
Q20	O114	O115	O116
Q21	O117	O118	O119
Q22	O120	O16	O32
Q23	O48	O64	O80
Q24	O96	O112	NA
Q25	NA	NA	NA
Q26	O128	NA	NA
Q27	NA	NA	NA
Q28	NA	NA	NA
Q29	NA	NA	NA
Q30	O136	NA	NA

Note: NA indicates no assignment.

compression processor (that computes pixel values of compressed image) requires 73 registers is that the design uses the maximum 73 independent primary inputs break-up of which is as follows:

73 = 64 (pixel values in the 8×8 block of input image) + 8 (elements of first row/column of 8×8 DCT coefficients matrix) + 1 (quantization coefficient t)

Demonstration of implanting constraints into JPEG codec

We demonstrate how a JPEG compression hardware accelerator can be secured using the IP vendor's biometric fingerprint. The demonstration is presented using the following steps:

Table 6.4 Scheduling and allocation of addition operations of JPEG compression processor

Control steps	Operations (O) assigned to *A*1	Operations (O) assigned to *A*2	Operations (O) assigned to *A*3
Q1	NA	NA	NA
Q2	O9	NA	NA
Q3	O10	O11	NA
Q4	O12	O13	NA
Q5	O25	O26	O14
Q6	O27	O29	O15
Q7	O28	O41	NA
Q8	O42	O30	NA
Q9	O43	O44	O45
Q10	O31	O57	O46
Q11	O58	O59	O47
Q12	O60	O61	NA
Q13	O73	O74	O62
Q14	O75	O77	O63
Q15	O76	O89	NA
Q16	O90	O78	NA
Q17	O91	O92	O93
Q18	O79	O105	O94
Q19	O106	O107	O95
Q20	O108	O109	NA
Q21	O121	O122	O110
Q22	O123	O125	O111
Q23	O124	O129	NA
Q24	O130	O126	NA
Q25	O131	O133	O127
Q26	NA	NA	NA
Q27	O132	NA	NA
Q28	O134	NA	NA
Q29	O135	NA	NA
Q30	NA	NA	NA

Note: NA indicates no assignment.

1. For demonstration, a digital template corresponding to the fingerprint image shown in Figure 6.4 is generated using the digital template generation process discussed earlier. The generated digital template has been shown in Figure 6.5.
2. Further, '0' and '1' bits in the digital template are mapped to the hardware security constraints using the mapping rules shown in Figure 6.6. In order to do so, first, an ordered list of storage variables is prepared in their increasing order (*S*0–*S*208). Next, based on the mapping rules of bits '0' and '1', storage variable pairs (*Si*, *Sj*) are sequentially picked up from the ordered list for converting each bit of the digital template into the corresponding hardware security constraints (artificial edges to be added into the CIG). Thus, obtained

Table 6.5 Fingerprint constraints (artificial edges in the CIG) corresponding to two hundred twenty-four 0s in the digital template

*S*0,*S*2	*S*0,*S*52	*S*0,*S*102	*S*0,*S*152	*S*0,*S*202	*S*2,*S*46	*S*2,*S*96	*S*2,*S*146	*S*2,*S*196
*S*0,*S*4	*S*0,54	*S*0,*S*104	*S*0,*S*154	*S*0,*S*204	*S*2,*S*48	*S*2,*S*98	*S*2,*S*148	*S*2,*S*198
*S*0,*S*6	*S*0,*S*56	*S*0,*S*106	*S*0,*S*156	*S*0,*S*206	*S*2,*S*50	*S*2,*S*100	*S*2,*S*150	*S*2,*S*200
*S*0,*S*8	*S*0,*S*58	*S*0,*S*108	*S*0,*S*158	*S*0,*S*208	*S*2,*S*52	*S*2,*S*102	*S*2,*S*152	*S*2,*S*202
*S*0,*S*10	*S*0,*S*60	*S*0,*S*110	*S*0,*S*160	*S*2,*S*4	*S*2,*S*54	*S*2,*S*104	*S*2,*S*154	*S*2,*S*204
*S*0,*S*12	*S*0,*S*62	*S*0,*S*112	*S*0,*S*162	*S*2,*S*6	*S*2,*S*56	*S*2,*S*106	*S*2,*S*156	*S*2,*S*206
*S*0,*S*14	*S*0,*S*64	*S*0,*S*114	*S*0,*S*164	*S*2,*S*8	*S*2,*S*58	*S*2,*S*108	*S*2,*S*158	*S*2,*S*208
*S*0,*S*16	*S*0,*S*66	*S*0,*S*116	*S*0,*S*166	*S*2,*S*10	*S*2,*S*60	*S*2,*S*110	*S*2,*S*160	*S*4,*S*6
*S*0,*S*18	*S*0,*S*68	*S*0,*S*118	*S*0,*S*168	*S*2,*S*12	*S*2,*S*62	*S*2,*S*112	*S*2,*S*162	*S*4,*S*8
*S*0,*S*20	*S*0,*S*70	*S*0,*S*120	*S*0,*S*170	*S*2,*S*14	*S*2,*S*64	*S*2,*S*114	*S*2,*S*164	*S*4,*S*10
*S*0,*S*22	*S*0,*S*72	*S*0,*S*122	*S*0,*S*172	*S*2,*S*16	*S*2,*S*66	*S*2,*S*116	*S*2,*S*166	*S*4,*S*12
*S*0,*S*24	*S*0,*S*74	*S*0,*S*124	*S*0,*S*174	*S*2,*S*18	*S*2,*S*68	*S*2,*S*118	*S*2,*S*168	*S*4,*S*14
*S*0,*S*26	*S*0,*S*76	*S*0,*S*126	*S*0,*S*176	*S*2,*S*20	*S*2,*S*70	*S*2,*S*120	*S*2,*S*170	*S*4,*S*16
*S*0,*S*28	*S*0,*S*78	*S*0,*S*128	*S*0,*S*178	*S*2,*S*22	*S*2,*S*72	*S*2,*S*122	*S*2,*S*172	*S*4,*S*18
*S*0,*S*30	*S*0,*S*80	*S*0,*S*130	*S*0,*S*180	*S*2,*S*24	*S*2,*S*74	*S*2,*S*124	*S*2,*S*174	*S*4,*S*20
*S*0,*S*32	*S*0,*S*82	*S*0,*S*132	*S*0,*S*182	*S*2,*S*26	*S*2,*S*76	*S*2,*S*126	*S*2,*S*176	*S*4,*S*22
*S*0,*S*34	*S*0,*S*84	*S*0,*S*134	*S*0,*S*184	*S*2,*S*28	*S*2,*S*78	*S*2,*S*128	*S*2,*S*178	*S*4,*S*24
*S*0,*S*36	*S*0,*S*86	*S*0,*S*136	*S*0,*S*186	*S*2,*S*30	*S*2,*S*80	*S*2,*S*130	*S*2,*S*180	*S*4,*S*26
*S*0,*S*38	*S*0,*S*88	*S*0,*S*138	*S*0,*S*188	*S*2,*S*32	*S*2,*S*82	*S*2,*S*132	*S*2,*S*182	*S*4,*S*28
*S*0,*S*40	*S*0,*S*90	*S*0,*S*140	*S*0,*S*190	*S*2,*S*34	*S*2,*S*84	*S*2,*S*134	*S*2,*S*184	*S*4,*S*30
*S*0,*S*42	*S*0,*S*92	*S*0,*S*142	*S*0,*S*192	*S*2,*S*36	*S*2,*S*86	*S*2,*S*136	*S*2,*S*186	*S*4,*S*32
*S*0,*S*44	*S*0,*S*94	*S*0,*S*144	*S*0,*S*194	*S*2,*S*38	*S*2,*S*88	*S*2,*S*138	*S*2,*S*188	*S*4,*S*34
*S*0,*S*46	*S*0,*S*96	*S*0,*S*146	*S*0,*S*196	*S*2,*S*40	*S*2,*S*90	*S*2,*S*140	*S*2,*S*190	*S*4,*S*36
*S*0,*S*48	*S*0,*S*98	*S*0,*S*148	*S*0,*S*198	*S*2,*S*42	*S*2,*S*92	*S*2,*S*142	*S*2,*S*192	*S*4,*S*38
*S*0,*S*50	*S*0,*S*100	*S*0,*S*150	*S*0,*S*200	*S*2,*S*44	*S*2,*S*94	*S*2,*S*144	*S*2,*S*194	–

hardware security constraints corresponding to two hundred twenty-four 0s and three hundred two 1s in the digital template are shown in Tables 6.5 and 6.6, respectively. As shown in Table 6.5, the hardware security constraint corresponding to the first 0 of the digital template is an edge between node pair (*S*0, *S*2). This is because the first two even numbers in the ordered list of storage variables are *S*0 and *S*2; hence, they correspond to the (*S*0, *S*2) node pair (based on mapping rule of bit '0') for implanting the constraint edge.

3. For securing the JPEG compression hardware accelerator, security constraints corresponding to the biometric fingerprint are implanted in the CIG of the design. Security constraints corresponding to each '0' and '1' bit, shown in Tables 6.5 and 6.6, are added as artificial edges into the CIG. Because of embedding of constraint edges into the CIG, the impact on register allocation of storage variables (*S*0–*S*208) has been shown using Tables 6.7 and 6.8, where Table 6.7 shows the register allocation before embedding security constraints, and Table 6.8 shows the register allocation post embedding security constraints. As shown in the table, storage variables *S*138, *S*196, *S*202 and *S*208 are subjected to reallocation of registers in order to enable the implantation of all constraint edges. The changes in register allocation of storage variables

Table 6.6 Fingerprint constraints (artificial edges in the CIG) corresponding to three hundred two 1s in the digital template

S1,S3	S1,S83	S1,S163	S3,S39	S3,S119	S3,S199	S5,S77	S5,S157
S1,S5	S1,S85	S1,S165	S3,S41	S3,S121	S3,S201	S5,S79	S5,S159
S1,S7	S1,S87	S1,S167	S3,S43	S3,S123	S3,S203	S5,S81	S5,S161
S1,S9	S1,S89	S1,S169	S3,S45	S3,S125	S3,S205	S5,S83	S5,S163
S1,S11	S1,S91	S1,S171	S3,S47	S3,S127	S3,S207	S5,S85	S5,S165
S1,S13	S1,S93	S1,S173	S3,S49	S3,S129	S5,S7	S5,S87	S5,S167
S1,S15	S1,S95	S1,S175	S3,S51	S3,S131	S5,S9	S5,S89	S5,S169
S1,S17	S1,S97	S1,S177	S3,S53	S3,S133	S5,S11	S5,S91	S5,S171
S1,S19	S1,S99	S1,S179	S3,S55	S3,S135	S5,S13	S5,S93	S5,S173
S1,S21	S1,S101	S1,S181	S3,S57	S3,S137	S5,S15	S5,S95	S5,S175
S1,S23	S1,S103	S1,S183	S3,S59	S3,S139	S5,S17	S5,S97	S5,S177
S1,S25	S1,S105	S1,S185	S3,S61	S3,S141	S5,S19	S5,S99	S5,S179
S1,S27	S1,S107	S1,S187	S3,S63	S3,S143	S5,S21	S5,S101	S5,S181
S1,S29	S1,S109	S1,S189	S3,S65	S3,S145	S5,S23	S5,S103	S5,S183
S1,S31	S1,S111	S1,S191	S3,S67	S3,S147	S5,S25	S5,S105	S5,S185
S1,S33	S1,S113	S1,S193	S3,S69	S3,S149	S5,S27	S5,S107	S5,S187
S1,S35	S1,S115	S1,S195	S3,S71	S3,S151	S5,S29	S5,S109	S5,S189
S1,S37	S1,S117	S1,S197	S3,S73	S3,S153	S5,S31	S5,S111	S5,S191
S1,S39	S1,S119	S1,S199	S3,S75	S3,S155	S5,S33	S5,S113	S5,S193
S1,S41	S1,S121	S1,S201	S3,S77	S3,S157	S5,S35	S5,S115	S5,195
S1,S43	S1,S123	S1,S203	S3,S79	S3,S159	S5,S37	S5,S117	S5,S197
S1,S45	S1,S125	S1,S205	S3,S81	S3,S161	S5,S39	S5,S119	S5,S199
S1,S47	S1,S127	S1,S207	S3,S83	S3,S163	S5,S41	S5,S121	–
S1,S49	S1,S129	S3,S5	S3,S85	S3,S165	S5,S43	S5,S123	
S1,S51	S1,S131	S3,S7	S3,S87	S3,S167	S5,S45	S5,S125	–
S1,S53	S1,S133	S3,S9	S3,S89	S3,S169	S5,S47	S5,S127	–
S1,S55	S1,S135	S3,S11	S3,S91	S3,S171	S5,S49	S5,S129	–
S1,S57	S1,S137	S3,S13	S3,S93	S3,S173	S5,S51	S5,S131	–
S1,S59	S1,S139	S3,S15	S3,S95	S3,S175	S5,S53	S5,S133	–
S1,S61	S1,S141	S3,S17	S3,S97	S3,S177	S5,S55	S5,S135	–
S1,S63	S1,S143	S3,S19	S3,S99	S3,S179	S5,S57	S5,S137	–
S1,S65	S1,S145	S3,S21	S3,S101	S3,S181	S5,S59	S5,S139	–
S1,S67	S1,S147	S3,S23	S3,S103	S3,S183	S5,S61	S5,S141	–
S1,S69	S1,S149	S3,S25	S3,S105	S3,S185	S5,S63	S5,S143	–
S1,S71	S1,S151	S3,S27	S3,S107	S3,S187	S5,S65	S5,S145	–
S1,S73	S1,S153	S3,S29	S3,S109	S3,S189	S5,S67	S5,S147	–
S1,S75	S1,S155	S3,S31	S3,S111	S3,S191	S5,S69	S5,S149	–
S1,S77	S1,S157	S3,S33	S3,S113	S3,S193	S5,S71	S5,S151	–
S1,S79	S1,S159	S3,S35	S3,S115	S3,S195	S5,S73	S5,S153	–
S1,S81	S1,S161	S3,S37	S3,S117	S3,S197	S5,S75	S5,S155	–

have been marked grey as shown in Table 6.8. The register reallocations of storage variables *S*138, *S*196, *S*202 and *S*208 (marked in red) have been performed in order to enable the implantation of constraint edges between node pairs ⟨*S*2,*S*138⟩, ⟨*S*0,*S*196⟩, ⟨*S*0,*S*202⟩, and ⟨*S*0,*S*208⟩, respectively. This is because initially register/colour 'R1' was used to store both storage variables/nodes of the pairs ⟨*S*0,*S*196⟩, ⟨*S*0,*S*202⟩ and ⟨*S*0,*S*208⟩, and register 'R3' was

Table 6.7 Register allocation to storage variables in JPEG compression processor before implanting fingerprint constraints

CS	Q0	Q1	Q2	Q3	Q4	...	Q22	Q23	Q24	Q25	Q26	Q27	Q28	Q29	Q30
R1	S0	S73	S137	S137	S141		S187	S196	S196	S202	S202	S202	S202	S207	S208
R2	S1	S74	NA	NA	NA	. . .	NA	NA	NA	NA	NA	NA	NA	NA	NA
R3	S2	S75	S75	S138	NA	. . .	NA	NA	NA	NA	NA	NA	NA	NA	NA
R4	S3	S3	S76	NA	NA	. . .	NA	NA	NA	NA	NA	NA	NA	NA	NA
.	.	.	.	.	.	.	.	.	.	.	.	.	.	.	.
.	.	.	.	.	.	.	.	.	.	.	.	.	.	.	.
.	.	.	.	.	.	.	.	.	.	.	.	.	.	.	.
R36	S35	S35	S35	S35	S35	. . .	NA	NA	NA	NA	NA	NA	NA	NA	NA
R37	S36	S36	S36	S36	S36	. . .	NA	NA	NA	NA	NA	NA	NA	NA	NA
.	.	.	.	.	.	.	.	.	.	.	.	.	.	.	.
.	.	.	.	.	.	.	.	.	.	.	.	.	.	.	.
.	.	.	.	.	.	.	.	.	.	.	.	.	.	.	.
R72	S71	S71	S71	S71	S71	. . .	S71	S71	S71	NA	NA	NA	NA	NA	NA
R73	S72	S72	S72	S72	S72	. . .	S72	S72	S72	S72	S72	S72	S72	S72	NA

Note: *NA indicates No Allocation*

used to store both storage variables of the node pairs ⟨*S*2,*S*138⟩ as shown in Table 6.7. To enable the implantation of constraint edges between these node pairs, the corresponding storage variables/nodes are required to be allocated to distinct registers/colours. Therefore, storage variables *S*196, *S*202 and *S*208 have been reallocated from register R1 to R2, and the storage variable *S*138 has been reallocated from register R3 to R2 post embedding hardware security constraints as shown in Table 6.8 (shaded).

As evident from the previous discussion, implantation of fingerprint constraints into the JPEG compression hardware accelerator design does not result into register overhead. Hence, the biometric fingerprinting approach is a highly robust and efficient security approach which provides very high security without incurring significant design overhead. Once the fingerprint constraints are implanted during the register allocation phase of HLS process, data path and controller synthesis phases are performed to generate a secure RTL circuit of JPEG compression hardware accelerator with embedded biometric fingerprint of the IP vendor. Therefore, a biometric fingerprint of the IP vendor can be used to secure the designs. The secretly embedded fingerprint information can be used to resolve conflict of IP ownership, in case a fraudulent user claims the ownership. Additionally, the embedded fingerprint information can be detected in the hardware designs to identify counterfeiting.

Table 6.8 Register allocation to storage variables in JPEG compression processor after implanting fingerprint constraints

CS	Q0	Q1	Q2	Q3	Q4	...	Q22	Q23	Q24	Q25	Q26	Q27	Q28	Q29	Q30
R1	S0	S73	S137	S137	S141	. . .	S187	NA	NA	NA	NA	NA	NA	S207	NA
R2	S1	S74	NA	S138	NA	. . .	NA	S196	S196	S202	S202	S202	S202	NA	S208
R3	S2	S75	S75	NA	NA	. . .	NA	NA	NA	NA	NA	NA	NA	NA	NA
R4	S3	S3	S76	NA	NA	. . .	NA	NA	NA	NA	NA	NA	NA	NA	NA
.	.	.	.	.	.	.	.	.	.	.	.	.	.	.	.
.	.	.	.	.	.	.	.	.	.	.	.	.	.	.	.
.	.	.	.	.	.	.	.	.	.	.	.	.	.	.	.
R36	S35	S35	S35	S35	S35	. . .	NA	NA	NA	NA	NA	NA	NA	NA	NA
R37	S36	S36	S36	S36	S36	. . .	NA	NA	NA	NA	NA	NA	NA	NA	NA
.	.	.	.	.	.	.	.	.	.	.	.	.	.	.	.
.	.	.	.	.	.	.	.	.	.	.	.	.	.	.	.
.	.	.	.	.	.	.	.	.	.	.	.	.	.	.	.
R72	S71	S71	S71	S71	S71	. . .	S71	S71	S71	S71	NA	NA	NA	NA	NA
R73	S72	S72	S72	S72	S72	. . .	S72	S72	S72	S72	S72	S72	S72	S72	NA

Note: *NA indicates No Allocation*

6.6.3 Detection and verification process of biometric fingerprint in a hardware accelerator design

IP vendor's biometric fingerprint embedded into a hardware accelerator design can be used to detect counterfeiting and nullifying false claim of IP ownership threat. The detection and verification process of an embedded biometric fingerprint in a hardware accelerator design is shown in Figure 6.8 and discussed in the following subsection.

6.6.3.1 Counterfeit detection

The counterfeit detection process is highlighted in Figure 6.8(a). To detect the counterfeiting or authenticity of a hardware accelerator design, the presence of embedded biometric fingerprint of a genuine IP vendor can be detected into the design. In order to do so, first the digital template of IP vendor's biometric fingerprint is converted into hardware security constraints (artificial edges into the CIG) using the vendor's mapping rules. This would determine the allocation of storage variables to the registers post implanting the constraints. Further, as shown in the lower part of Figure 6.8(a), inspection of register allocation of storage variables (through Mux/Demux inputs/outputs) in the RTL data path of design under test is performed. Thus, the information of register allocation of storage variables is matched against the hardware security constraints (corresponding to the digital

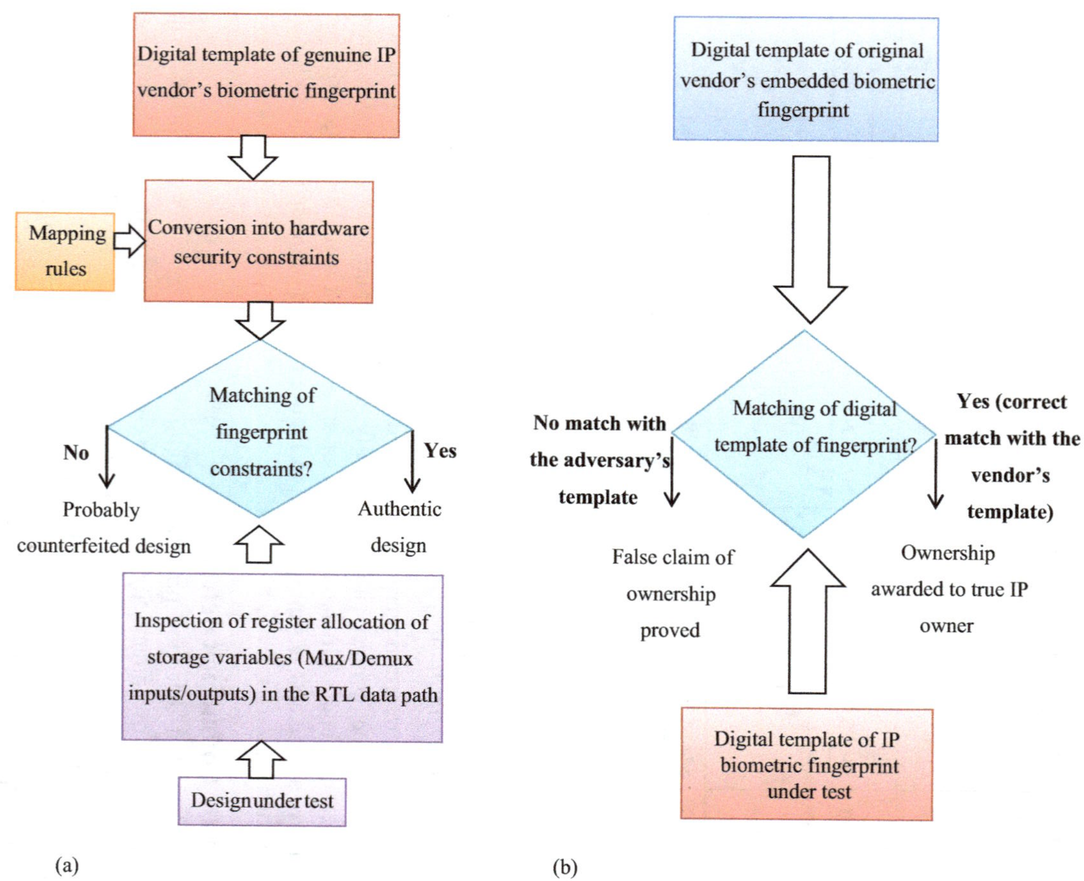

Figure 6.8 *(a) Counterfeit detection and (b) proving true IP ownership (nullifying false claim) using biometric fingerprint (Sengupta and Rathor, 2020a)*

template) of an original IP vendor's biometric fingerprint. If the extracted information from the design matches with the security constraints corresponding to the digital template of the biometric fingerprint, then the design is proved to be authentic, as the presence of vendor's biometric fingerprint is found in the design. However, if the presence of fingerprint constraints is not found, then the design may probably be a counterfeit.

6.6.3.2 Nullifying false claim of IP ownership

The verification process of biometric fingerprint for nullifying the ownership claim by a fraudulent user is highlighted in Figure 6.8(b). In order to discern the true ownership of the design, the embedded digital template of fingerprint is matched against the digital template of the incoming fingerprint. Only the digital template of a genuine IP owner will match with the embedded digital template. Hence, the digital template corresponding to the adversary's (fraudulent claimer) biometric fingerprint cannot match with embedded digital template, as fingerprint digital template is also naturally unique. This nullifies the fraudulent claim of IP ownership by an adversary and proves the ownership of a genuine owner. It is important to note that although two digital templates of two distinct fingerprints may contain the same number of 0s and 1s, the respective bit positions in two digital templates are always different. This is because the digital template contains information of all the minutiae points of a biometric fingerprint; further, the positions of '0' and '1' bits in a digital template encapsulate the information of all four attributes of each minutiae point. This renders the digital template corresponding to a biometric fingerprint unique and thus enables the seamless verification process.

6.7 Analysis on case studies

This section discusses the analysis of the biometric-fingerprint-based hardware security approach for different fingerprint images and various resource constraints. For demonstration, JPEG codec application is used. First, the relationship between generated hardware security constraints and the respective biometric fingerprint is discussed. Second, discussions on security analysis in terms of *Pc* and comparison with a contemporary approach (Sengupta and Rathor, 2020c) have been presented. Subsequently, the impact of the biometric-fingerprint-based hardware security approach on the design cost of JPEG codec hardware accelerator has been discussed for different resource constraints/configurations (Sengupta and Rathor, 2020a).

6.7.1 Analysing the relationship between biometric fingerprint and strength of hardware security constraints

The relationship of a biometric fingerprint and corresponding strength of hardware security constraints depends on the following factors: (i) the number of minutiae points extracted from the biometric fingerprint, (ii) attributes of each minutiae point, (iii) the algorithm of digital template generation and (iv) mapping rules for converting digital template to the corresponding hardware security constraints. This subsection analyses the relationship of various biometric fingerprints with the number of corresponding hardware security constraints generated for implanting into a target design (DSP or multimedia hardware accelerator) for security (Sengupta and Rathor, 2020a). In order to do so, different fingerprint images have been chosen from a publicly available database (Alilou, 2019) of biometric fingerprints. The relationship of various biometric fingerprints with the hardware security constraints has been depicted in Figure 6.9. As shown in the figure, the number of minutiae points extracted from the biometric fingerprint images 101_1, 101_2 and 101_8 are 22, 15 and 24, respectively. Further, their corresponding digital templates are of size 526, 350 and 555 bits, respectively, as shown in Figure 6.9. In the digital template of biometric fingerprint 101_1 of size 526 bits, the number of secret biometric constraints corresponding to bits '0' and '1' are 224 and 302, respectively. These secret biometric constraints represent the number of hardware security constraints to be embedded into the target design. Similarly, Figure 6.9 also shows the relationship of the number of hardware security constraints corresponding to the fingerprint images 101_2 and 101_8, respectively. In the digital template of biometric fingerprint 101_2 of size 350 bits, the number of secret biometric constraints corresponding to bits '0' and '1' are 148 and 202, respectively. Further, in the digital template of biometric fingerprint 101_8 of size 555 bits, the number of secret biometric constraints corresponding to bits '0' and '1' are 242 and 313, respectively. It can be observed from Figure 6.9 that the number of secret biometric constraints corresponding to bits '0' and '1' vary according to the fingerprint image. It is highly unlikely that the number of secret biometric constraints (0s and 1s) corresponding to two different fingerprint images exactly matches. Nevertheless, in a rare case, in case the number of 0s and 1s in the digital template corresponding to two biometric fingerprints matches, then both the templates can be distinguished on the basis of the respective positions of 0 and 1 bits (the respective positions of 0 and 1 bits can never match due to the natural uniqueness of minutiae points of a fingerprint).

Further, the relationship of fingerprints of different fingers from the same individual, with the generated hardware security constraints, has been depicted in Figure 6.10. For the fingerprint image of little finger, the number of secret biometric constraints corresponding to bits '0' and '1' are 274 and 366, respectively, as shown in Figure 6.10. Further, for the fingerprint image of middle finger, the numbers of secret biometric constraints corresponding to bits '0' and '1' are 292 and 429, respectively, and for the fingerprint image of ring finger, the numbers of secret biometric constraints corresponding to bits '0' and '1' are 386 and 509, respectively. It is to be noted that the corresponding hardware security constraints of biometric fingerprints of different fingers vary because of variation in the number of minutiae points and the corresponding digital template. Therefore, an IP vendor can use fingerprint of different fingers (of the same person) to vary the number of hardware security constraints to be embedded for securing the designs (Sengupta and Rathor, 2020a).

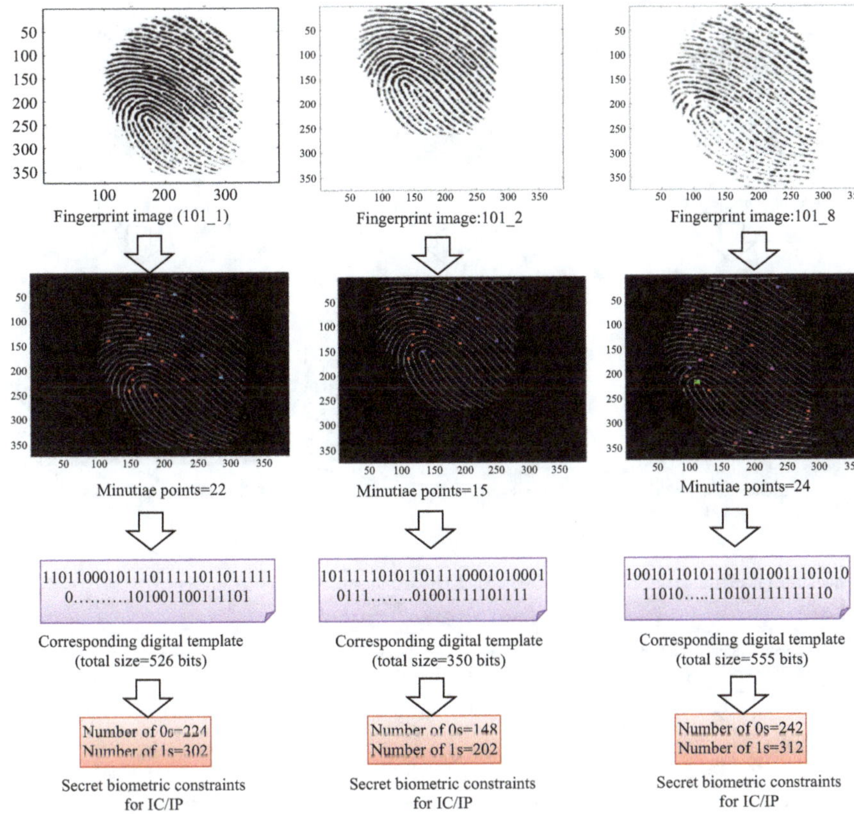

Figure 6.9 Secret biometric constraints for various fingerprint images (Sengupta and Rathor, 2020a)

In contrast to the biometric-fingerprint-based security approach (Sengupta and Rathor, 2020a), the variation in secret constraints to be embedded in contemporary security approaches (Koushanfar *et al.*, 2005; Le Gal and Bossuet, 2012; Sengupta *et al.*, 2016) is performed by varying the signature size or combination. In the contemporary approaches, it is hard to make prior estimation of the impact of varying signature size, strength of the encoding algorithm used and signature combination on the generated secret constraints. The underlying reason is that a larger signature size may not always correspond to higher number of secret constraints, as some signature digits may not be implanted into the target design (Sengupta and Rathor, 2020a).

6.7.2 Security analysis

This subsection analyses the security achieved using the biometric fingerprint approach (Sengupta and Rathor, 2020a) in terms of *Pc* metric.

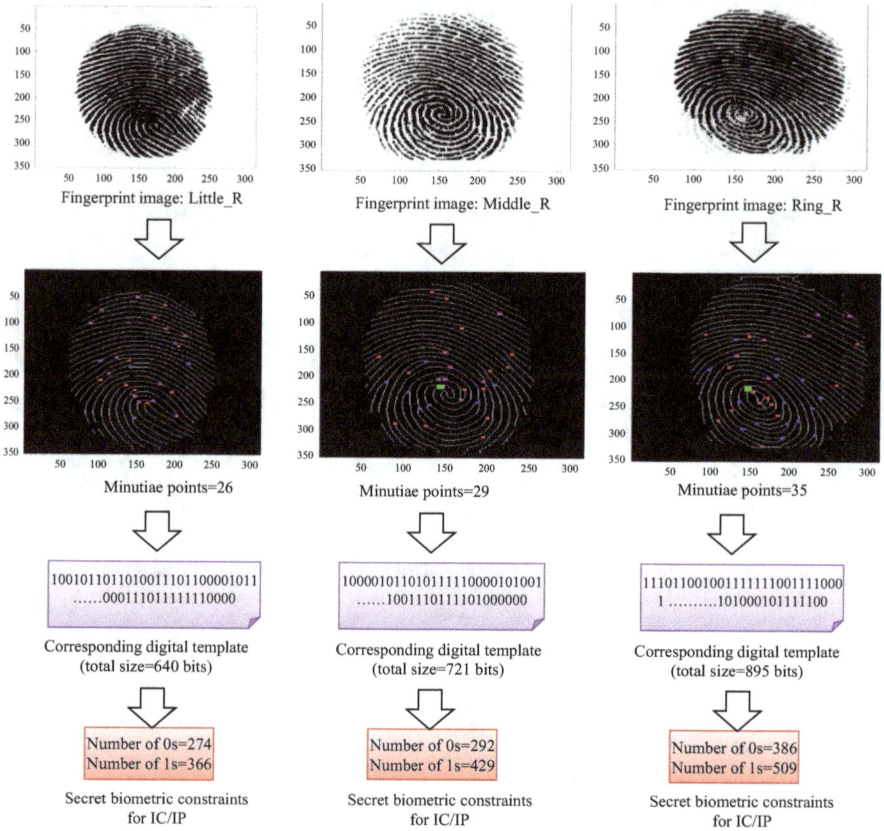

Figure 6.10 Secret biometric constraints for fingerprints of different fingers of an individual (Sengupta and Rathor, 2020a)

Here, the *Pc* achieved using biometric fingerprint approach indicates the strength of ownership proof for nullifying the false claim of ownership. Further, the *Pc* value also indicates the robustness of embedded fingerprint constraints (in the form of digital evidence) into a hardware accelerator design for detecting the counterfeiting. Furthermore, this subsection also discusses the comparison of *Pc* value of biometric fingerprint approach (Sengupta and Rathor, 2020a) with a contemporary security approach (Sengupta and Rathor, 2020c).

The *Pc* metric has been evaluated using the following function (Sengupta and Rathor, 2020a):

$$Pc = \left(1 - \frac{1}{h}\right)^{k1} * \left(1 - \frac{1}{\pi_{j=1}^{m} N(U_j)}\right)^{k2} \tag{6.8}$$

where h indicates the number of colours/registers in the CIG of target DSP hardware accelerator design before embedding constraints and $k1$ indicates the number of secret constraints embedded during the register allocation phase. Further, $k2$ indicates the number of secret constraints embedded during the resource allocation phase, $N(U_j)$ indicates the number of resources of FU type U_j and m indicates the total types of FU resources present in the design. A lower Pc value is desirable for achieving high robustness of embedded secret information and high strength of ownership proof. It can be achieved by embedding a larger number of secret constraints as evident from (6.8). In the case of biometric fingerprint approach, the secret hardware security constraints have been embedded during register allocation phase only. Hence, the value of $k1$ in the formula of Pc indicates the total number of hardware security constraints (secret biometric constraints corresponding to bits '0' and '1') embedded into the design and the value of $k2$ remains 0. The Pc using biometric fingerprint approach has been analysed for various fingerprint images. Figure 6.11 reports the variation in minutiae points corresponding to various fingerprint images available in publicly accessible database (Alilou, 2019). The number of hardware security constraints ($k1$) corresponding to the minutiae points of these fingerprint images has been reported in Figure 6.12. Based on the number of hardware security constraints ($k1$), the variation in Pc on JPEG codec hardware accelerator design for the different fingerprint images has been shown in Figure 6.13. As shown in the figure, very low Pc value is achieved for JPEG codec hardware accelerator design indicating the higher strength of ownership proof (Sengupta and Rathor, 2020a).

Further, the Pc using biometric fingerprint approach for JPEG codec has been analysed for the fingerprints of different fingers of the same individual (Sengupta and Rathor, 2020a). Figure 6.14 reports the variation in minutiae points corresponding to the fingerprints of different fingers of an individual. The hardware security constraints ($k1$) corresponding to the minutiae points of these fingerprints have been reported in Figure 6.15. Based on the number of hardware security constraints ($k1$), the variation in Pc for JPEG codec hardware design for the fingerprint images of different fingers has been shown in Figure 6.16. As shown in the figure, an IP vendor can achieve low Pc by embedding fingerprint of a specific finger that has a greater number of minutiae points (equivalent to a larger number of security constraints).

Further, the biometric-fingerprint-based approach (Sengupta and Rathor, 2020a) has been compared with a contemporary approach (Sengupta and Rathor, 2020c) that also secures the JPEG compression hardware accelerator against false claim of ownership and piracy threats. The contemporary approach (Sengupta and Rathor, 2020c) implants secret constraints in the register allocation and FU vendor allocation phases. The implanted constraints in the register allocation phase are indicated by $k1$ and that in FU vendor allocation phase are indicated by $k2$ (shown in the formula of Pc shown in (6.8)). The total implanted hardware security constraints are indicated by $G = k1+k2$. The comparative analysis of biometric fingerprinting approach and the contemporary approach (Sengupta and Rathor, 2020c) for JPEG compression hardware in terms of Pc has been shown in Figure 6.17. The

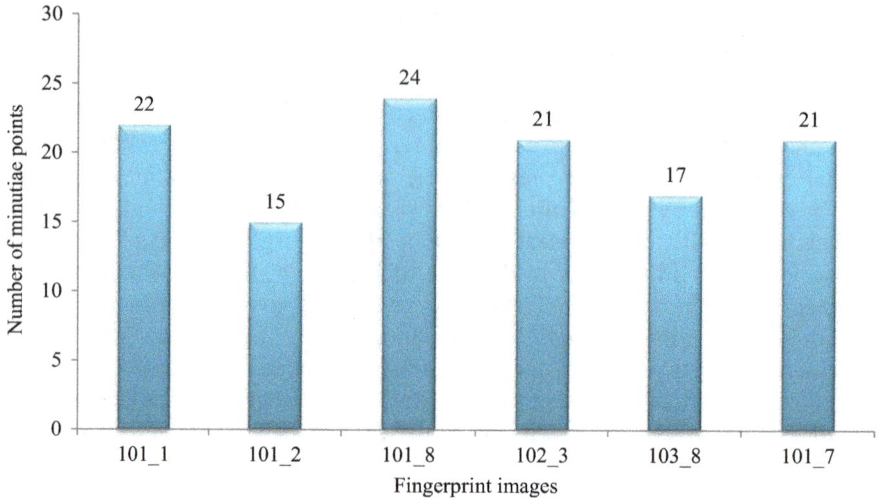

Figure 6.11 Variation in the number of minutiae points for different fingerprint images (Sengupta and Rathor, 2020a)

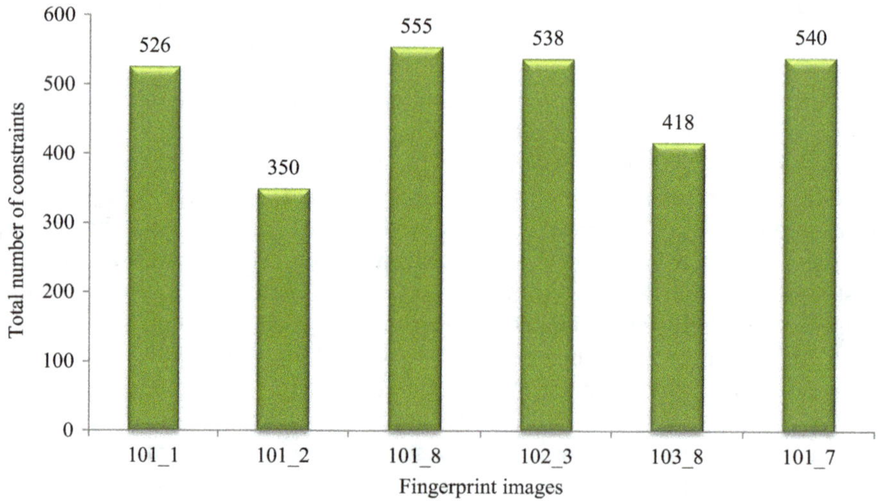

Figure 6.12 Variation in total number of constraints (k1) for different fingerprint images (Sengupta and Rathor, 2020a)

comparison of biometric fingerprinting approach with the contemporary approach (Sengupta and Rathor, 2020c) has been shown for different sizes of total secret constraints (G) generated from the contemporary approach. In this comparison, the Pc value corresponding to the constraints sizes $G = 20$, 60 and 100 in the

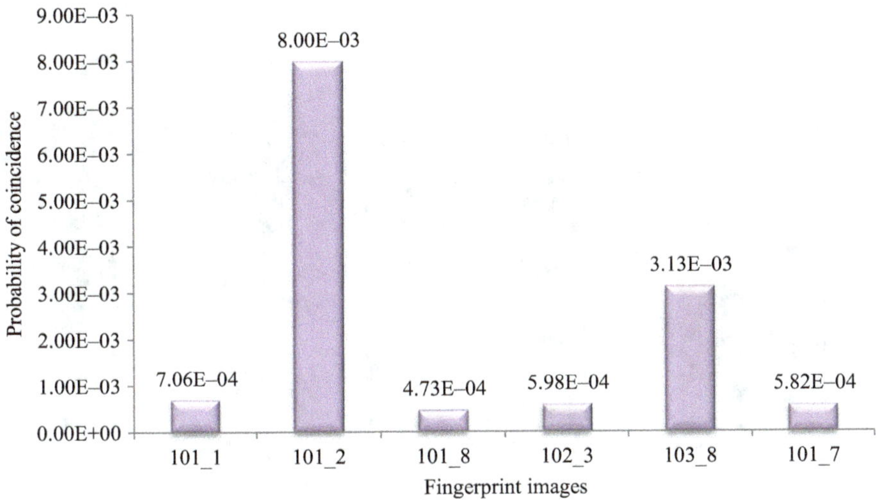

Figure 6.13 Variation in probability of coincidence for JPEG codec, for different fingerprint images

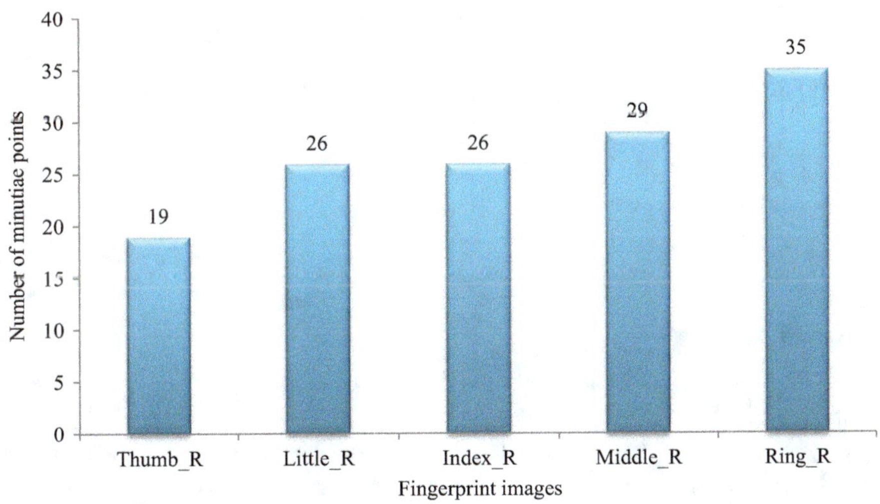

Figure 6.14 Variation in minutiae points for fingerprints of different fingers of an individual (Sengupta and Rathor, 2020a)

contemporary approach has been compared with the *Pc* value corresponding to the fixed number of minutiae points ($n = 35$) of a fingerprint in the biometric fingerprinting approach (Sengupta and Rathor, 2020a). As shown in Figure 6.17, the *Pc* value achieved using biometric fingerprinting approach is always less compared

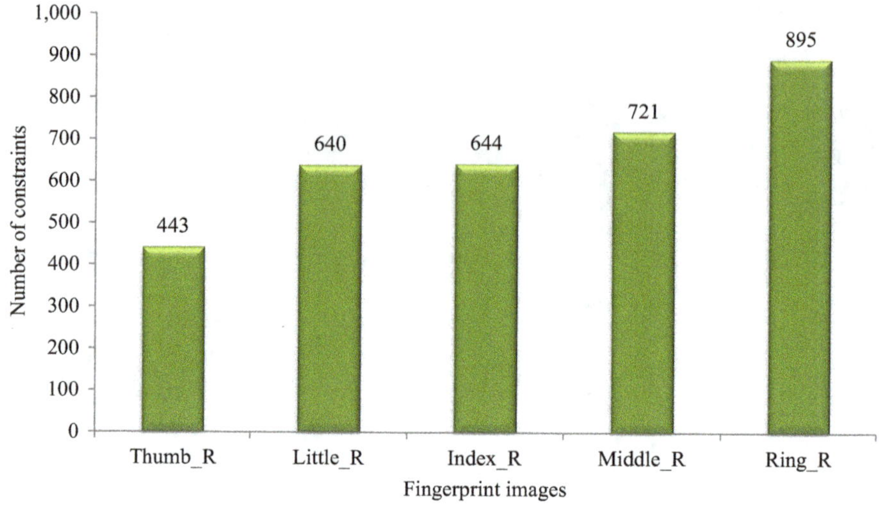

Figure 6.15 Variation in total number of constraints (k1) for fingerprints of different fingers of an individual (Sengupta and Rathor, 2020a)

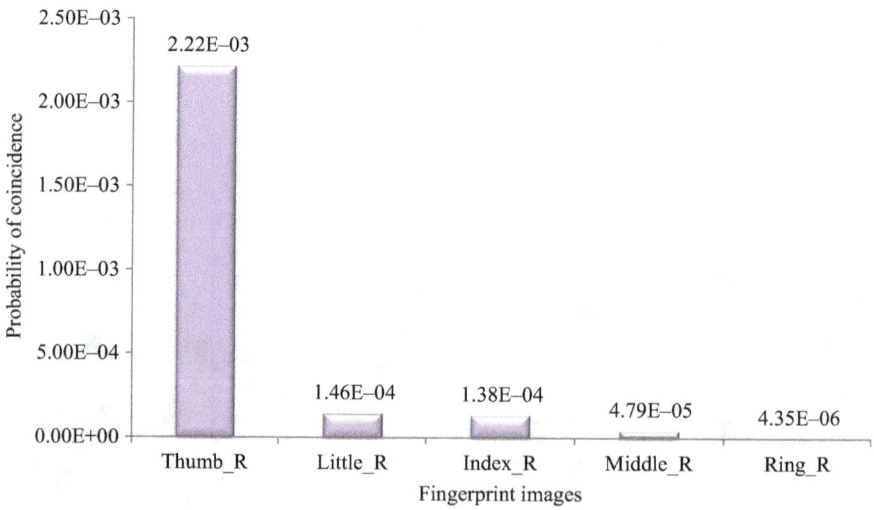

Figure 6.16 Variation in probability of coincidence for JPEG codec design, for fingerprints of different fingers of an individual

with the contemporary approach for different sizes of total secret constraints (G). This is because the biometric fingerprinting approach implants a higher number of secret constraints, i.e. $k1 = 895$ (corresponding to the minutiae points $n = 35$) compared with the contemporary approach. This indicates stronger digital evidence achievable using biometric fingerprinting approach.

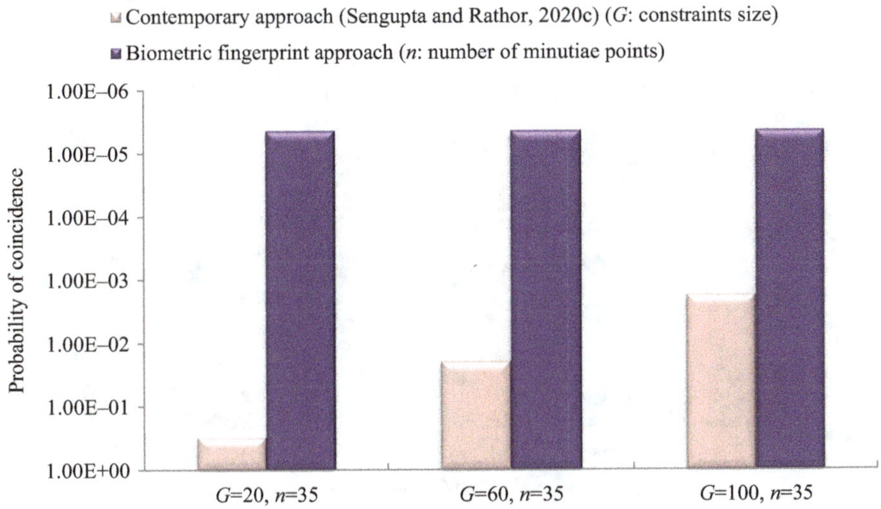

Figure 6.17 Comparison of probability of coincidence with contemporary approach for JPEG compression hardware

6.7.3 Design cost analysis

This subsection discusses the impact of employing biometric-fingerprinting-based security on design cost. The case study in terms of design cost has been performed on JPEG compression hardware accelerator design, for varying resource constraints. The following equation is used to evaluate the design cost (Sengupta and Rathor, 2020a):

$$C_d(U_i) = \rho_1 \frac{L_d}{L_m} + \rho_2 \frac{A_d}{A_m} \tag{6.9}$$

where $C_d(U_i)$ is the design cost of JPEG compression hardware accelerator for resource constraints U_i; further, L_d and L_m are the design latency at specified resource constraints and the maximum design latency, respectively; A_d and A_m are the design area at specified resource constraints and maximum area, respectively, and ρ_1, ρ_2 are the weights which are kept fixed at 0.5.

Since in the biometric fingerprinting approach (Sengupta and Rathor, 2020a) the security constraints are implanted into the register allocation phase of HLS process, the impact of the employed security on the design cost depends on whether the implanting of constraints incurs register overhead or not. The case study analysis of biometric-fingerprint-based security approach (Sengupta and Rathor, 2020a) on JPEG compression design shows that the number of registers before and after implanting security constraints remains the same. The comparison of registers before and after implanting security constraints for various resource constraints of the JPEG compression hardware accelerator design has been shown in Figure 6.18 (Sengupta and Rathor, 2020a). The reason why the register overhead is not incurred

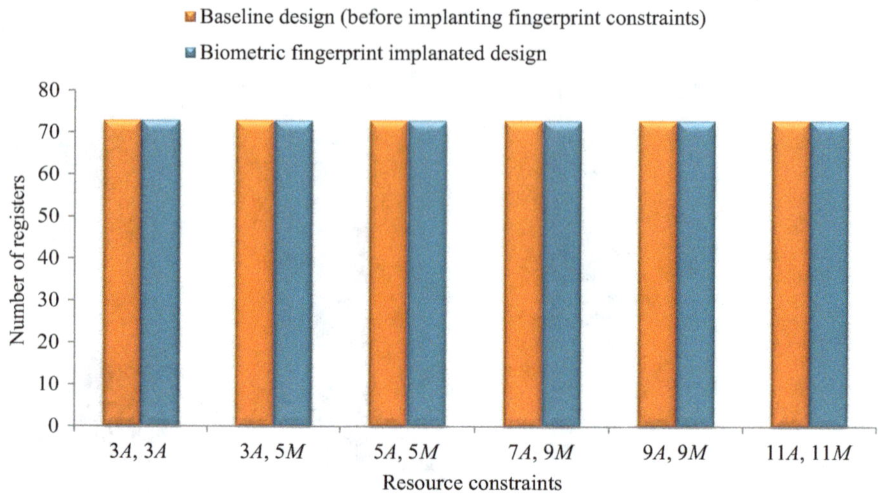

Figure 6.18 Comparison of the number of registers before and after implanting fingerprint constraints into JPEG compression hardware accelerator design (Sengupta and Rathor, 2020a)

is because embedding of security constraints into the CIG of the JPEG design is possible, by performing local alterations in the register allocation of storage variables. Since the biometric fingerprinting approach (Sengupta and Rathor, 2020a) does not require extra registers to accommodate all fingerprint constraints (during the register allocation phase), the design cost also remains the same (before and after implanting constraints). The design cost comparison for various resource constraints has been depicted in Figure 6.19. This comparison reveals that the biometric-fingerprint-based security approach (Sengupta and Rathor, 2020a) is capable of providing higher security at zero design cost overhead.

6.8 Benefits and advantages of biometric-fingerprint-based IP protection

This subsection highlights the benefits and advantages of biometric-fingerprint-based IP core protection (Sengupta and Rathor, 2020a).

1. *Resilient against vulnerability*: The hardware security constraints corresponding to a biometric fingerprint are non-vulnerable (Sengupta and Rathor, 2020a). This is because the hardware security constraints correspond to the minutiae features of a biometric fingerprint, which are always unique for an individual. Hence, the secret constraints cannot be compromised by an adversary. Even if the adversary becomes aware of the implanted secret biometric constraints into the design, he/she cannot fraudulently claim his/her right over them (Sengupta and Rathor, 2020a). However, in the case of contemporary

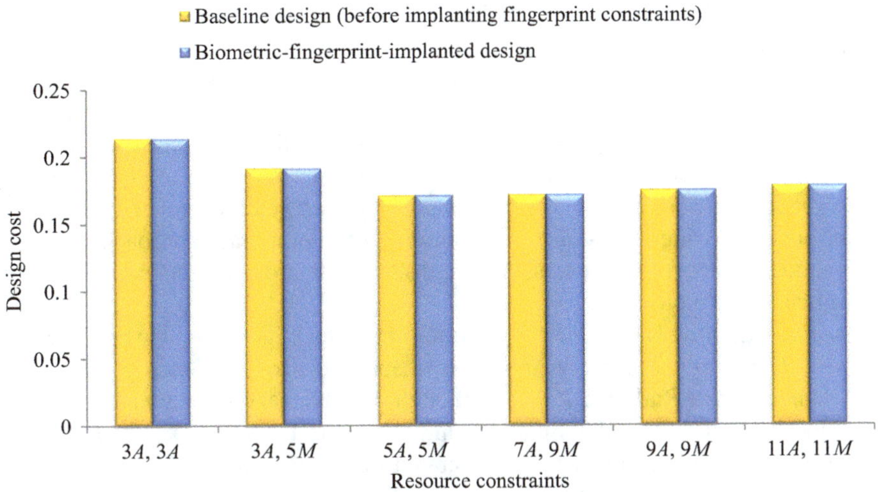

Figure 6.19 Comparison of design cost of JPEG compression hardware accelerator before and after implanting fingerprint constraints (Sengupta and Rathor, 2020a)

approaches (Koushanfar *et al.,* 2005; Le Gal and Bossuet, 2012; Sengupta *et al.,* 2016), the generation of security constraints depends on the vendor's chosen signature size, combination and encoding rules. Since the chosen signature and the encoding rules may be stolen/compromised by an attacker, the contemporary approaches (Koushanfar *et al.,* 2005; Le Gal and Bossuet, 2012; Sengupta *et al.,* 2016) are not highly resilient against vulnerability. This vulnerability of contemporary approaches (Koushanfar *et al.,* 2005; Le Gal and Bossuet, 2012; Sengupta *et al.,* 2016) may defeat the goal of securing against piracy and false claim of ownership threats.

2. *Impossible to replicate*: The hardware security constraints corresponding to a biometric fingerprint cannot be replicated for reuse by an adversary (Sengupta and Rathor, 2020a). This is because of the natural uniqueness of a biometric fingerprint. An adversary can never prove someone else's biometric fingerprint as his/her own. Therefore, the adversary cannot replicate the hardware security constraints corresponding to the IP vendor's biometric fingerprint for realizing wrong intentions of counterfeiting and false claiming ownership. However, the hardware security constraints generated from contemporary approaches (Koushanfar *et al.,* 2005; Le Gal and Bossuet, 2012; Sengupta *et al.,* 2016) can be replicated by an adversary once she/he succeeds in compromising or leaking information such as the number of variables, size and combination of the vendor's signature and encoding rules of signature variables.
3. *No dependence on storage*: The hardware security constraints corresponding to a biometric fingerprint do not require to be necessarily stored. This is because the security constraints can always be generated from the respective biometric

fingerprint (Sengupta and Rathor, 2020a) of the same individual. For implanting fingerprint constraints into the design, the biometric fingerprint of a highly trustful insider in the IP vendor's house can be chosen. However, in the case of contemporary approaches (Koushanfar *et al.,* 2005; Le Gal and Bossuet, 2012; Sengupta *et al.,* 2016), the generation of security constraints depends on some secret information which are required to be stored for regenerating the implanted constraints during verification process. The secret information which is required to be stored in contemporary approaches may be vendor's chosen signature size (length), combination and encoding rules of signature variables.

4. *Security against brute force*: The brute-force attack does not help an adversary in determining secret constraints corresponding to a biometric fingerprint (Sengupta and Rathor, 2020a). Although an adversary may attempt to determine the exact digital template using the brute-force attack, she/he cannot prove the relationship of the regenerated digital template (using brute force) with his/her own biometric fingerprint. The original digital template and the corresponding security constraints will always only belong to the genuine IP vendor's biometric fingerprint (Sengupta and Rathor, 2020a).
5. *Tamper tolerant*: The fingerprint constraints corresponding to a biometric fingerprint are highly tamper tolerant. This is because to perform tampering, an adversary needs to obtain the exact digital template corresponding to the genuine IP owner's biometric fingerprint. However, an adversary would fail to do so owing to the huge size of digital template and its dependence on unique minutiae points corresponding to the original IP vendor's biometric fingerprint (Sengupta and Rathor, 2020a).
6. *Zero dependence on secret key*: The biometric-fingerprint-based security approach does not involve any secret key to generate hardware security constraints corresponding to a biometric fingerprint. This makes the fingerprinting approach free from adversary's attempt of determining security constraints by compromising the secret key (Sengupta and Rathor, 2020a).
7. *Security in terms of point matching difference function (PMDF)*: The PMDF is also a measure of indicating the security achieved using biometric fingerprinting approach. The PMDF function is given as follows (Sengupta and Rathor, 2020a):

$$\text{PMDF} = \sum_{i,j=1}^{n} |MP_i - MP_j| \tag{6.10}$$

where n indicates the total number of minutiae points corresponding to the implanted biometric fingerprint. Further, MP_i indicates decimal representation of a minutia point of the implanted fingerprint, and MP_j indicates the decimal representation of a minutia point of the vendor's fingerprint. The 0 value of PMDF indicates the matching of minutiae points of implanted fingerprint and the vendor's incoming fingerprint. Since all the minutiae points of the incoming fingerprint of a genuine IP vendor would match with the minutiae points corresponding to the implanted biometric fingerprint, the PMDF would result into 0. However, in the case of false IP vendor (or adversary), PMDF results into a non-zero value as all the minutiae points of his/her incoming fingerprint would not match with the minutiae points of the implanted fingerprint. Hence, the non-zero value of PMDF, in the case of false IP vendor, nullifies/corroborates the false claim of IP ownership (Sengupta and Rathor, 2020a).

6.9 Conclusion

This chapter discussed a biometric-fingerprint-based security approach (Sengupta and Rathor, 2020a) which is a paradigm shift in the security techniques used against IP piracy and false claim of IP ownership threat. Strength of the biometric-fingerprint-based security approach (Sengupta and Rathor, 2020a) lies in the fact that the hardware security constraints for securing hardware accelerator designs are generated from the IP vendor's biometric fingerprint, which is naturally unique. Therefore, the corresponding hardware security constraints are also unique and can seamlessly be verified in the design.

This chapter further demonstrated the process of obtaining corresponding hardware security constraints from a biometric fingerprint and the process of implanting them into a real-life design, viz. JPEG compression hardware accelerator. The biometric-fingerprint-based security approach (Sengupta and Rathor, 2020a) implanted the security constraints during the register allocation phase of HLS process. This approach can be utilized for securing all kinds of DSP and multimedia hardware accelerators whose design process exploits HLS framework.

Additionally, this chapter discussed the impact of employing biometric-fingerprint-based security approach (Sengupta and Rathor, 2020a) on the design cost. The analysis on a real-life case study revealed that the biometric fingerprinting approach does not incur design cost overhead. Furthermore, the biometric fingerprinting approach offers many benefits and advantages such as resiliency against vulnerability, infeasibility against replication, tamper tolerance, security against brute-force attack and security in terms of PMDF. The biometric-fingerprint-based security approach (Sengupta and Rathor, 2020a) is a promising alternative to the contemporary security approaches used against false claim of IP ownership and piracy threat.

At the end of this chapter, a reader gains knowledge about the following:

- Hardware threats to DSP and multimedia hardware accelerators.
- Summary of contemporary security techniques against hardware threats.
- Introduction to biometric-fingerprint-based security approach for DSP and multimedia hardware accelerators.
- Background on biometric fingerprint, minutiae points and pre-processing of fingerprint image.
- Minutiae point extraction from a biometric fingerprint.
- Digital template generation corresponding to a biometric fingerprint.

- Robust correlation or mapping of fingerprint's digital template into corresponding hardware security constraints.
- Implantation process of hardware security constraints into hardware accelerator design during HLS process.
- Demonstration of securing a JPEG codec hardware accelerator using the vendor's biometric fingerprint.
- Detection and verification process of implanted biometric fingerprint constraints.
- Relationship between various fingerprints and corresponding strength of hardware security constraints.
- Case study on security analysis of biometric fingerprinting approach.
- Case study on design cost analysis of biometric fingerprinting approach.
- Benefits and advantages of biometric fingerprinting approach.

6.10 Questions and exercise

1. What is a heterogeneous SoC?
2. What are some of the major applications of a hardware accelerator?
3. What is the difference between IP piracy and false claim of ownership threat?
4. State the major differences between biometric-fingerprinting-based IP protection and hardware watermarking.
5. State the similarities between biometric-fingerprinting-based IP protection and hardware watermarking.
6. State the major differences between biometric-fingerprinting-based IP protection and crypto-digital signature techniques.
7. What information is required during the detection process of cryptographic digital signature IP protection methodology?
8. What is a digital template in biometric fingerprinting?
9. What is ridge bifurcation?
10. What is a minutiae point?
11. What is a ridge termination?
12. State the attributes of a minutiae point.
13. Explain the pre-processing steps used in biometric fingerprint.
14. What is the role of FFT enhancement?
15. What is segmentation in biometric fingerprint?
16. What is binarization in biometric fingerprint?
17. What is thinning in biometric fingerprint?
18. Explain the digital template generation process.
19. What function is used to form digital templates of individual minutiae points in biometric-fingerprint-based hardware security?
20. Is a digital template always unique? If so why?
21. During conversion process of a digital template to its equivalent hardware security constraints, what factors is this dependent upon?
22. What is a 2D-DCT? Explain its role in JPEG compression.

23. What is the computation function used to perform quantization on each DCT-transformed pixel value?
24. Explain the counterfeit detection and IP ownership verification process using biometric fingerprint.
25. What is the relationship between fingerprints of an individual with generated hardware security constraints?
26. What is the role of digital bits of generated digital template during embedding process in a hardware accelerator?
27. Explain the probability of coincidence (*Pc*) metric.
28. Explain the benefits and advantages of biometric-fingerprint-based security approach.
29. How does biometric-fingerprint-based security approach secure against brute-force attack?
30. What is point matching difference function and how is it useful in nullifying/corroborating the false claim of IP ownership?

References

V. K. Alilou (2019), "FingerPrint Matching: A Simple Approach," MATLAB Central File Exchange, https://www.mathworks.com/matlabcentral/fileexchange/44369-fingerprint-matching-a-simple-approach, retrieved Dec. 2019.

F. Koushanfar, I. Hong, and M. Potkonjak (2005), 'Behavioral synthesis techniques for intellectual property protection,' *ACM Trans. Des. Autom. Electron. Syst.*, vol. 10(3), pp. 523–545.

B. Le Gal and L. Bossuet (2012), 'Automatic low-cost IP watermarking technique based on output mark insertions,' *Des. Autom. Embedded Syst.*, vol. 16(2), pp. 71–92.

H. R. Mahdiany, A. Hormati and S. M. Fakhraie (2001), 'A hardware accelerator for DSP system design,' *Proc. ICM*, pp. 141–144.

C. Pilato, S. Garg, K. Wu, R. Karri and F. Regazzoni (2018), 'Securing hardware accelerators: a new challenge for high-level synthesis,' *IEEE Embedded Syst. Lett.*, vol. 10(3), pp. 77–80.

S. M. Plaza and I. L. Markov (2015), 'Solving the third-shift problem in IC piracy with test-aware logic locking,' *IEEE Trans. Comput. Aided Des. Integr. Circuits Syst.*, vol. 34(6), pp. 961–971.

D. Saha and S. Sur-Kolay (2011), 'SoC: a real platform for IP reuse, IP infringement, and IP protection,' *CAD for Gigascale SoC Design and Verification Solutions*, vol. 2011, doi:10.1155/2011/731957.

R. Schneiderman (2010), 'DSPs evolving in consumer electronics applications,' *IEEE Signal Process. Mag.*, vol. 27(3), pp. 6–10.

A. Sengupta and S. Bhadauria (2016), 'Exploring low cost optimal watermark for reusable IP cores during high level synthesis,' *IEEE Access*, vol. 4, pp. 2198–2215.

A. Sengupta, S. Bhadauria and S. P. Mohanty (2016), 'Embedding low cost optimal watermark during high level synthesis for reusable IP core protection,' *Proc. ISCAS*, Montreal, QC, pp. 974–977.

A. Sengupta, E. R. Kumar and N. P. Chandra (2019), 'Embedding digital signature using encrypted-hashing for protection of DSP cores in CE,' *IEEE Trans. Consum. Electron.*, vol. 65(3), pp. 398–407.

A. Sengupta and S. P. Mohanty (2019), 'IP core and integrated circuit protection using robust watermarking,' *IP Core Protection and Hardware-Assisted Security for Consumer Electronics*, e-ISBN: 9781785618000, pp. 123–170.

A. Sengupta and M. Rathor (2020a), 'Securing hardware accelerators for CE systems using biometric fingerprinting,' *IEEE Trans. Very Large Scale Integr. VLSI Syst.*, Accepted, June 2020, doi: 10.1109/TVLSI.2020.2999514.

A. Sengupta and M. Rathor (2020b), 'Enhanced security of DSP circuits using multi-key based structural obfuscation and physical-level watermarking for consumer electronics systems,' *IEEE Trans. Consum. Electron.*, doi: 10.1109/TCE.2020.2972808.

A. Sengupta (2017), 'Hardware security of CE devices [hardware matters],' *IEEE Consum. Electron. Mag.*, vol. 6(1), pp. 130–133.

A. Sengupta (2016), 'Intellectual property cores: protection designs for CE products,' *IEEE Consum. Electron. Mag.*, vol. 5(1), pp. 83–88.

A. Sengupta and M. Rathor (2019), 'Crypto-based dual-phase hardware steganography for securing IP cores,' *Lett. IEEE Comput. Soc.*, vol. 2(4), pp. 32–35.

A. Sengupta and M. Rathor (2020c), 'Structural obfuscation and crypto-steganography-based secured JPEG compression hardware for medical imaging systems,' *IEEE Access*, vol. 8, pp. 6543–6565.

A. Sengupta, D. Roy, S. P. Mohanty and P. Corcoran (2018), 'Low-cost obfuscated JPEG CODEC IP core for secure CE hardware,' *IEEE Trans. Consum. Electron.*, vol. 64(3), pp. 365–374.

A. Sengupta and D. Roy (2017), 'Antipiracy-aware IP chipset design for CE devices: a robust watermarking approach,' *IEEE Consum. Electron. Mag.*, vol. 6(2), pp. 118–124.

A. Sengupta (2020), 'Frontiers in securing IP cores – forensic detective control and obfuscation techniques', The Institute of Engineering and Technology (IET), ISBN-10: 1-83953-031-6, ISBN-13: 978-1-83953-031-9.

F. Zhao and X. Tang (2007), 'Preprocessing and postprocessing for skeleton-based fingerprint minutiae extraction,' *Pattern Recognit.*, vol. 40(4), pp. 1270–1281.

Chapter 7

Key-triggered hash-chaining-based encoded hardware steganography for securing DSP hardware accelerators

Anirban Sengupta[1]

This chapter describes a multi-encoding-driven key-triggered hash-chaining-based hardware steganography approach for securing digital signal processing (DSP) hardware accelerators which uses multiple layers of encoding and key-based parallel switch blocks (SBs) to drive multiple secure hash-chaining blocks in the algorithm. The presented approach is highly robust against fraud ownership claim and piracy threats.

The chapter is organized as follows: Section 7.1 provides some introduction to the research problem, followed by some discussion on other selected approaches in Section 7.2; Section 7.3 describes the presented hash-chaining-based hardware steganography approach. Section 7.4 presents the design process of securing finite impulse response (FIR) filter using this hardware steganography process; Section 7.5 presents the KHC-stego tool of this corresponding key-triggered hash-chaining hardware steganography approach; Section 7.6 discusses the analysis on case studies; Section 7.7 concludes the chapter, while Section 7.8 presents some exercise for readers.

7.1 Introduction

The rage of development is increasing the demand of electronics systems in the market. To cope up with the increasing demand, the design of reusable intellectual property (IP) cores is automated through design automation tools to reduce the design time and effort (Sengupta *et al.,* 2019a). Moreover, the design of highly complex IP cores such as DSP and multimedia makes the role of automated high-level synthesis (HLS) process more significant (Schneiderman, 2010; Sengupta *et al.,* 2010; Plaza and Markov; 2015). The HLS process accepts the algorithmic description of an application and generates its equivalent register transfer level or Verilog/VHDL description (which is referred to as soft IP

[1]Computer Science and Engineering, Indian Institute of Technology Indore, Indore, India

core) (Chakraborty and Bhunia, 2009). Further, the journey of an electronics design from its algorithmic description to an end integrated circuit (IC) involves various design houses forming a design chain. More explicitly, an IP core design process is outsourced to a separate design house in the supply chain for SoC integration later. The various design houses involved in the design supply chain are situated globally and may not be trustworthy (Yasin *et al.,* 2016). This is because an attacker in the untrustworthy design house may pirate the IP core and sell the counterfeited/cloned IPs/ICs in the market to earn illegal revenue. Moreover, a dishonest IP buyer may fraudulently claim the IP ownership, deceiving the genuine owner. This renders an IP core design vulnerable against 'fraudulent claim of ownership' and piracy threats (Sengupta, 2016, 2017; Zhang, 2016; Sengupta *et al.,* 2019b).

To nullify the false ownership claim and to enable the detection of counterfeiting/cloning, designer's secret-mark can be embedded into an IP core design (Newbould, 2002; Colombier and Bossuet, 2015; Sengupta *et al.,* 2018, 2019c; Sengupta and Mohanty, 2019). However, the embedded secret constraints (secret-mark) are also required to be highly secured so that an attacker would not be able to nullify the purpose of embedding secret-mark by extracting it (Sengupta, 2020). In order to enhance security of DSP cores, Sengupta and Rathor (2020) proposed an *encoding and key-driven hash-chaining-based hardware steganography* approach that produces a highly robust stego-mark which is arduous to be extracted or regenerated by an adversary. Hence, the adversary fails to evade counterfeit detection process by embedding stego-mark of genuine author post regeneration or extraction of it. The encoding and hash-chaining-driven hardware steganography approach (Sengupta and Rathor, 2020) ensures robustness of the secret-mark by generating it through a large-size secret-key, designer-chosen encoding rules, encoded bitstreams and the number of iterations of round function in each hash unit (HU) of the chaining process.

7.2 Discussion on selected approaches

In the field of protection of reusable IP cores for DSP hardware accelerators, there are two categories of approaches available in the literature, first, hardware (IP) watermarking approaches (Koushanfar *et al.,* 2005; Le Gal and Bossuet, 2012; Sengupta and Bhadauria, 2016; Sengupta and Roy, 2017) which are signature-based mechanisms and second, hardware (IP) steganography approaches (Sengupta and Rathor, 2019a, 2019b) which are signature-free techniques. For example, in hardware watermarking approaches (Sengupta and Bhadauria, 2016; Sengupta and Roy, 2017), multivariable signature encoding algorithm was employed to generate the watermarking constraints for implanting into HLS phase of DSP hardware accelerator design, while Koushanfar *et al.* (2005) employed binary digit combination to form vendor signature, followed by implanting watermarking constraints during HLS. Further, Le Gal and Bossuet (2012) use mathematical relationships between numeric values as inputs and outputs at specified times to generate watermarking constraints. On the other hand, hardware steganography approaches (Sengupta and Rathor, 2019a, 2019b) derive secret relationships from cover design data (coloured interval graph (CIG) of architectural synthesis phase) of DSP hardware design process and construct encoding algorithms to form secret stego-constraints for implanting, without requiring IP vendor signature at all. The threat model for both the categories of techniques is the same and intends to primarily prove fraudulent claim of ownership as well as detect pirated IP cores such as counterfeited/cloned ICs/IPs in the design chain.

However, it is well established that in the context of DSP hardware accelerators, steganography provides better designer control in terms of secretly implanted design constraints and incurs lesser design overhead than hardware watermarking approaches (Sengupta, 2020). This is vastly due to the inherent nature of the IP steganography algorithms which depends neither on the strength of the vendor signature chosen nor its combination. Instead, it is dependent on the stego-constraints converter algorithm and its equivalent secret encoding algorithm. Nevertheless, in both existing IP steganography (Sengupta and Rathor, 2019a, 2019b) and watermarking approaches (Koushanfar *et al.,* 2005; Le Gal and Bossuet, 2012; Sengupta and Bhadauria, 2016; Sengupta and Roy, 2017), proof of ownership could still be relatively easier from an attacker's perspective due to the non-existence of robust secret keys and multilayered encoding algorithms. On the contrary, in the switch-based encoding-driven hardware steganography approach (Sengupta and Rathor, 2020), the secret stego-constraints are generated through interweaving of hash-chaining process, switch-based HUs, robust stego keys, multiple encoding algorithms and steganography embedder algorithm. The complexity of encoding and key-driven hash-chaining-based steganography (Sengupta and Rathor, 2020) algorithm makes it infeasible for an attacker to generate/extract the implanted secret stego-constraints to prove IP ownership. The switch-based encoding-driven steganography (Sengupta and Rathor, 2020) provides stronger security than hardware watermarking approaches in terms of lower probability of coincidence (reflecting more digital evidence), greater designer control (due to no dependency on signature) and lower design overhead.

7.3 Encoding and key-driven hash-chaining-based hardware steganography methodology

7.3.1 Threat model

A vendor's IP core is susceptible to fraudulent claim of ownership by a dishonest buyer. The encoding and key-driven hash-chaining-based hardware steganography approach (Sengupta and Rathor, 2020) handles the 'fraudulent claim of ownership' threat for IP cores by embedding vendor's secret stego-mark (secret steganography information) into the design. The embedded robust stego-mark also enables the detection of counterfeited/cloned IPs/ICs.

7.3.2 High-level description

The encoding and key-driven hash-chaining-based steganography methodology (Sengupta and Rathor, 2020) generates secret stego-constraints (steganography information) through key-triggered hash-chaining steganography process. The generated stego-constraints are embedded during register and functional unit (FU) allocation phase of the HLS process. This results into a stego-embedded DSP hardware design which is secured against the fraudulent claim of ownership and counterfeiting/cloning threats.

The overview of the switch-based encoding-driven steganography approach (Sengupta and Rathor, 2020) is shown in Figure 7.1. As shown in the figure, the following are the inputs to the approach: (i) DSP application in the form of data-flow graph (DFG), (ii) resource constraints and (iii) module library. These inputs are fed to the HLS block, output of which is a scheduled DFG of the DSP application. The role of this block is to schedule the DFG based on designer-specified resource constraints. Further, the scheduled DFG is fed as input to the parallel encoding block. The output of this block is a number of bitstream representations of the respective DSP application. The role of this block is to encode the scheduled DFG of DSP application into a number of unique bitstreams based on designer-chosen encoding rules. Thereafter, all encoded bitstreams are fed to the stego-constraints generator block, output of which is the stego-constraints to be implanted into the DSP core, as secret steganography information. The role of this block is to generate stego-constraints through two sub-blocks, parallel SB and hash-chaining block as shown in Figure 7.1. Another input to the stego-constraints generator is the output of stego-key block which provides keys required to generate the stego-constraints. These stego-keys are fed to the parallel SB along with the encoded bitstreams generated from the parallel encoding block. The role of each switch in the parallel SB is to select an encoded bitstream (out of all available) based on specific stego-key value. The output of parallel SB is the selected bitstreams (based on vendor stego-key values) which are fed to the hash-chaining block as input. All the encoded bitstreams generated from parallel encoding block are also directly fed to the hash-chaining block. The role of hash-chaining block is to generate stego-constraints through a chain of HUs employed inside. The final output of hash-chaining block is the stego-constraints which are fed to the stego-embedder block as input. The stego-embedder block embeds the stego-constraints during the register and FU allocation phase of HLS. The output of stego-embedder block is the steganography-embedded DSP hardware accelerator.

7.3.3 In-depth description of key-triggered hash-chaining-based hardware steganography

This section discusses the key-triggered hash-chaining steganography approach (Sengupta and Rathor, 2020) in details. Figure 7.2 depicts the details of generating stego-embedded DSP core through key-triggered hash-chaining-based steganography. As shown in the figure, the entire process is divided into the following blocks.

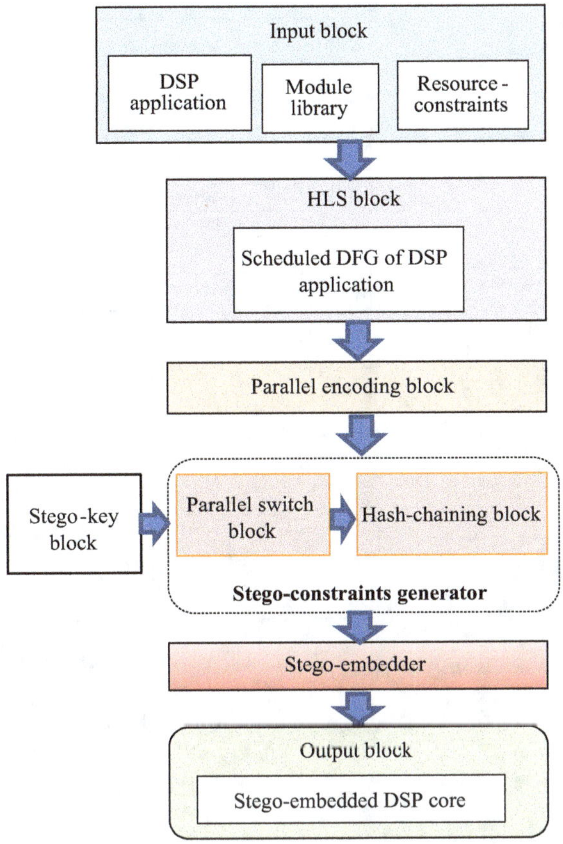

Figure 7.1 Overview of key-triggered hash-chaining-based steganography (Sengupta and Rathor, 2020)

7.3.3.1 Input block

This block contains the inputs fed to the key-triggered hash-chaining-based steganography approach as shown in Figure 7.2. The DSP application to be secured can be fed as input in the following forms: C/C++ code, transfer function or intermediate representation such as DFG. Here, DFG representation of DSP application is used as input to the key-triggered hash-chaining steganography methodology (Sengupta and Rathor, 2020). Besides DFG, module library and resource constraints are also fed as inputs to the HLS block.

7.3.3.2 HLS block

This block performs HLS process on the DFG of the DSP application. First, the operations present in the DFG are scheduled using a scheduling algorithm into control steps (Q), based on designer-specified resource/FU constraints. This results

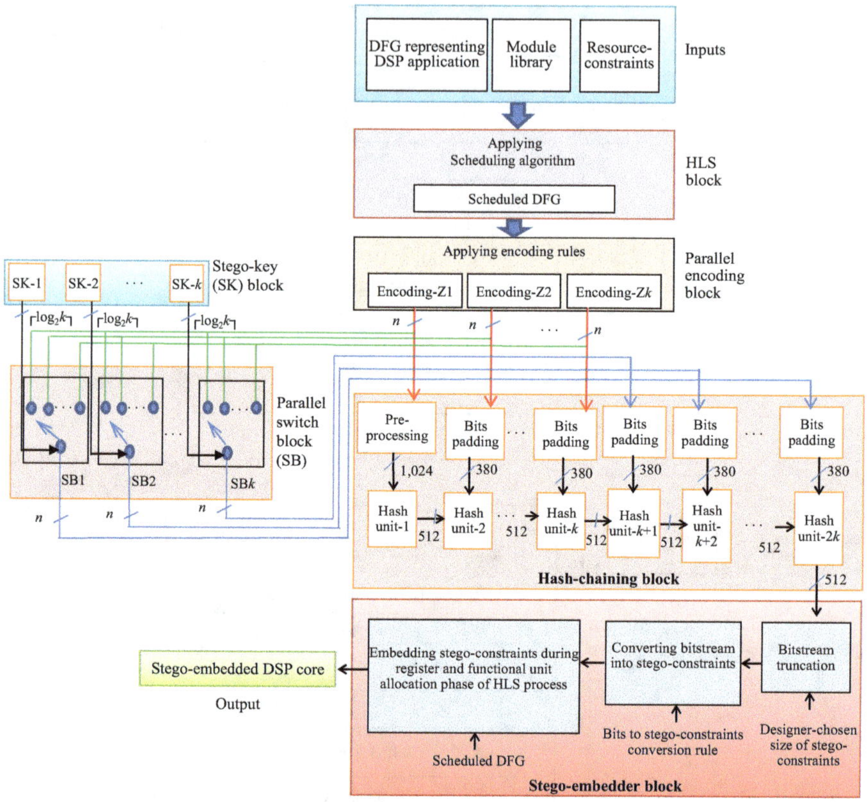

Figure 7.2 Details of securing DSP cores using key-triggered hash-chaining-based steganography (Sengupta and Rathor, 2020)

into a scheduled DFG of the DSP application, which is fed as input to the next block.

7.3.3.3 Parallel encoding block

This block encodes the scheduled DFG of the DSP application into manifold bitstream representations by applying encoding rules (Sengupta and Rathor, 2020). Each encoding rule is capable of encoding the DSP application into a unique bitstream. Length of each bitstream is equal to the number of operations in the DSP application. For a DSP application having n operations, total possible encoded bitstreams are 2^n and the length of each bitstream is n. Out of 2^n, a designer can generate k number of encoded bitstreams (for implementation) by applying k-encoding rules (where $1 \leq k \leq 2^n$) as shown in Figure 7.2. These k encoded bitstreams are fed as input to the parallel SB (these inputs are represented by green lines in Figure 7.2) and hash-chaining block (these inputs are represented by red

Table 7.1 Encoding rules to encode a DSP application into bitstreams

Encoding	Encoding rules (operation (O), control step (Q))	Encoded bit
Z1	If O# and corresponding Q# are both even	0
	Otherwise	1
Z2	If O# and corresponding Q# are of the same parity (both even or both odd parity)	0
	If O# and corresponding Q# are of different parity	1
Z3	If O# and corresponding Q# are both odd	0
	Otherwise	1
Z4	If O# and corresponding Q# are of different parity	0
	If O# and corresponding Q# are of the same parity	1
Z5	If O# and corresponding Q# are both prime	0
	Otherwise	1
Z6	If O# and corresponding Q# are both prime	1
	Otherwise	0
Z7	If GCD of O# and corresponding Q# is 1	0
	If GCD of O# and corresponding Q# is not 1	1
Z8	If (O#) mod (corresponding Q#) is 0	0
	If (O#) mod (corresponding Q#) is not 0	1
Z9	If Q# is equal to second odd sequence of O#	0
	Otherwise	1

lines in the same figure). The nine (k=9) encoding rules (shown in Table 7.1) to generate nine encoded bitstreams of the DSP application have been proposed by Sengupta and Rathor (2020). Because of total 2^n possible bitstream representations of the DSP application, an attacker is required to put extremely high effort to find out the actual encoded bitstreams used in generating the secret stego-constraints (or stego-mark).

Further, the encoding rules applied by the designer also remain unknown to the attacker. This renders almost infeasible for an attacker to find the encoded bitstreams used in subsequent blocks for generating stego-constraints.

7.3.3.4 Parallel switch block

For k number of encodings chosen by the designer, total k-parallel switches are used inside the parallel SB as shown in Figure 7.2. Each SB selects one out of k encoded bitstreams based on the stego-key value. Each switch operates on a distinct stego-key value; hence, total k stego-keys are required. Each switch can be implemented using a multiplexer of size k:1, where a stego-key acts as the select lines/bits and the size of select line/stego-key is $\lceil \log_2 k \rceil$ bits. The selected k-bitstreams through all k-switches are fed to the hash-chaining block (these inputs are represented by blue lines in Figure 7.2).

7.3.3.5 Stego-key block

This block provides stego-keys to operate the k-switches in the parallel SB. As discussed in the previous step, the size of each stego-key used to operate single switch is $\lceil \log_2 k \rceil$ bits. Since there are total k-parallel switches, the total size of stego-key becomes $k \times \lceil \log_2 k \rceil$ bits. Since the maximum possible encoded bit-streams are $k=2^n$ (as discussed earlier in parallel encoding block), a parallel SB can have maximum 2^n switches and the size of each stego-key becomes $\lceil \log_2 2^n \rceil = n$. Hence, the maximum possible key size (Max key size) is given by the following equation:

$$\begin{aligned} &\text{Max key size} = (\text{Max number of switches}) \times (\text{size of each stego key}) \\ &\text{Max key size} = 2^n \times \lceil \log_2 2^n \rceil = n \times 2^n \text{bits} \end{aligned} \tag{7.1}$$

Because of very large size stego-key, it is very hard for an attacker to find out the correct key used to generate the stego-constraints (or stego-mark). This strengthens the security achieved through key-triggered hash-chaining-based steganography approach.

7.3.3.6 Hash-chaining block

This block is responsible for generating the stego-constraints in the form of hashed bitstream. A chain of total $2k$ HUs is employed inside the hash-chaining block, where k is the total number of encodings chosen by the designer. Each HU is fed with the input of 1,024 bit and generates output of 512 bits. Each HU employs 'round function computation (RFC)' process (Sengupta and Rathor, 2019c) to generate 512-bit hash output. The number of rounds the RFC process is run for each HU is specified by the designer. The number of rounds varies from 1 to 80. Because of designer-specified number of rounds, the hardness of finding the hash output and consequently stego-constraints enhances from an attacker's perspective.

Further, in the key-triggered hash-chaining steganography approach (Sengupta and Rathor, 2020), two different processes are used to feed input to the HUs:

1. *Pre-processing*: This process is used to feed 1,024-bit input to the HU-1 (HU_1) only. The rule of constructing 1,024 bits through pre-processing is given by the following equation:

$$\begin{aligned} &(1{,}024 \text{ bits})_{HU_1} = \\ &(n \text{ bits of encoded bitstream obtained from encoding Z1}) \\ &\&'1'\&(\text{sequence of}(896 - n - 1)\text{zeros}) \\ &\&(128 \text{ bit representation of the length } n) \end{aligned} \tag{7.2}$$

2. *Bit padding*: This is the process to form 380-bit input to the remaining HUs (HU_2–HU_{2k}). To form the 380 bits, the designer-chosen combination of $(380-n)$ bits is padded before the n bits of the encoded bitstream. Thus, the size of encoded bitstreams (corresponding to encoding Z2–Zk) fed to the HUs (HU_2–HU_{2k}) becomes 380 bits which are only known to the designer and unknown to an attacker.

Further, the total 1,024-bit input to the remaining HUs (HU_2–HU_{2k}) is formed using the following equation:

$$\begin{aligned}&(1,024\ \text{bits})_{\text{HU}_{2-2k}} = \\&(512\ \text{bit output of previous hash unit})\\&\&\text{`1000'}\&(380\ \text{bit output of bit padding})\\&\&(128\ \text{bit representation of the length of previous hash output})\end{aligned} \tag{7.3}$$

The $2k$ HUs employed in the hash-chaining block are categorized into two types, regular HUs and key-triggered HUs.

1. *Regular HUs (HU_1–HU_k)*: The first k HUs in the hash-chaining block are the regular HUs. These HUs (HU_1–HU_k) exploit encodings Z1–Zk, respectively, to generate hash output. The 1,024-bit input for HU_1 is formed using (7.2) and 1,024-bit input for HU_2–HU_k is formed using (7.3).
2. *Key-triggered HUs (HU_{k+1}–HU_{2k})*: The last k HUs in the hash-chaining block are the key-triggered HUs. This is because a particular encoding (out of k-encodings) used by a key-triggered HU is selected on the basis of the specific value of stego-key. The 1,024-bit input for HU_{k+1}–HU_{2k} is formed using (7.3).

It is important to note here that the total HUs ($2k$) employed in the hash-chaining block are two times of the number of encodings (k) chosen by the designer, whereas total key-triggered hash-blocks (k) are equal to the number of encodings (k) chosen by the designer. The 512-bit hash output of the hash-chaining block is fed as input to the next block.

7.3.3.7 Stego-embedder block

This block is responsible for embedding stego-constraints into the design during HLS process. The entire process is explained using the following three subprocesses:

1. *Bitstream truncation*: This process truncates the 512-bit hash bitstream (obtained from hash-chaining block) to the designer-specified size of stego-constraints.
2. *Conversion into stego-constraints*: All ‘0’ and ‘1’ bits in the truncated bitstream are converted into corresponding stego-constraints by applying conversion rules as shown in Table 7.2. Thus, obtained stego-constraints are fed to the ‘embedding stego-constraints’ process.
3. *Embedding stego-constraints*: Inputs to this process are scheduled DFG and stego-constraints to be embedded. This process embeds stego-constraints into the DSP core design during register allocation and FU allocation phase of HLS process. The stego-constraints corresponding to bit ‘0’ are embedded during register allocation, and those corresponding to bit ‘1’ are embedded during FU allocation. To embed stego-constraints corresponding to bit ‘0’ during register allocation phase, a CIG framework is leveraged. A CIG represents graphically the assignments of all storage variables of the design to the minimum possible registers, where each register is denoted using a distinct colour and storage

Table 7.2 Rules to convert a hash bitstream into corresponding stego-constraints

Bit	Conversion rule
'0'	An edge is embedded between node pair ⟨even, even⟩ in to CIG
'1'	If operation (O)# is odd, then it is allocated to FU of vendor type-1 (V1) and if O# is even, then it is allocated to FU of vendor type-2 (V2)

variables are denoted using nodes in the CIG (Sengupta and Bhadauria, 2016). An edge between two nodes in the CIG indicates that the respective storage variables/nodes are essentially executed through two distinct registers/colours. The storage variables of the design can be extracted out from the scheduled DFG, where all primary and intermediate inputs/outputs of the design are assigned to the variables. The stego-constraints corresponding to bit '0' are embedded as extra edges (constraint edges) in the CIG complying with the condition that two nodes connected through an edge should be of different colours. To satisfy this condition, alteration of node colours in the CIG may be required to perform. The added constraint edges into the CIG are reflected in the scheduled DFG through the reallocation of storage variables to the registers/colours.

Further, the stego-constraints corresponding to bit '1' are embedded by reallocating the operations in scheduled DFG to a particular vendor type, which is determined on the basis of the rule mentioned in Table 7.2. Thus, each embedded stego-constraint corresponding to bit '1' indicates the allocation of an operation to a fixed vendor type.

The embedding stego-constraints (stego-mark) result into a stego-embedded DSP core. By detecting the embedded steganography during forensic detection, false claim of ownership of the IP core can be nullified and counterfeited/cloned IPs/ICs can be detected.

7.3.4 Detection of steganography

To nullify the fraudulent claim of ownership, the stego-constraints (or secret steganography information) are detected in the design under test. To do so, the RTL data path of the design is inspected, and the information about the inputs of multiplexers associated with each register (which tells about the assignment of storage variables to the registers) and vendor-type allocation of each operation is collected. This information is matched with the designer's stego-constraints generated through the key-triggered hash-chaining steganography approach (Sengupta and Rathor, 2020). If both matches, then the presence of stego-constraints of the genuine owner is verified in the design; thus, ownership is awarded to the genuine owner. Figure 7.3 depicts the steganography detection process in detail.

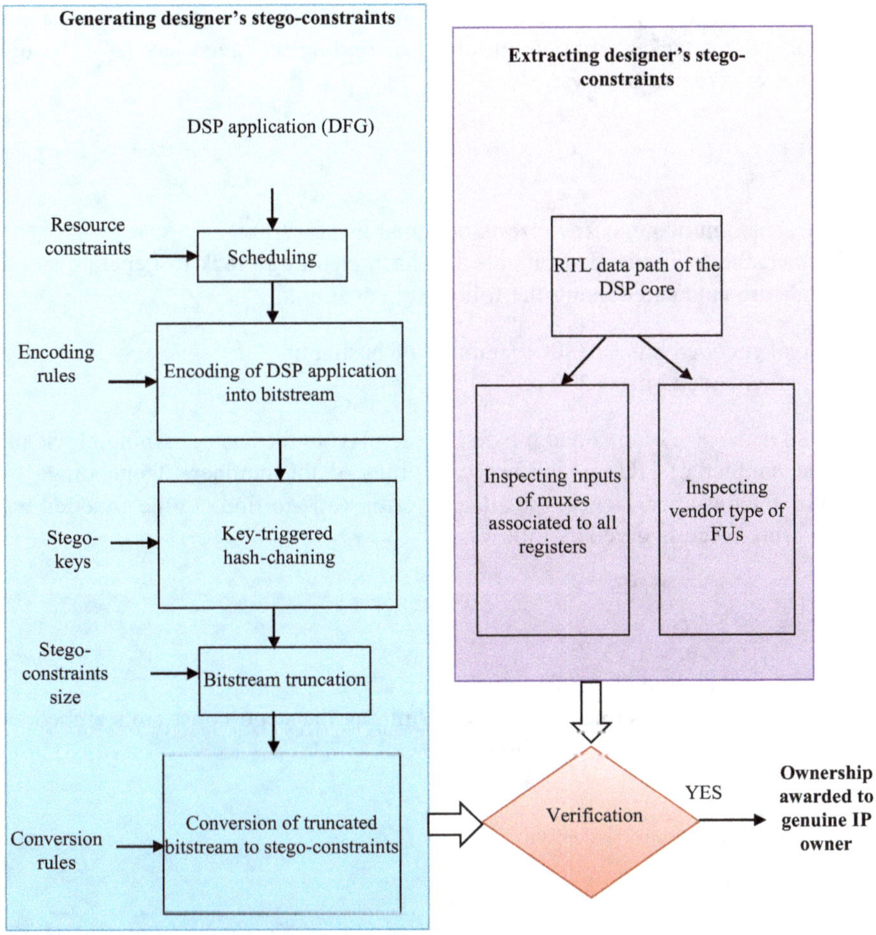

Figure 7.3 Detection process of steganography (Sengupta and Rathor, 2020)

7.3.5 Security from an attacker's perspective

A dishonest IP buyer/user (attacker) can claim the ownership of the design, deceiving the genuine owner. If secret information (stego-mark) of the genuine owner is not embedded into the design, then the attacker can easily realize his/her malicious intent of claiming IP ownership fraudulently. The embedded stego-mark thwarts an attacker to do so. However, an attacker may try to claim stego-mark (stego-constraints) as his/her own, defeating the purpose of steganography. In order to do so, the attacker is required to find the secret stego-constraints embedded into the design. The key-triggered hash-chaining-based steganography enhances the attacker's effort of finding stego-constraints manifold. The attacker needs to find out the valid stego-key and the encoded bitstreams used to generate the stego-constraints. The attacker's maximum effort of finding the stego-key (e_{sk}^{m}) through brute-force is given as follows:

$$\begin{aligned} e_{sk}^{m} &= 2^{\text{Maxkey size}} \\ e_{sk}^{m} &= 2^{n\times 2^{n}} \end{aligned} \tag{7.4}$$

where the maximum stego-key size is obtained using (7.1).

Further, the total encoded bits used in hash-chaining block to generate stego-constraints are calculated using the following equation:

$$\begin{aligned} \text{Total encoded bits} &= 380 \times (\text{number of hash units}) \\ \text{Total encoded bits} &= 380 \times (2k) \end{aligned} \tag{7.5}$$

where 380 is the size of encoded bits fed to each HU in the hash-chaining block and $2k$ is the number of HUs which are two times of the number of encodings (k) chosen by the designer. Hence, the attacker effort (e_{H}^{eb}) to find out the encoded bits through brute-force is given as follows:

$$\begin{aligned} e_{H}^{eb} &= 2^{\text{total encoded bits}} \\ e_{H}^{eb} &= 2^{380\times(2k)} \end{aligned} \tag{7.6}$$

where total encoded bits are calculated using (7.5).

Hence, an attacker's total effort in determining the stego-constraints embedded into the design is quantified as follows:

$$\begin{aligned} \text{Total effort} &= e_{sk}^{m} \times e_{H}^{eb} \\ \text{Total effort} &= 2^{n\times 2^{n}} \times 2^{380\times(2k)} \\ \text{Total effort} &= 2^{(n\times 2^{n}+(380\times 2k))} \end{aligned} \tag{7.7}$$

It is evident from (7.7) that an attacker requires a huge effort in order to find out the stego-constraints embedded into the design. This renders the generated stego-mark using key-triggered hash-chaining steganography approach (Sengupta and Rathor, 2020) highly robust and secure.

7.4 Design process of securing FIR filter using encoding and key-driven hash-chaining steganography

Digital FIR filters are widely used in electronics applications such as digital communication for de-noising. Here, an FIR filter core is chosen to illustrate the hash-chaining-based steganography approach (Sengupta and Rathor, 2020). The FIR filter application in the form of DFG is fed as input along with module library and resource (FU) constraints of four multipliers (*M*) and four adders (*A*). The steps of generating a stego-embedded FIR filter core using key-triggered hash-chaining-based steganography are as follows:

1. *Generating scheduled DFG of FIR application*: The DFG of FIR application is scheduled on the basis of designer-chosen resource (FU) constraints of four M and four A. The scheduled DFG is shown in Figure 7.4. As shown in the figure, 23 operations of the FIR applications are executed through 9 control steps (Q). Operations have been allocated to FUs from two different vendors (V1 and V2). Note: for an FU instance (M_j^i or A_j^i), instance number and vendor type are denoted by the subscripts j and i, respectively. Since chosen constraints of multiplier are four, two instances of multiplier of vendor V1 and another two instances of multiplier of vendor V2 have been used to execute multiplication operations in a control step. Similarly, chosen constraints of adder are four; therefore, two instances of adder of vendor V1 and another two instances of adder of vendor V2 have been used to execute addition operations in a control step. Further, the primary and intermediate inputs and outputs of the design are stored using total 31 storage variables ($S0$–$S30$). The corresponding CIG of the FIR application is shown in Figure 7.5, where nodes $S0$–$S30$ indicate the storage variables of the design. Table 7.3 captures the allocation of storages variables to registers (colours) into different control steps.
2. *Generating encoded bitstreams*: By applying nine (k=9) encoding rules (Sengupta and Rathor, 2020), the scheduled DFG of FIR application is encoded into nine unique bitstreams as shown in Table 7.4. As there are n=23 operations in the FIR application, the length of each bitstream is 23.
3. *Pre-processing and bit padding of encoded bitstreams*: The first bitstream generated using encoding rule Z1 is preprocessed to form 1,024 bits using (7.2). The remaining eight bitstreams (each of size '23') generated from encoding rules Z2–Z9 are converted to 380 bits by padding designer-chosen combination of 357 bits before them. Thus, obtained eight bitstreams of 380-bit each are converted to 1,024 bits using (7.3). Now, all the nine bitstreams are of length 1,024 bits and ready to be fed to the hash-chaining block.
4. *Feeding bitstreams to hash-chaining block*: The first bitstream of 1,024 bit (corresponding to encoding-Z1) is fed to the HU-1 (HU_1). The remaining eight bitstreams (corresponding to encoding-Z2–Z9) are fed to the regular and key-triggered HUs. Please note that as 9 encodings are chosen by the designer, total 9×2=18 HUs are employed in the hash-chaining block, where HU_1–HU_9 are the regular HUs and HU_{10}–HU_{18} are the key-triggered HUs. All the nine bitstreams are directly fed to the nine regular HUs (HU_1–HU_9), respectively. However, the selection of a specific bitstream (out of nine) fed to a particular key-triggered HU is based on a stego-key value of size $\lceil \log_2 9 \rceil = 4$ bits. Overall the all stego-keys used (in this demonstration) to select the bitstreams fed to the all nine key-triggered HUs are as follows: '1000-0111-0110-0101-0100-0011-0010-0001-0000'.
5. *Generating final hash bitstream of 512-bit long*: Based on the designer-chosen stego-key values and the 1,024-bit inputs fed to the HUs, the hash-chaining block generates 512-bit-long hash bitstream which is as follows:

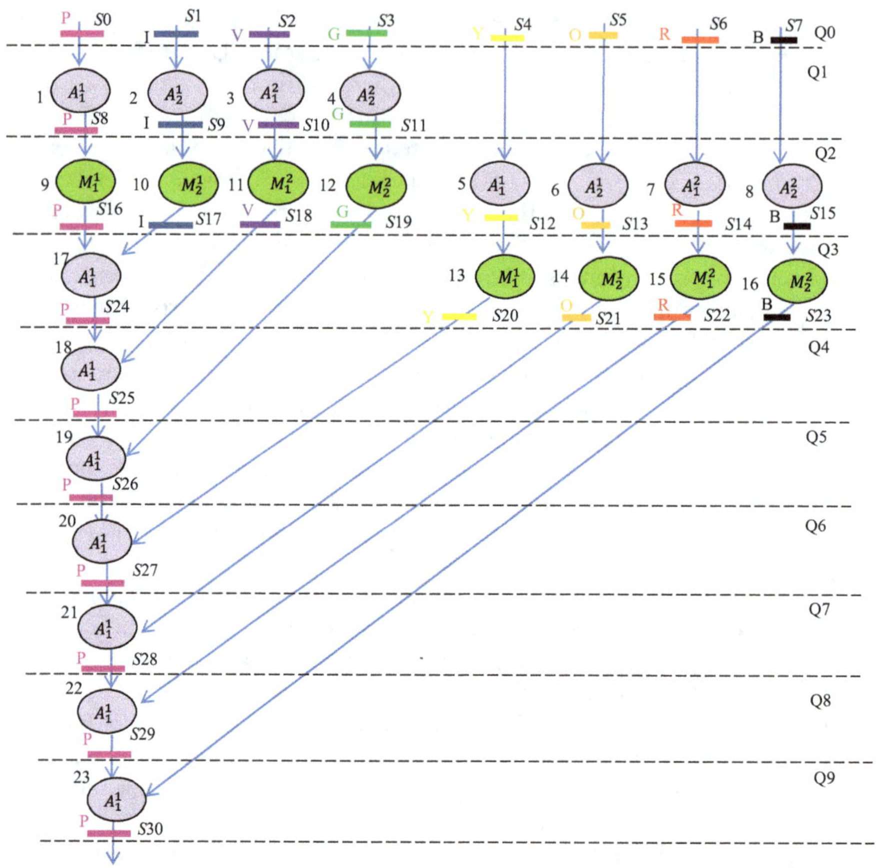

Figure 7.4 Scheduled and resource-allocated DFG of FIR filter application

'01000011001000100001000000001001011001011101010101100101010
1001001011001110101001101011110001100100110011010011110000
0100000101011011111101001011001011110111001110111110101010100
0101101000011000101001100100010101000011010111101001101000
0101111101111101101100101001101101101111110001110110011110
1111001011101001110001100111101011100101010101101001010000
1100000011111001001101011010010010101001001010100001110101
1111110000001000000000101100110000011011111111011100010110
10011101001001111100111010000101000110000'

6. *Hash-bitstream truncation*: The hash bitstream is truncated to the designer-chosen size of stego-constraints. For the stego-constraints size (W) of 26, the truncated bitstream is as follows: '01000011001000100001000000'.

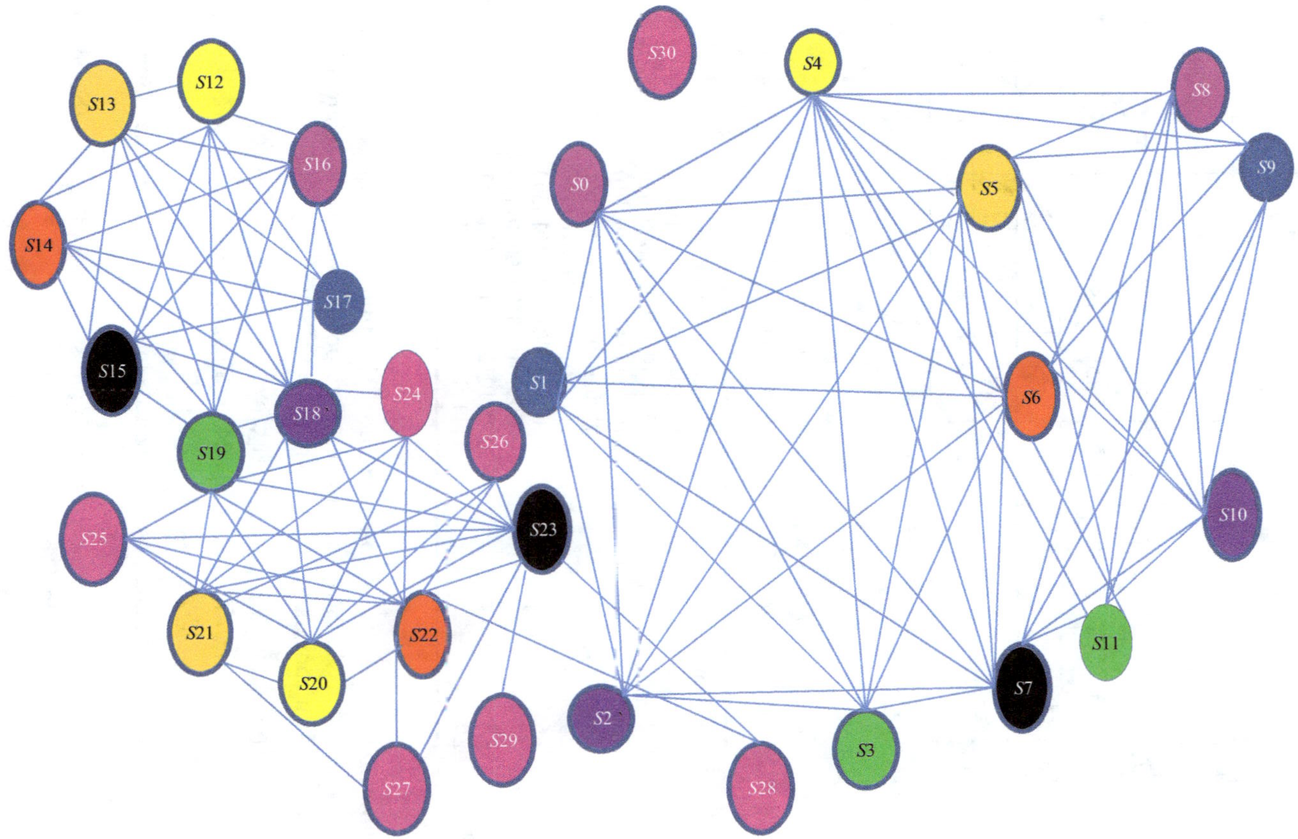

Figure 7.5 CIG of FIR filter hardware accelerator (IP core) before steganography

Table 7.3 Register allocation of FIR before implanting steganography

Q	Q0	Q1	Q2	Q3	Q4	Q5	Q6	Q7	Q8	Q9
P	S0	S8	S16	S24	S25	S26	S27	S28	S29	S30
I	S1	S9	S17	–	–	–	–	–	–	–
V	S2	S10	S18	S18	–	–	–	–	–	–
G	S3	S11	S19	S19	S19	–	–	–	–	–
Y	S4	S4	S12	S20	S20	S20	–	–	–	–
O	S5	S5	S13	S21	S21	S21	S21	–	–	–
R	S6	S6	S14	S22	S22	S22	S22	S22	–	–
B	S7	S7	S15	S23	S23	S23	S23	S23	S23	–

Table 7.4 Encoded bitstreams corresponding to FIR filter core using nine encoding rules

Encodings	Encoded bitstream
Z1	1111101010101111101010 1
Z2	0101101010100101000000 0
Z3	0101111111110101010101 0
Z4	1010010101011010111111 1
Z5	1111010111010111010111 1
Z6	0000101000101000101000 0
Z7	0000010101010010010111 0
Z8	0000101010101101111101 1
Z9	1111111111111111111111 1

7. *Conversion of bits to stego-constraints*: The truncated bitstream contains twenty 0s and six 1s. Stego-constraints corresponding to twenty 0s are as follows (based on conversion rule shown in Table 7.2):

$\langle S0,S2\rangle$, $\langle S0,S4\rangle$, $\langle S0,S6\rangle$, $\langle S0,S8\rangle$, $\langle S0,S10\rangle$, $\langle S0,S12\rangle$, $\langle S0,S14\rangle$, $\langle S0,S16\rangle$, $\langle S0,S18\rangle$, $\langle S0,S20\rangle$, $\langle S0,S22\rangle$, $\langle S0,S24\rangle$, $\langle S0,S26\rangle$, $\langle S0,S28\rangle$, $\langle S0,S30\rangle$, $\langle S2,S4\rangle$, $\langle S2,S6\rangle$, $\langle S2,S8\rangle$, $\langle S2,S10\rangle$, $\langle S2,S12\rangle$

These stego-constraints are embedded into the CIG in the form of additional edges.

Further, stego-constraints corresponding to six 1s are as follows (based on conversion rule shown in Table 7.2):

$$O1 \rightarrow V1, O2 \rightarrow V2, O3 \rightarrow V1, O4 \rightarrow V2, O5 \rightarrow V1, O6 \rightarrow V2$$

i.e. odd operations (O1, O3, O5) are assigned to vendor type-1 (V1) and even operations (O2, O4, O6) are assigned to the vendor type-2 (V2).

8. *Embedding stego-constraints during HLS process*: The constraint edges (stego-constraints corresponding to bit '0') are embedded into the CIG one-by-one. During the embedding of an edge as a stego-constraint, the condition that the respective two nodes should be of different colour must be taken care of. If it is not so, then the colour of one of the node is modified in order to accommodate that edge. For example, before embedding the constraint edge $\langle S0,S8\rangle$, both nodes ($S0$ and $S8$) are of the same colour as shown in Figure 7.5. However, embedding the edge $\langle S0,S8\rangle$ results into change in the colour of node S8 from pink (P) to indigo (I) as shown in Figure 7.6. The same figure shows the CIG after embedding all constraint edges. The embedded constraint edges have been drawn as red dotted lines in Figure 7.6. As shown in the figure, the colour of storage variables/ nodes $S8$, $S9$, $S10$, $S11$, $S16$, $S17$, $S24$, $S26$, $S28$ and $S30$ is modified to embed all the constraint edges. Table 7.5 captures the register allocation of storage variables into different control steps post embedding stego-constraints. The storage variables subjected to register reallocation post embedding constraints have been marked grey in the table. Further, to embed the constraints corresponding to all six 1s, reallocation of operations (O1–O6) to the particular vendor type is performed on the basis of the rule shown in Table 7.2. The scheduled DFG of FIR application with embedded stego-constraints (corresponding to both 0s and 1s) is shown in Figure 7.7. As shown in the figure, the FUs (corresponding to O1–O6) highlighted in red colour indicate the embedded stego-constraints corresponding to bit '1'. Embedding stego-constraints produces a stego-embedded FIR filter core which is secure with designer's robust stego-mark.

7.5 Key-triggered hash-chaining-driven steganography tool for securing hardware accelerators

Authors have developed a ***KHC-stego tool*** (key-triggered hash-chaining-driven steganography tool) to simulate and analyse the functionality of key-triggered hash-chaining-based steganography approach (Sengupta and Rathor, 2020) for DSP hardware accelerators. *This tool provides a friendly graphical interface to users and available for free download publicly at*: http://www.anirban-sengupta.com/Hardware_Security_Tools.php.

A snapshot of the graphical user interface of the tool is shown in Figure 7.8. The left portion of the tool shows the panel for providing required inputs to the tool, whereas the right portion shows the panel with output buttons to see the intermediate and final outputs of the key-triggered hash-chaining-based steganography approach. The panel in the middle shows the status of the intermediate steps, viz. control data flow graph (CDFG) scheduling of DSP application, encoding selection, executing hash-chaining process and insert constraints, as shown in Figure 7.8. Initially, these status bars remain red. Upon applying the inputs, the respective status bar turns green. The KHC-stego tool accepts the DSP application

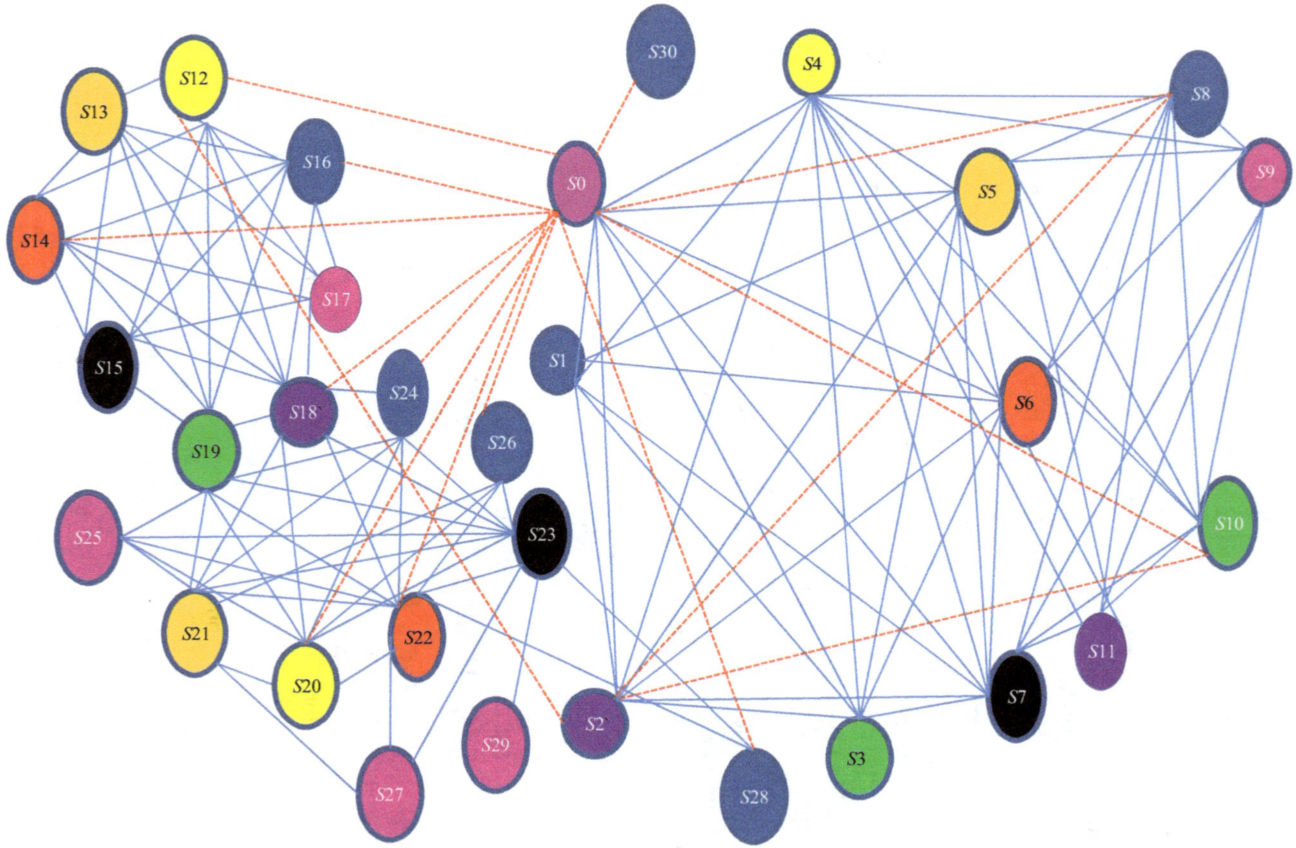

Figure 7.6 CIG of FIR filter hardware accelerator (IP core) after steganography

Table 7.5 Register allocation of FIR after implanting steganography

Q	Q0	Q1	Q2	Q3	Q4	Q5	Q6	Q7	Q8	Q9
P	S0	**S9**	**S17**	–	S25	–	S27	–	S29	–
I	S1	**S8**	**S16**	**S24**	–	**S26**	–	**S28**	–	**S30**
	S2	**S11**	S18	S18	–	–	–	–	–	–
G	S3	**S10**	S19	S19	S19	–	–	–	–	–
Y	S4	S4	S12	S20	S20	S20	–	–	–	–
O	S5	S5	S13	S21	S21	S21	S21	–	–	–
R	S6	S6	S14	S22	S22	S22	S22	S22	–	–
B	S7	S7	S15	S23	S23	S23	S23	S23	S23	–

input in the form of CDFG along with module library and resource constraints. The tool shows the output of intermediate steps of key-triggered hash-chaining-based steganography, finally generated stego-constraints, the security metric in terms of probability of coincidence, pre and post-steganography design cost. Further, it also shows scheduling and registers allocation pre and post embedding steganography constraints, onto the output window.

Let us generate all the intermediate and final outputs of key-triggered hash-chaining-based steganography approach for FIR filter core using the KHC-stego tool. We will provide the same inputs and stego-keys used during the demonstration of FIR filter core discussed in Section 7.4. Here, we can match the output generated with the tool and that obtained in the demonstration. First of all, inputs CDFG of FIR filter core, resource constraints of four adder and four multipliers and module library, are fed to the tool as shown in Figure 7.9. On clicking on the button 'Initial scheduled DFG' on output panel, the scheduled DFG becomes available onto the output window shown at the bottom of the tool. As the DFG gets scheduled, the respective status bar ('DFG scheduled' shown in Figure 7.9) turns green, whereas the remaining status bar are still red as shown in Figure 7.9. Further, the number of encodings is entered in the tool. The tool pops up a window which shows the definitions of nine different encodings, as shown in Figure 7.10. Further, the encoding numbers are entered in the tool as shown in Figure 7.11. Upon clicking on the output button 'Encoded Bitstream' at output panel, the nine different encoded bitstreams of scheduled DFG of FIR filter are available onto the output window as shown in the same figure. Further, stego-key value for each key-driven hash block and the number of rounds for all the 18 hash blocks in the hash-chain are entered in the tool as shown in Figure 7.12. The final hash output is also shown in the same

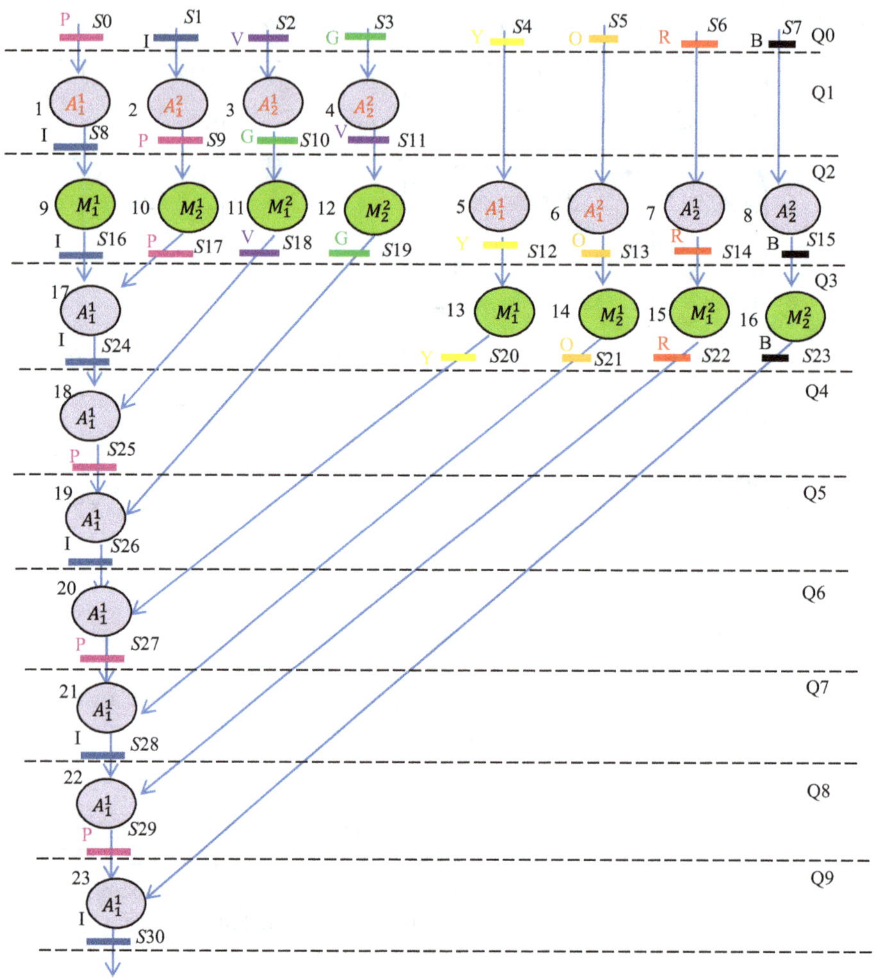

Figure 7.7 Scheduled DFG of FIR filter application after embedding steganography

figure on to the output window. Further, Figure 7.13 shows that the constraints size=26 has been fed as input and the final truncated bitstream is made available on to the output window by clicking on the button 'truncated bitstream'. The number of 0s and 1s in the truncated bitstream can also be shown at output window by clicking on the respective buttons on output panel. As shown in all the snapshots of the tool, the same inputs and stego-keys that are used in the demonstration of FIR filter core in Section 7.4 have been fed here. The tool produces the desired outputs that match with the demonstration of FIR filter core in Section 7.4. Further, Figure 7.14 shows the register allocation post embedding stego-constraints corresponding to '0' bits. Here, values under the column headings 1–8 show the storage

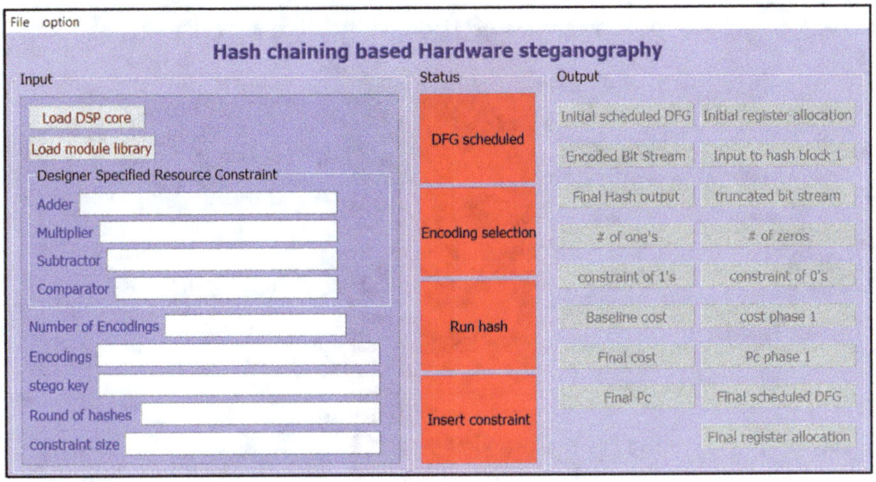

Figure 7.8 A snapshot of key-triggered hash-chaining-driven steganography tool

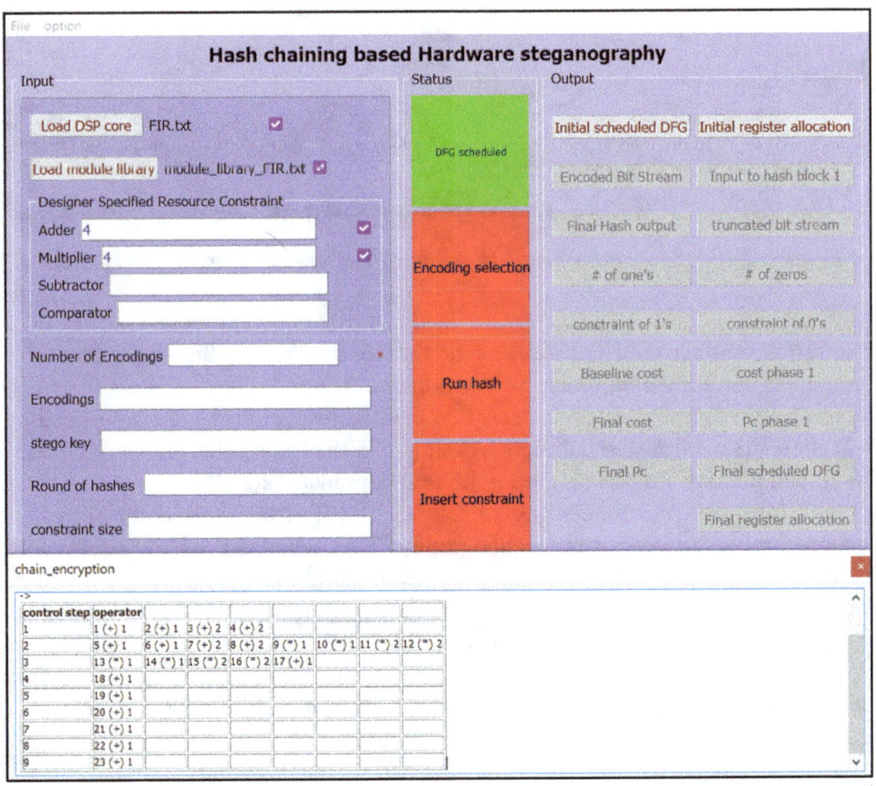

Figure 7.9 Snapshot post entering input DSP application, library and constraints size

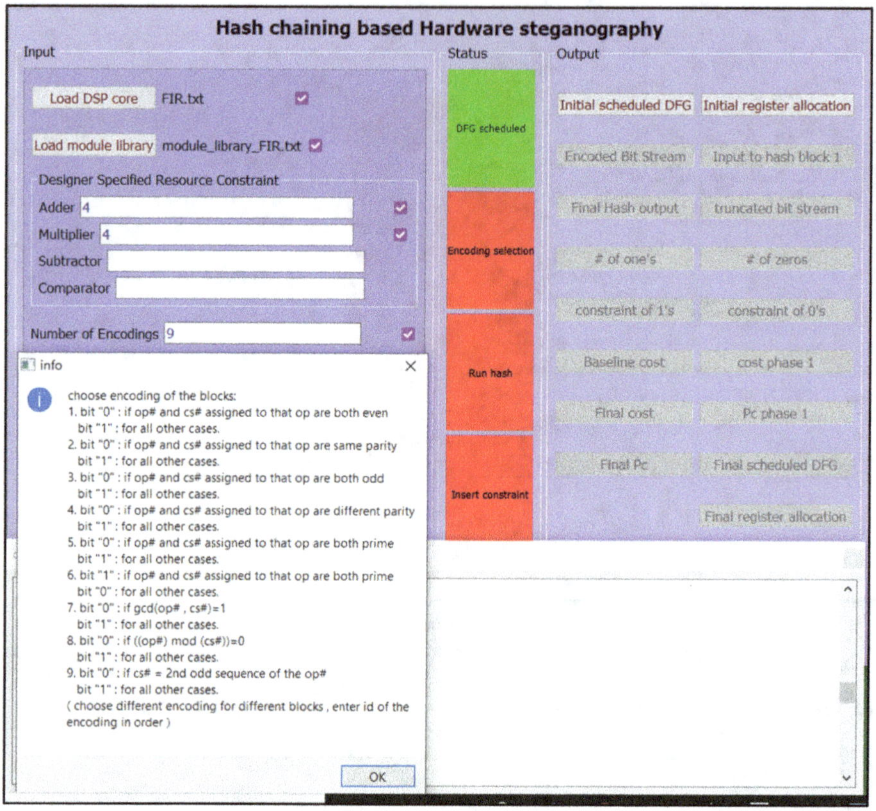

Figure 7.10 Snapshot of tool showing different encodings to be chosen

variable (S) number, and the heading of the column shows the register number, where pink, indigo, violet, green, yellow, orange, red and black registers have been denoted by the number 1, 2, 3, 4, 5, 6, 7 and 8, respectively. The row headings (0–9) show the control step numbers. As shown in the figure, this register allocation post embedding constraints matches with that demonstrated in Section 4.7. Further, Figure 7.15 shows the scheduling of CDFG of FIR filter post embedding key-triggered hash-chaining-based steganography. This scheduled CDFG captures the impact of embedding stego-constraints corresponding to bit '1' in the form of FU vendor reallocation. Additionally, design cost pre and post embedding steganography can be seen at output window by clicking on the respective buttons on output panel as shown in Figure 7.16. Further, the value of probability of coincidence metric can also be seen at output window of the tool as shown in Figure 7.17.

Thus, the key-triggered hash-chaining-based steganography approach can be simulated and analysed using the KHC-stego tool developed by the authors. This tool is useful for analysing various kinds of DSP hardware accelerator applications.

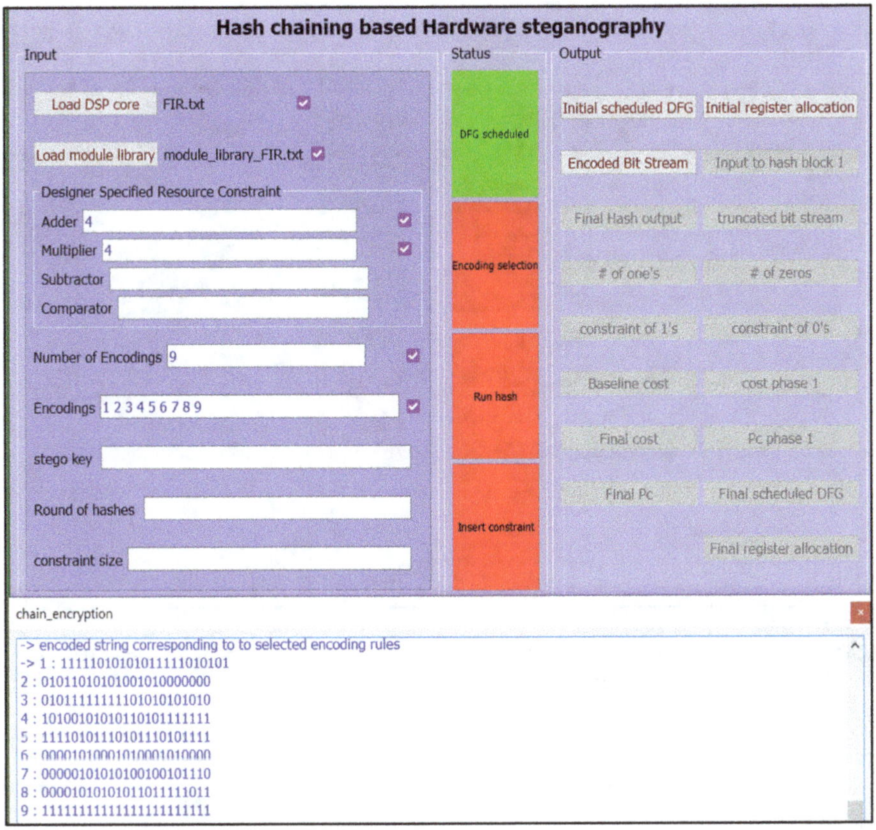

Figure 7.11 *Snapshot of tool showing encoded bitstreams of scheduled DFG of FIR filter*

7.6 Analysis on case studies

The case studies of different DSP applications have been analysed to show the security achieved through key-triggered hash-chaining steganography approach (Sengupta and Rathor, 2020). Further, a security comparison, in terms of key strength, has been shown with respect to contemporary watermarking and steganography approaches. Furthermore, the impact of key-triggered hash-chaining steganography approach on design cost has been assessed by comparing it with the baseline design cost (before steganography). The security and design cost analysis along with comparative assessment are discussed in the following subsections.

7.6.1 Security analysis (Sengupta and Rathor, 2020)

The key-triggered hash-chaining steganography approach (Sengupta and Rathor, 2020) targets the security of DSP cores against false claim of ownership and

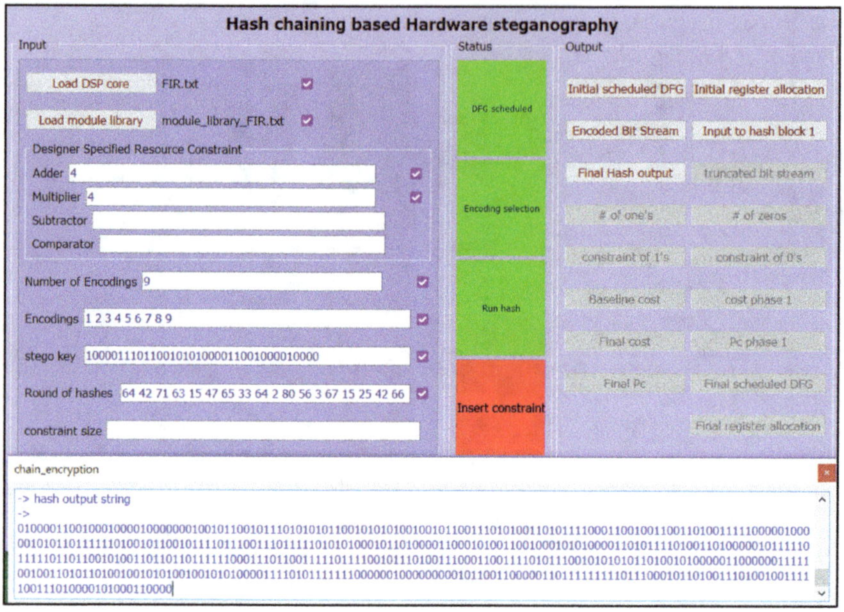

Figure 7.12 *Snapshot of tool post entering encoding numbers, stego-key and rounds of hash*

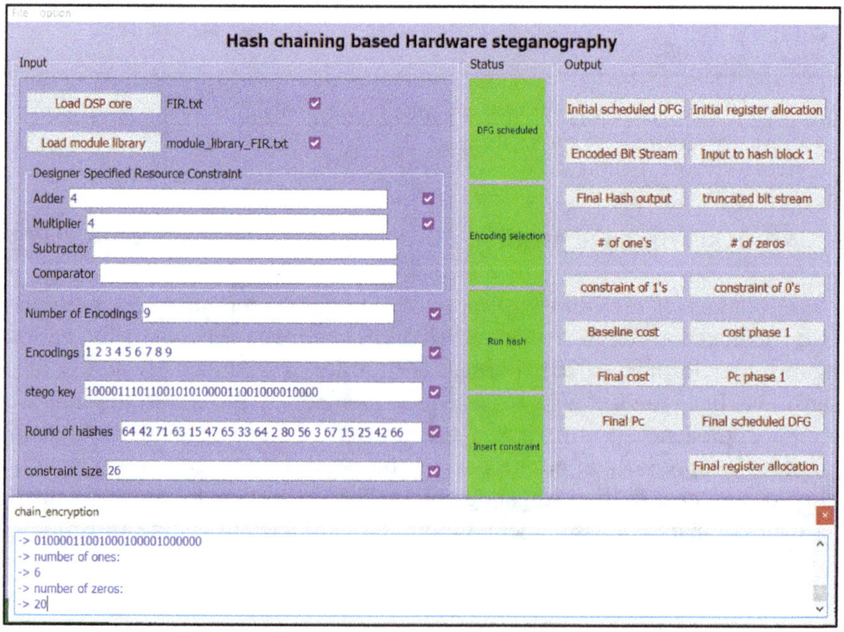

Figure 7.13 *Snapshot of tool post entering constraint size*

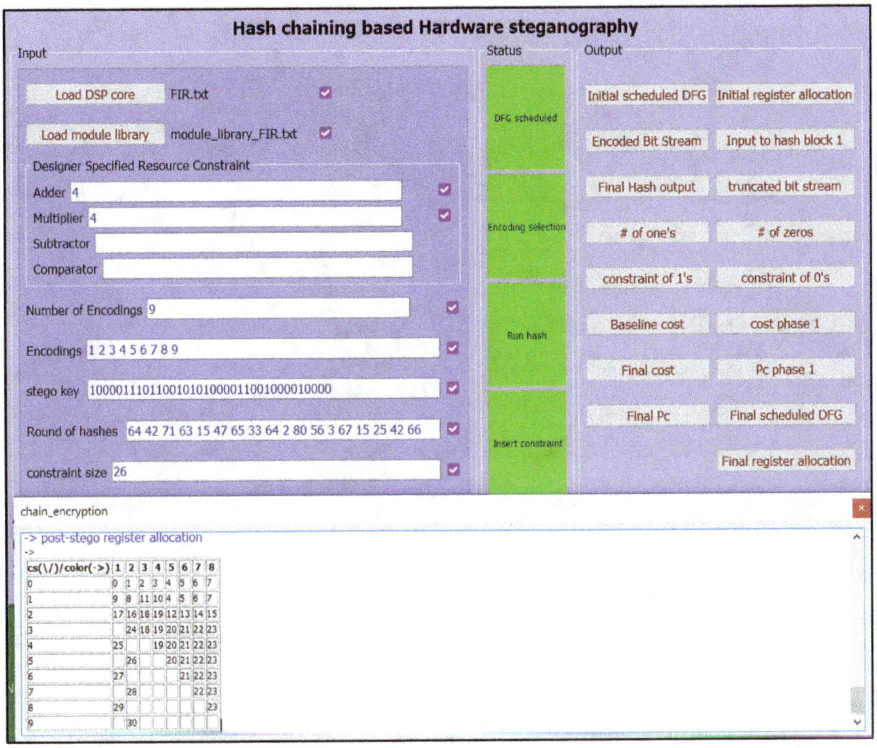

Figure 7.14 Snapshot of tool showing register allocation post embedding constraints

counterfeiting/cloning threats. The false claim of ownership can be nullified by embedding significant amount of digital evidence, which shows the proof of ownership. The strength of ownership proof is assessed in terms of probability of coincidence (*Pc*) which represents the probability of coincidently finding designer's stego-mark in a non-steganography-embedded design. The *Pc* metric is evaluated using the following formula (Sengupta and Rathor, 2019b):

$$Pc = \left(1 - \frac{1}{h}\right)^{k1} \times \left(1 - \frac{1}{\pi_{j=1}^{m} N\left(U_j\right)}\right)^{k2} \tag{7.8}$$

where h indicates the number of colours/registers in the CIG of DSP hardware accelerator design before steganography, and $k1$ indicates the number of stego-constraints embedded during the register allocation phase (i.e. the number of 0s embedded). Further, $k2$ indicates the number of stego-constraints embedded during the FU resource allocation phase (i.e. effective number of 1s embedded), $N(U_j)$ indicates the number of resources of FU-type U_j and m indicates the total types of FU resources present in the DSP hardware accelerator design. Here, $k1$ (the number

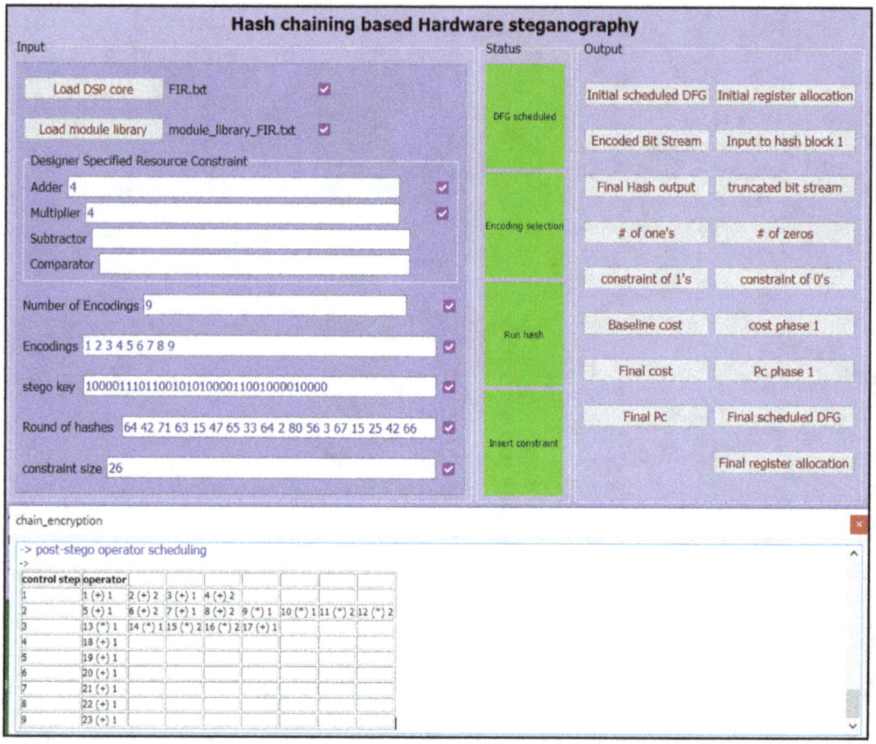

Figure 7.15 Snapshot of tool showing scheduling and allocation post embedding constraints

of stego-constraints embedded during register allocation) and *k*2 (the number of stego-constraints embedded FU allocation phase) indicate the amount of digital evidence hidden within the design. Higher the value of *k*1 and *k*2 (i.e. higher digital evidence), lower is the *Pc* value which, in turn, indicates the higher strength of ownership proof. Table 7.6 shows the number of stego-constraints *k*1 and *k*2 and total constraints size $W=k1+k2$ for various DSP applications. Further, Figure 7.18 shows the probability of coincidence achieved through key-triggered hash-chaining steganography approach (Sengupta and Rathor, 2020) post embedding stego-constraints (*k*1) corresponding to bit '0'. Figure 7.19 shows the final probability of coincidence achieved post embedding stego-constraints ($k1+k2$) corresponding to bits '0' and '1'. Further, variation in *Pc* for increasing number of constraints is shown in Figure 7.20. As shown in the figure, the *Pc* value decreases on increasing the size of stego-constraints ($W=k1+k2$), indicating higher strength of ownership proof. In addition, counterfeiting can be detected by inspecting the presence of stego-mark in a counterfeited IP/IC. The counterfeited IP/IC does not contain the authentic stego-mark of original vendor. Further, cloned IPs/ICs are detected by verifying the presence of authentic stego-mark in the IPs/ICs of different brands.

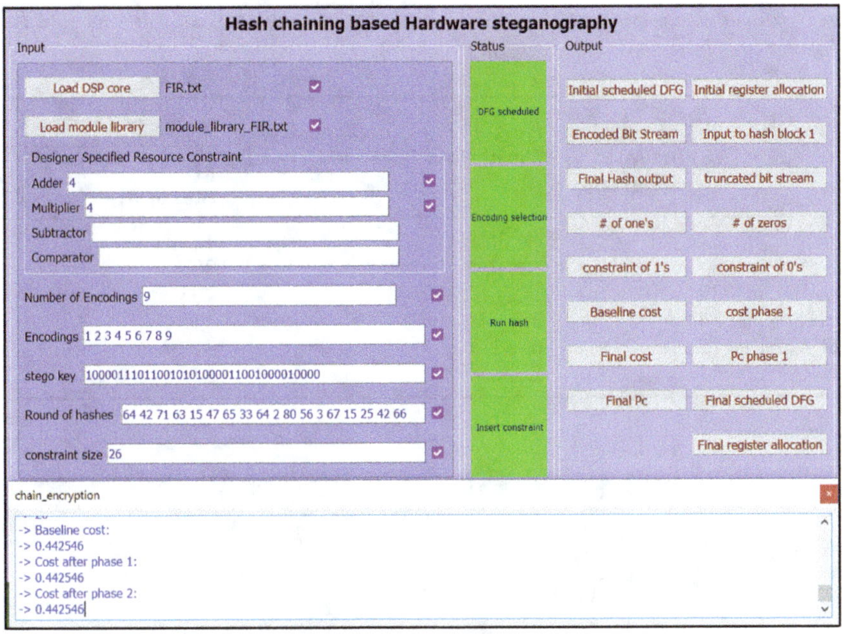

Figure 7.16 Snapshot of tool showing design cost pre and post embedding stego-constraints

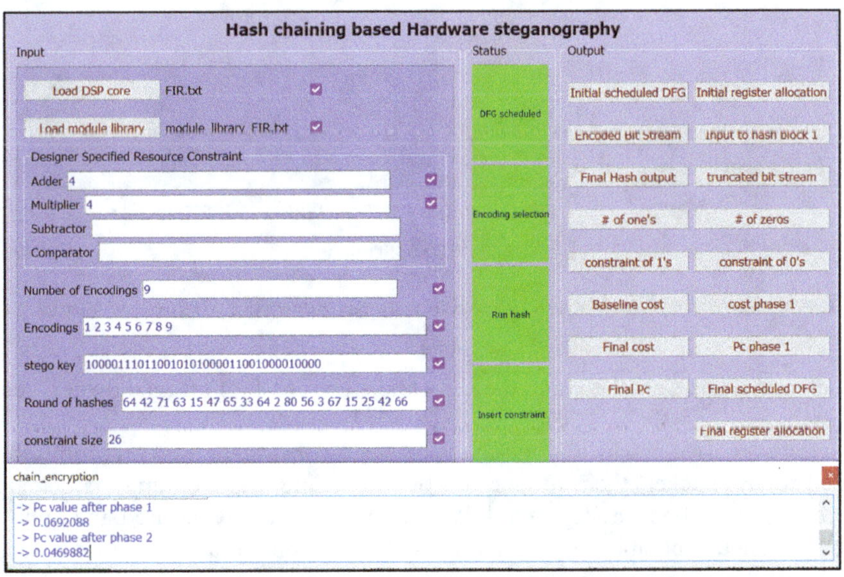

Figure 7.17 Snapshot of tool showing probability of coincidence (Pc) value post embedding stego-constraints corresponding to bits '0' and '1'

Table 7.6 Stego-constraints k1 and k2 and total constraints size W=k1+k2 for various DSP applications

DSP applications	No. of registers (*h*) before steganography	Resource constraints	Total constraint size (*W*=*k*1+*k*2)	No. of constraints	
				*k*1 (no. of 0s)	*k*2 (no. of 1s)
DCT	8	1*A*, 4*M*	23	13	10
FIR	8	4*A*, 4*M*	26	20	6
JPEG_IDCT	29	12*A*, 12*M*	312	203	109
MPEG	14	3*A*, 7*M*	37	21	16
JPEG_sample	12	6*A*, 1*M*	51	30	21
EWF	7	2*A*, 1*M*	52	34	18

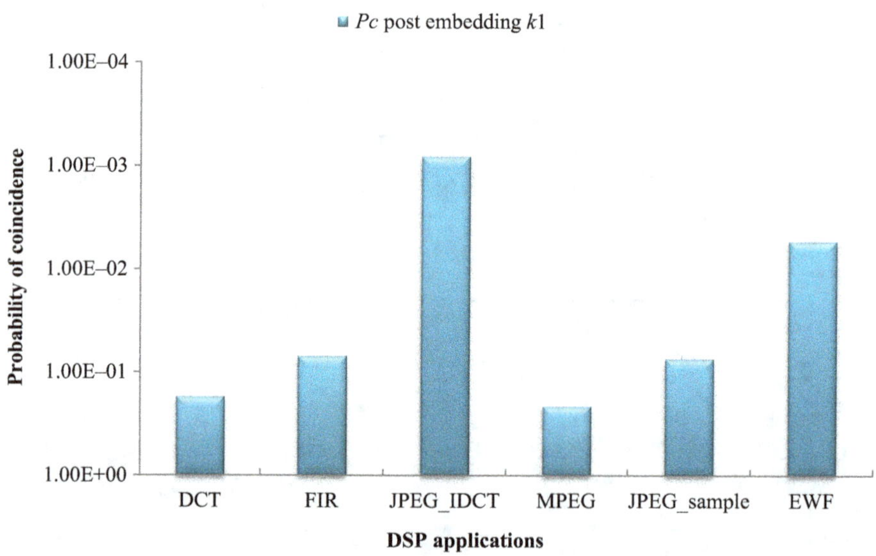

Figure 7.18 Probability of coincidence post embedding stego-constraints k1

The cloned IPs/ICs shall contain the authentic stego-mark of original owner, hence detected.

However, if embedded stego-mark or stego-constraints are compromised or leaked to an attacker, then the purpose of embedding steganography is defeated. Therefore, an embedded stego-mark is required to be highly secured and only the IP owner should be able to generate it. In order to obtain a highly robust stego-mark, the key-triggered hash-chaining steganography approach (Sengupta and Rathor, 2020) generates it using designer-specified stego-key. Moreover, the encoded bitstreams used in the stego-constraints generation process also enhance

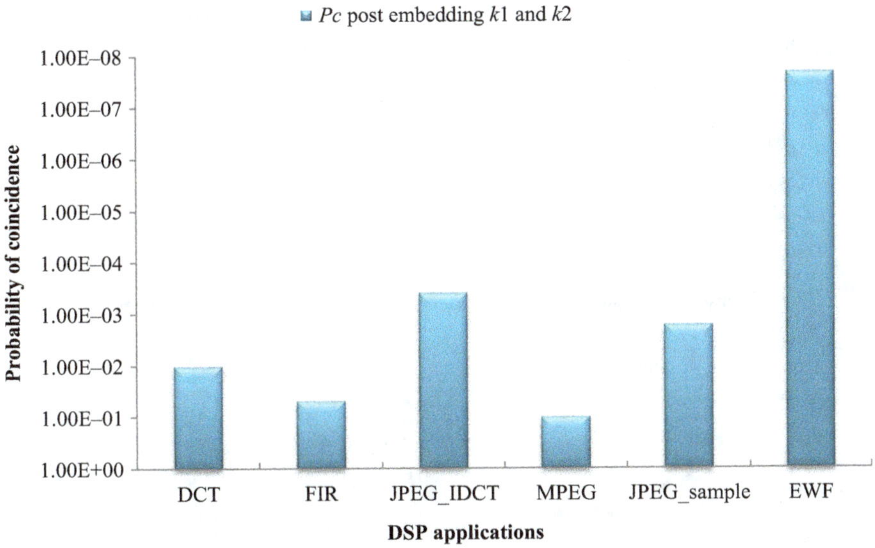

Figure 7.19 Final probability of coincidence post embedding stego-constraints k1 and k2

the robustness of the stego-mark. Both the stego-keys and the encoded bitstreams are only known to the designer. For an attacker, to find the stego-constraints, both the stego-keys and encoded bitstreams need to be determined. From an attacker's perspective, the maximum possible stego-key size required to generate the stego-constraints is reported in Table 7.7. Further, the table also shows the required attacker's maximum effort to find out the valid key value. Additionally, an attacker is also required to find out the total encoded bits (in the bitstreams) used during stego-constraints generation. The attacker's effort of deriving actual encoded bits through brute force is reported in Table 7.7. The overall security in terms of attacker's total effort of finding stego-constraints is also shown in the same table. As evident from the table, determining stego-constraints becomes almost infeasible for an attacker because of huge effort required to find out the stego-key value and encoded bits.

Tamper tolerance: The key-triggered hash-chaining steganography approach (Sengupta and Rathor, 2020) is not vulnerable to tampering in contrast to watermarking-based approaches (Koushanfar *et al.*, 2005; Sengupta and Bhadauria, 2016; Sengupta and Roy, 2017). This is because, unlike watermarking, the steganography is free from signature digits combinations.

Security comparison with contemporary approaches: Watermarking and steganography are the contemporary approaches that secure an IP core against ownership abuse and counterfeiting/cloning. However, the purpose of watermarking approach fails if the designer signature gets leaked to an attacker. This is because, the watermarking approaches do not use secret key to generate the watermarking

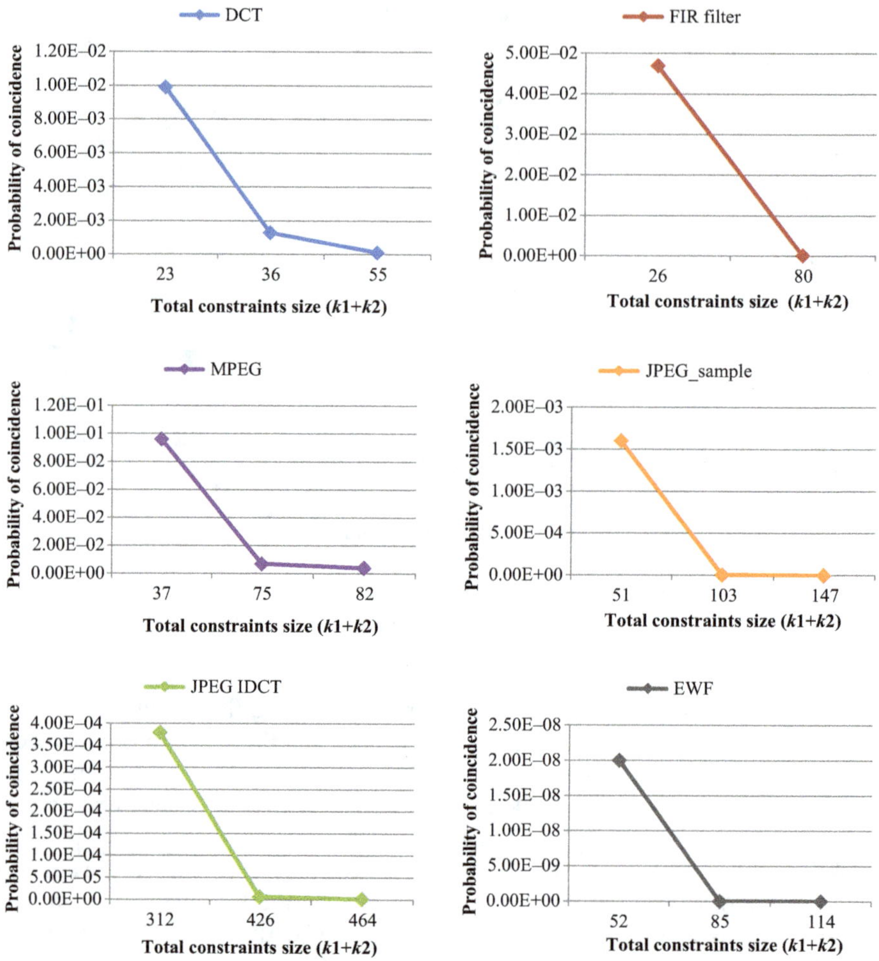

Figure 7.20 Impact of increasing stego-constraints size on probability of coincidence (Pc)

constraints. Similarly, steganography approach proposed by Sengupta and Rathor (2019a) does not employ a secret key to generate stego-constraints. Although the steganography approach proposed by Sengupta and Rathor (2019b) uses stego-key, the key size is very small compared to the key-triggered hash-chaining steganography approach (Sengupta and Rathor, 2020). The comparison of key size with respect to Sengupta and Bhadauria (2016) and Sengupta and Rathor (2019a, 2019b) is shown in Table 7.8. As shown in the table, the key size in the watermarking (Sengupta and Bhadauria, 2016) and steganography approaches (Sengupta and Rathor, 2019a) is zero because of non-involvement of secret key to generate the secret constraints. Further, it is evident from the table that the key-triggered

Table 7.7 Security analysis of key-triggered hash-chaining-based steganography in terms of stego-key strength and attacker's total effort of finding stego-constraints

DSP applications	**Maximum key size using (7.1)**	**Attacker's maximum effort in terms of finding valid key (using (7.4))**	**Attacker's maximum effort in terms of finding encoded bits (using (7.6))**	**Attacker's total effort (using (7.7))**
DCT	491,520	$>10^{147603}$	10^{2059}	$>10^{149662}$
FIR	192,937,984	$>10^{57939334}$	10^{2059}	$>10^{57941393}$
JPEG_IDCT	5.8153×10^{35}	$>10^{1.74\times10^{35}}$	10^{2059}	$>10^{1.74\times10^{35}}$
MPEG	7,516,192,768	$>10^{2277634172}$	10^{2059}	$>10^{2277636231}$
JPEG_sample	283,467,841,536	$>10^{85899345920}$	10^{2059}	$>10^{85899347979}$
EWF	584,115,552,256	$>10^{177004712804}$	10^{2059}	$>10^{177004714863}$

Table 7.8 Key size comparison of key-triggered hash-chaining-based steganography with contemporary approaches

DSP applications	**Maximum key size in bits**		
	Key-triggered hash-chaining steganography approach	**Sengupta and Rathor (2019b)**	**Sengupta and Bhadauria (2016) and Sengupta and Rathor (2019a)**
DCT	491,520	610	0
FIR	192,937,984	625	0
JPEG_IDCT	5.8153×10^{35}	785	0
MPEG	7,516,192,768	620	0
JPEG_sample	283,467,841,536	665	0
EWF	584,115,552,256	640	0

hash-chaining-based steganography offers very high security in terms of robustness of the stego-mark (security of stego-constraints). This renders an attacker almost infeasible to find out the stego-constraints embedded into the design.

7.6.2 Design cost analysis (Sengupta and Rathor, 2020)

This subsection discusses the impact of employing key-triggered hash-chaining-driven steganography-based security on design cost. The following equation is used to evaluate the design cost (Sengupta and Rathor, 2019b):

$$C_d(U_i) = \rho_1 \frac{L_d}{L_m} + \rho_2 \frac{A_d}{A_m} \tag{7.9}$$

where $C_d(U_i)$ is the design cost of DSP cores for resource constraints U_i; further, L_d and L_m are the design latency at specified resource constraints and maximum design latency, respectively, A_d and A_m are the design area at specified resource constraints and maximum area, respectively, and ρ_1, ρ_2 are the weights which are fixed at 0.5. Variation in the design cost for increasing size of stego-constraints is shown in Figure 7.21. As shown in the figure, the impact of increasing stego-constraint size on design cost is either zero or nominal.

Design cost comparison with baseline: Design cost comparison with the baseline is shown in Figure 7.22 for a particular constraint size (*W*). As shown in the figure, the design cost may increase by a marginal value because of the possibility of increment in the number of registers required to embed the stego-constraints. However, no extra register is required for most of the DSP applications. This

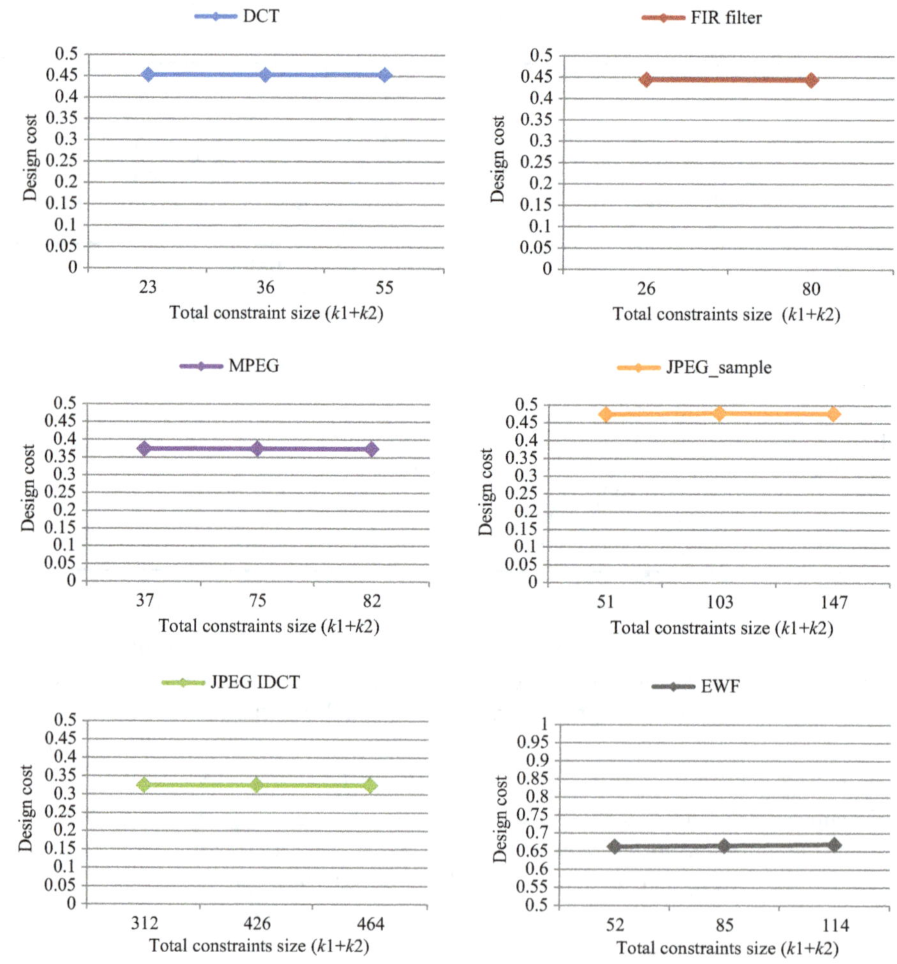

Figure 7.21 Impact of increasing stego-constraints size on design cost of key-triggered hash-chaining steganography

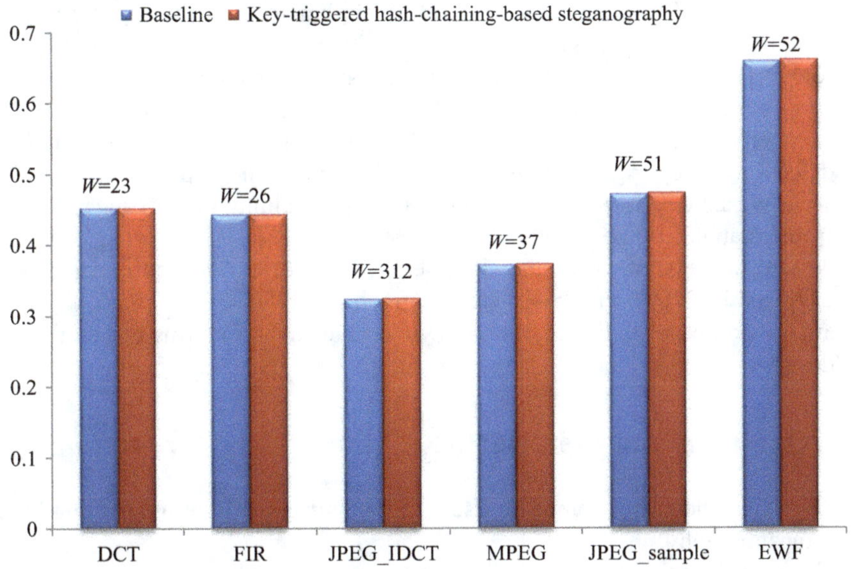

Figure 7.22 Design cost comparison of key-triggered hash-chaining steganography approach with baseline. Note: W indicates the stego-constraints size

indicates that the key-triggered hash-chaining steganography approach (Sengupta and Rathor, 2020) achieves very high security at almost zero overhead.

7.7 Conclusion

This chapter discusses a key-triggered hash-chaining-based hardware steganography approach (Sengupta and Rathor, 2020) which offers very high security against false claim of IP ownership threat. Additionally, the key-triggered hash-chaining steganography approach is also capable of detecting counterfeited/cloned IPs/ICs. The stego-mark generated through the key-triggered hash-chaining steganography approach is highly robust as it is produced using secret stego-key of very large size, designer-selected encoded bitstreams and the number of iterations of round function in each HU of the chaining process. The robustness of the stego-mark has been evaluated in terms of key size, attacker's total effort of finding stego-constraints and probability of coincidence. These case studies show that the key-triggered hash-chaining steganography approach provides higher security than contemporary approaches at trivial design overhead.

At the end of this chapter, a reader gains the following concepts:

- need of security of DSP hardware accelerators;
- key-triggered hash-chaining-based steganography methodology;
- various encodings of a DSP application;

- role of encoded bitstreams of a DSP application in key-triggered hash-chaining-based steganography;
- role of HUs in key-triggered hash-chaining-based steganography;
- concept of regular and key-triggered HUs;
- stego-embedder block in key-triggered hash-chaining-based steganography;
- detection of key-triggered hash-chaining-based steganography;
- security achieved using key-triggered hash-chaining-based steganography from an attacker's perspective;
- design process of obtaining stego-embedded FIR filter core using key-triggered hash-chaining-based steganography and
- analysis on case studies of various DSP applications, in terms of security and design cost.

7.8 Questions and exercise

1. Explain the threat model used in key-triggered hash-chaining-based steganography.
2. Explain the role of encoded bitstreams of a DSP application in key-triggered hash-chaining-based steganography.
3. Explain the role of hash units in key-triggered hash-chaining-based steganography.
4. Explain the concept of regular and key-triggered hash units.
5. Explain the function of stego-embedder block in key-triggered hash-chaining-based steganography.
6. Explain the function of the detection of key-triggered hash-chaining-based steganography.
7. What is the significance of parallel switch blocks?
8. How many encoding algorithms can be used in the key-triggered hash-chaining-based steganography?
9. What is the output bit size of the bit padding block? How is this size determined?
10. What is the output bit size of the parallel switch blocks?
11. What is the maximum key size of the stego-key block?
12. What is the rule of constructing 1,024 bits through pre-processing?
13. Explain the different encoding rules used to encode a DSP application into bitstreams.
14. What is the attacker's maximum effort of finding the stego-key?
15. Determine the total encoded bits used in hash-chaining block to generate stego-constraints.
16. How to determine an attacker's total effort in determining the stego-constraints embedded into the design?
17. What is a KHC-stego tool?
18. How many phases are used for embedding stego-constraints into the design?
19. Compare any hardware watermarking with key-triggered hash-chaining-based steganography, in terms of security achieved and design overhead.

References

R. S. Chakraborty and S. Bhunia (2009), 'HARPOON: an obfuscation-based SoC design methodology for hardware protection,' *IEEE Trans. Comput.-Aided Des. Integr. Circuits Syst.*, vol. 28(10), pp. 1493–1502.

B. Colombier and L. Bossuet (2015), 'Survey of hardware protection of design data for integrated circuits and intellectual properties,' *IET Comput. Digit. Tech.*, vol. 8(6), pp. 274–287.

F. Koushanfar, I. Hong, and M. Potkonjak (2005), 'Behavioral synthesis techniques for intellectual property protection,' *ACM Trans. Des. Autom. Electron. Syst.*, vol. 10(3), pp. 523–545.

B. Le Gal and L. Bossuet (2012), 'Automatic low-cost IP watermarking technique based on output mark insertions,' *Des. Autom. Embedded Syst.*, vol. 16(2), pp. 71–92.

R. D. Newbould, J. D. Carothers and J. J. Rodriguez (2002), 'Watermarking ICs for IP protection,' *IET Electron. Lett.*, vol. 38(6), pp. 272–274.

S. M. Plaza and I. L. Markov (2015), 'Solving the third-shift problem in IC piracy with test-aware logic locking,' *IEEE Trans. Comput.-Aided Des. Integr. Circuits Syst.*, vol. 34(6), pp. 961–971.

R. Schneiderman (2010), 'DSPs evolving in consumer electronics applications,' *IEEE Signal Process. Mag.*, vol. 27(3), pp. 6–10.

A. Sengupta and S. Bhadauria (2016), 'Exploring low cost optimal watermark for reusable IP cores during high level synthesis,' *IEEE Access*, vol. 4, pp. 2198–2215.

A. Sengupta, D. Kachave and D. Roy (2019a), 'Low cost functional obfuscation of reusable IP cores used in CE hardware through robust locking,' *IEEE Trans. Comput.-Aided Des. Integr. Circuits Syst.*, vol. 38(4), pp. 604–616.

A. Sengupta (2016), 'Intellectual property cores: protection designs for CE products,' *IEEE Consum. Electron. Mag.*, vol. 5(1), pp. 83–88.

A. Sengupta (2017), 'Hardware security of CE devices [hardware matters],' *IEEE Consum. Electron. Mag.*, vol. 6(1), pp. 130–133.

A. Sengupta, E. R. Kumar and N. P. Chandra (2019c), 'Embedding digital signature using encrypted-hashing for protection of DSP cores in CE,' *IEEE Trans. Consum. Electron.*, vol. 65(3), pp. 398–407.

A. Sengupta and S. P. Mohanty (2019), 'IP core and integrated circuit protection using robust watermarking,' *IP Core Protection and Hardware-Assisted Security for Consumer Electronics*, e-ISBN: 9781785618000, pp. 123–170.

A. Sengupta and M. Rathor (2019a), 'IP core steganography for protecting DSP kernels used in CE systems,' *IEEE Trans. Consum. Electron.*, vol. 65(4), pp. 506–515.

A. Sengupta and M. Rathor (2019b), 'Crypto-based dual-phase hardware steganography for securing IP cores,' *Lett. IEEE Comput. Soc.*, vol. 2(4), pp. 32–35.

A. Sengupta and M. Rathor (2019c), 'Security of functionally obfuscated DSP core against removal attack using SHA-512 based key encryption hardware,' *IEEE Access*, vol. 7, pp. 4598–4610.

A. Sengupta and M. Rathor (2020), 'IP core steganography using switch based key-driven hash-chaining and encoding for securing DSP kernels used in CE systems,' *IEEE Trans. Consum. Electron*, vol. 66(3), pp. 251–260.

A. Sengupta, D. Roy and S. P. Mohanty (2019b), 'Low-overhead robust RTL signature for DSP core protection: new paradigm for smart CE design,' *Proc. 37th IEEE International Conference on Consumer Electronics (ICCE)*, pp. 1–6.

A. Sengupta and D. Roy (2017), 'Antipiracy-aware IP chipset design for CE devices: a robust watermarking approach [hardware matters],' *IEEE Consum. Electron. Mag.*, vol. 6(2), pp. 118–124.

A. Sengupta, D. Roy and S. P. Mohanty (2018), 'Triple-phase watermarking for reusable IP core protection during architecture synthesis,' *IEEE Trans. Comput.-Aided Des. Integr. Circuits Syst.*, vol. 37(4), pp. 742–755.

A. Sengupta, R. Sedaghat and Z. Zeng (2010), 'A high level synthesis design flow with a novel approach for efficient design space exploration in case of multi-parametric optimization objective,' *Microelectron. Reliab.*, vol. 50(3), pp. 424–437.

A. Sengupta (2020), 'Frontiers in securing IP cores – forensic detective control and obfuscation techniques', The Institute of Engineering and Technology (IET), ISBN-10: 1-83953-031-6, ISBN-13: 978-1-83953-031-9.

M. Yasin, J. J. Rajendran, O. Sinanoglu, and R. Karri (2016), 'On improving the security of logic locking,' *IEEE Trans. Comput.-Aided Des. Integr. Circuits Syst.*, vol. 35(9), pp. 1411–1424.

J. Zhang (2016), 'A practical logic obfuscation technique for hardware security,' *IEEE Trans. Very Large Scale Integr. (VLSI) Syst.*, vol. 24(3), pp. 1193–1197.

Chapter 8

Designing a secured *N*-point DFT hardware accelerator using obfuscation and steganography

Anirban Sengupta[1] and Mahendra Rathor[1]

The chapter describes a design flow of a secured *N*-point discrete Fourier transform (DFT) hardware accelerator using obfuscation and steganography. The end goal of this chapter is to familiarize the reader about designing process of a secured *N*-point DFT hardware accelerator that can thwart against reverse engineering (RE) and detect intellectual property core piracy. The state-of-the-art methods have been used to employ security in the hardware design and synthesis process.

The chapter is organized as follows: Section 8.1 discusses about the introduction of the chapter; Section 8.2 presents the details of the secured *N*-point DFT hardware accelerator design methodology that includes the secured design flow, design process and other security details. Section 8.3 presents the analysis of the case study that includes design overhead analysis, security analysis, etc.; Section 8.4 presents conclusion and Section 8.5 concludes the chapter with important exercise for the readers.

8.1 Introduction

Digital signal processing (DSP) algorithms have wide utilization in electronics devices to facilitate several applications such as audio/video compression/decompression, denoising. DSP algorithms execute core function of these applications. DFT is an important DSP algorithm which is used for spectral analysis of signals, frequency response analysis of systems and so forth. Owing to computational intensiveness of DFT algorithm, it is efficient to be employed as an application-specific processor or a hardware accelerator. Realizing DFT algorithm as a hardware accelerator helps in satisfying stringent time and power constraints. However, utilization of DFT hardware accelerators in electronics systems also invites security risks because of popular hardware threats such as Trojan insertion, ownership abuse and counterfeiting/cloning (Sengupta, 2017, 2020; Zhang and Tehranipoor, 2011; Sengupta *et al.*, 2017a; Pilato *et al.*, 2018; Sengupta and Mohanty, 2019). Therefore,

[1]Computer Science and Engineering, Indian Institute of Technology Indore, Indore, India

ensuring security of hardware accelerator designs is also becoming an important part of the design process in this modern era of very large scale integration technology.

Structural obfuscation (Lao and Parhi, 2015; Sengupta *et al.*, 2017b, 2018) is a security mechanism that makes RE arduous for an attacker, hence prevents against Trojan insertion and counterfeiting/cloning threats. However, this preventive control can be made ineffective by an attacker if she/he succeeds in deducing original design structure/functionality by performing RE. This guides a designer to additionally employ a detective control to remain extra secured. The detective control can be enabled using hardware watermarking (Sengupta and Bhadauria, 2016; Sengupta *et al.*, 2019) or hardware steganography (Sengupta and Rathor, 2019a, 2019b) techniques.

Rathor and Sengupta (2020) proposed a secured design flow of an *N*-point DFT hardware accelerator using both preventive and detective control against aforementioned threats. The security has been deployed by integrating structural obfuscation and crypto-steganography together to provide enhanced security to the design process of an *N*-point DFT hardware accelerator.

8.2 Secured *N*-point DFT hardware accelerator design methodology

DFT is a transformation of a discrete signal from its discrete-time representation to a discrete-frequency representation. The input (discreet time) and output (discreet frequency) data set of DFT are of the same length. Main applications of DFT include spectral analysis of signals and frequency response analysis of systems. This section discusses the design process of developing a secured *N*-point DFT hardware accelerator design (Rathor and Sengupta, 2020) using two security mechanisms viz. structural obfuscation and crypto-steganography. The overall methodology is discussed in the following subsections.

8.2.1 Secured design flow

The entire flow of designing a secured *N*-point DFT hardware accelerator is depicted in Figure 8.1. As shown in the figure, the *N*-point DFT algorithm in the form of mathematical relationship is subjected to security-aware high-level synthesis (HLS) framework proposed by Rathor and Sengupta (2020). The security-aware HLS framework employs security mechanism-1 based on structural obfuscation and security mechanism-2 based on crypto-steganography to generate a secured register transfer level (RTL) design of an *N*-point DFT hardware accelerator. Apart from algorithmic description of *N*-point DFT, the following inputs are also required to the security-aware HLS framework: module library, resources constraints and stego-keys.

The structural obfuscation-based security mechanism is performed on data-flow graph (DFG), an intermediate representation, of *N*-point DFT processor as shown in Figure 8.1. A tree height transformation (THT)-based structural transformation has been exploited to structurally obfuscate the design. Post-performing

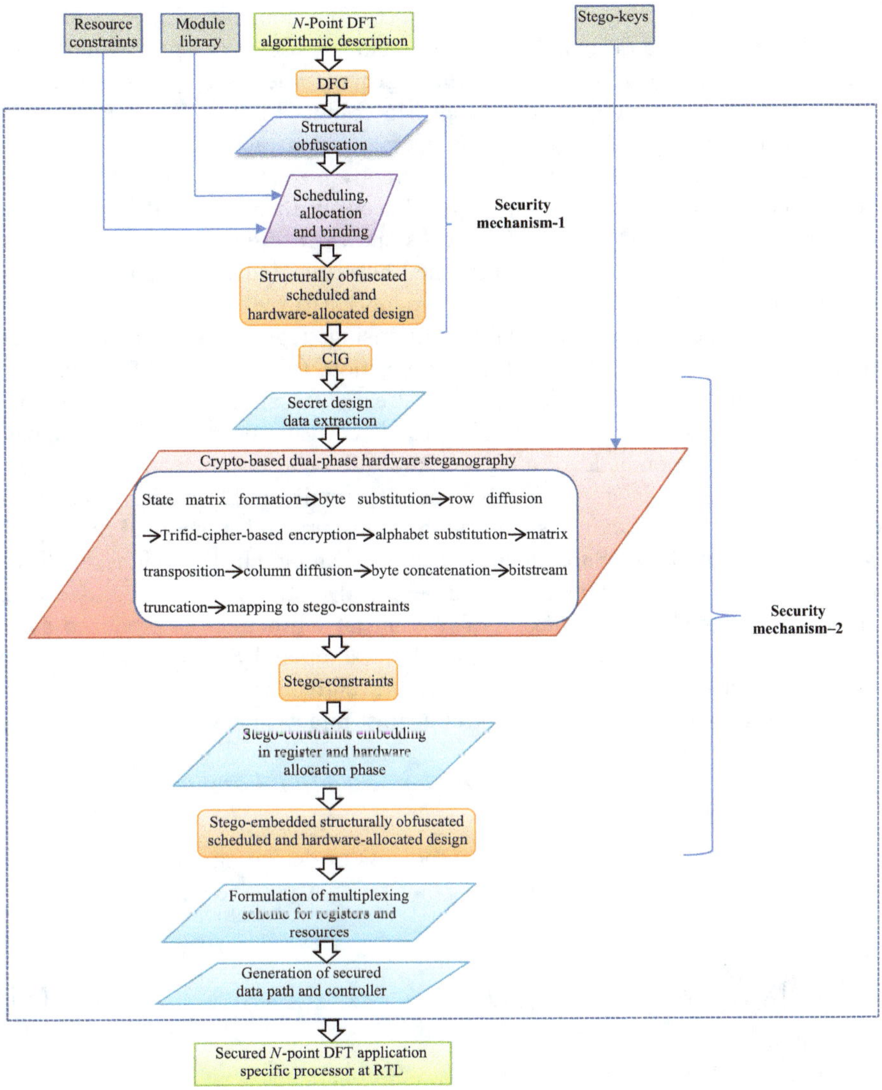

Figure 8.1 Design flow of generating secured N-point DFT hardware accelerator at RTL (Rathor and Sengupta, 2020)

the THT-based structural transformation, scheduling, allocation and binding steps of HLS are executed which result in a structurally obfuscated scheduled and resource-allocated DFT design. Thus, obtained structurally obfuscated design is subjected to security mechanism-2.

The crypto-steganography-based security mechanism-2 is performed on scheduled and resource-allocated form of structurally obfuscated DFT design. In order to do so, first, a coloured interval graph (CIG) is constructed using scheduled and hardware-allocated design. A CIG is a graphical process of showing allocation of storages variables of the design to the registers, where storages variables are mapped to nodes and registers are mapped to colours in the CIG (Sengupta and Bhadauria, 2016). Thus obtained CIG is utilized for extracting secret design data which is fed as input to the crypto-steganography mechanism. Apart from secret design data, the crypto-steganography-based security mechanism also uses secret stego-keys to generate stego-constraints. As shown in Figure 8.1, the process of stego-constraints generation is accomplished by sequentially executing following steps: (i) state matrix formation using stego-key1, (ii) byte substitution, (iii) row diffusion using stego-key2, (iv) Trifid cipher using stego-key3, (v) alphabet substitution using stego-key4, (vi) matrix transposition, (vii) column diffusion, (viii) byte concatenation using stego-key5, (ix) bitstream truncation and (x) mapping bits to stego-constraints based on designer's formulated mapping rules. Thus, obtained stego-constraints are embedded in register allocation and resource allocation phase of HLS. This results in a stego-embedded structurally obfuscated *N*-point DFT design. Further, multiplexing schemes for registers and functional unit (FU) resources are formulated. Subsequently, data path and controller synthesis phases of HLS are performed to generate a stego-embedded structurally obfuscated RTL design. Thus, a secured RTL design of *N*-point DFT processor is produced using security-aware HLS framework proposed by Rathor and Sengupta (2020). The employed security mechanisms enable (i) preventive control against hardware Trojan insertion and (ii) detective control against piracy threat.

8.2.2 *Design process of secured* N*-point DFT hardware accelerator*

This subsection discusses the elaborative process of designing secured DFT hardware accelerator using an example of 4-point DFT. A generic equation of 4-point DFT is given as follows:

$$W[k] = \sum_{n=0}^{3} w[n]e^{-j\pi nk/2} k = 0, 1, 2, 3 \tag{8.1}$$

where input discrete-data sequence is represented by $w[n]$ and output discrete-data sequence is represented by $W[k]$. Each discrete value of output sequence of 4-point DFT is computed as follows:

$$W[0] = w[0] * 1 + w[1] * 1 + w[2] * 1 + w[3] * 1 \tag{8.2}$$

$$W[1] = w[0] * 1 + w[1]e^{-j\pi/2} + w[2]e^{-j\pi} + w[3]e^{-j3\pi/2} \tag{8.3}$$

$$W[2] = w[0] * 1 + w[1]e^{-j\pi} + w[2]e^{-j2\pi} + w[3]e^{-j3\pi} \tag{8.4}$$

$$W[3] = w[0] * 1 + w[1]e^{-j3\pi/2} + w[2]e^{-j3\pi} + w[3]e^{-j9\pi/2} \tag{8.5}$$

The algorithmic description (mathematical relationship of input and output samples) of 4-point DFT is converted into a DFG representation as shown in Figure 8.2.

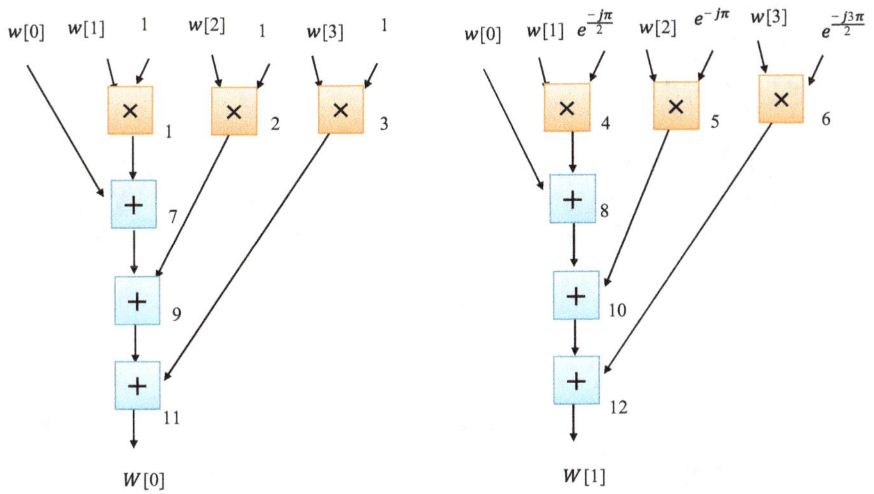

Figure 8.2 DFG of 4-point DFT for parallel dual output (Rathor and Sengupta, 2020)

It is important to note in Figure 8.2 that the DFG computes two output samples in parallel ($W(i)$ and $W(i+1)$) in order to obtain parallel dual output. This DFG is fed as input to the security aware HLS framework which produces a secured RTL design of 4-point DFT for computing parallel dual output. The two security mechanisms employed to design a secured 4-point DFT processor are illustrated in the following subsections.

8.2.2.1 THT-based structural obfuscation – the security mechanism-1

The DFG representing 4-point DFT processor for parallel dual output is subjected to THT-based structural obfuscation. This security mechanism structurally transforms the DFG which is further subjected to scheduling, allocation and binding phases of HLS process to generate structurally obfuscated scheduled and resource-allocated design. THT-based structural obfuscation is a structural transformation of DFG, where the sequential execution flow in the graph is broken and the execution of parallel sub-computations is enabled without affecting the functionality of the design. Figure 8.3 shows the DFG post-performing THT-based structural transformation. As shown in Figures 8.2 and 8.3, the sequential executions of operations 7, 9 and 11 (in Figure 8.2) are broken and parallel executions of operation 7 and 9 are enabled (in Figure 8.3). Similarly, the sequential executions of operation 8, 10 and 12 (in Figure 8.2) are broken and parallel executions of operation 8 and 10 are enabled (in Figure 8.3). Thus, obtained structurally transformed DFG is subjected to scheduling, allocation and binding phases based on designer's chosen resource constraints of three multipliers (M) and two adders (A). The scheduled and

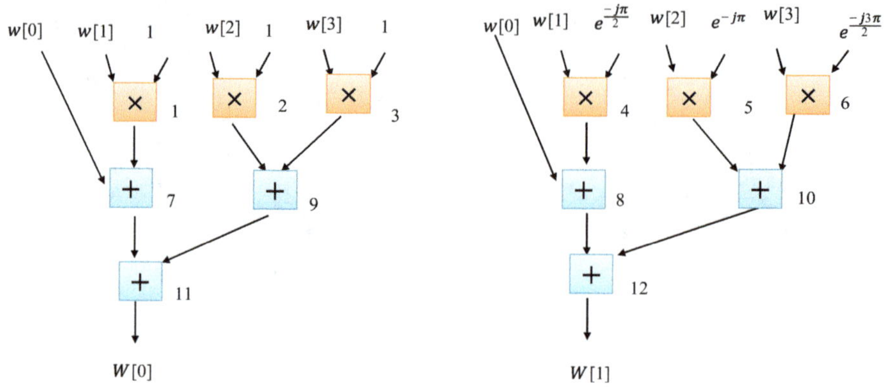

Figure 8.3 Structurally obfuscated DFG of 4-point DFT using THT (Rathor and Sengupta, 2020)

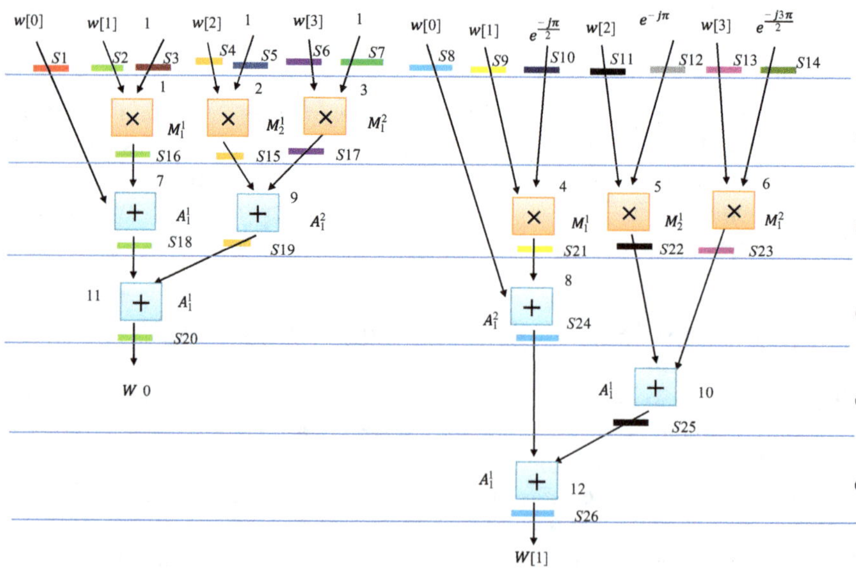

Figure 8.4 Scheduled and resource-allocated DFG of obfuscated 4-point DFT based on resources constraints of 3M and 2A (Rathor and Sengupta, 2020)

resource-allocated DFG is shown in Figure 8.4. As shown in the figure, 12 operations of the 4-point DFT (for parallel dual output) are scheduled in 5 control steps (Q). Multiplication and addition operations have been assigned to respective FUs from two different vendors V1 and V2. Since chosen constraint of multipliers is 3, the **two instances from vendor V1** and **one instance from vendor V2** are chosen for allocation of multiplier resources to the multiplication operations in a control step. Similarly, chosen constraint of adders is 2, **one instance from vendor V1** and **another instance from vendor V2** are chosen for allocation of adder resources to the addition operations in a control step.

The THT-based structural transformation employed in the DFG leads to manifold changes in the RTL structure of the design post-HLS, without affecting the functionality. The changes in the design structure due to structural obfuscation include changes in the size and count of multiplexers and de-multiplexers and changes in the I/O (inputs/outputs) connectivity of FU resources. The structurally obfuscated design is quite hard to interpret through RE by an attacker (as it becomes unobvious), hence preventing Trojan insertion. Next, this obtained structurally obfuscated 4-point DFT design is subjected to security mechanism-2, i.e. crypto-steganography to augment the security level against piracy threat.

8.2.2.2 Crypto-based steganography – the security mechanism-2

This security mechanism enables detective control over counterfeiting and cloning by embedding owner's secret stego-mark into the design during HLS process. The stego-mark (or stego-constraints) is generated using crypto-steganography process which requires following inputs: (i) secret design data and (ii) stego-keys. Further, the generated stego-constraints are embedded into the scheduled and resource-allocated DFG (cover design data) by performing register and resource reallocation during HLS process. The overall process of generating stego-embedded structurally obfuscated 4-point DFT design is discussed as follows:

1. A CIG is constructed from scheduled and allocated DFG (shown in Figure 8.4). As shown in Figure 8.4, 26 storage variables ($S1$–$S26$) are executed using 14 registers. In the CIG, 26 storage variables have been represented using 26 nodes and 14 registers have been represented using 14 distinct colours as shown in Figure 8.5. The register allocation of storage variables into different control steps (Q1–Q5) is captured in Table 8.1.
2. The secret design data is extracted from the CIG. It is represented by a set or collection of indices (i, j) of all node pairs (S_i, S_j) of the same colours in the CIG. The secret design data extracted from CIG of 4-point DFT processor is given as follows:

   A = {(2,16), (2,18), (2, 20), (16,18), (16,20), (18,20), (4,15), (4,19), (15,19), (6,17), (8,24), (8,26), (24,26), (9, 21), (11,22), (11,25), (22,25), (13,23)}

   Post-applying modulo 15 and converting into hexadecimal notation, the resultant secret design data is given as follows:

   A = {(2, 1), (2, 3), (2, 5), (1, 3), (1, 5), (3, 5), (4, F), (4, 4), (F, 4), (6, 2), (8, 9), (8, B), (9, B), (9, 6), (B, 7), (B, A), (7, A), (D, 8)}

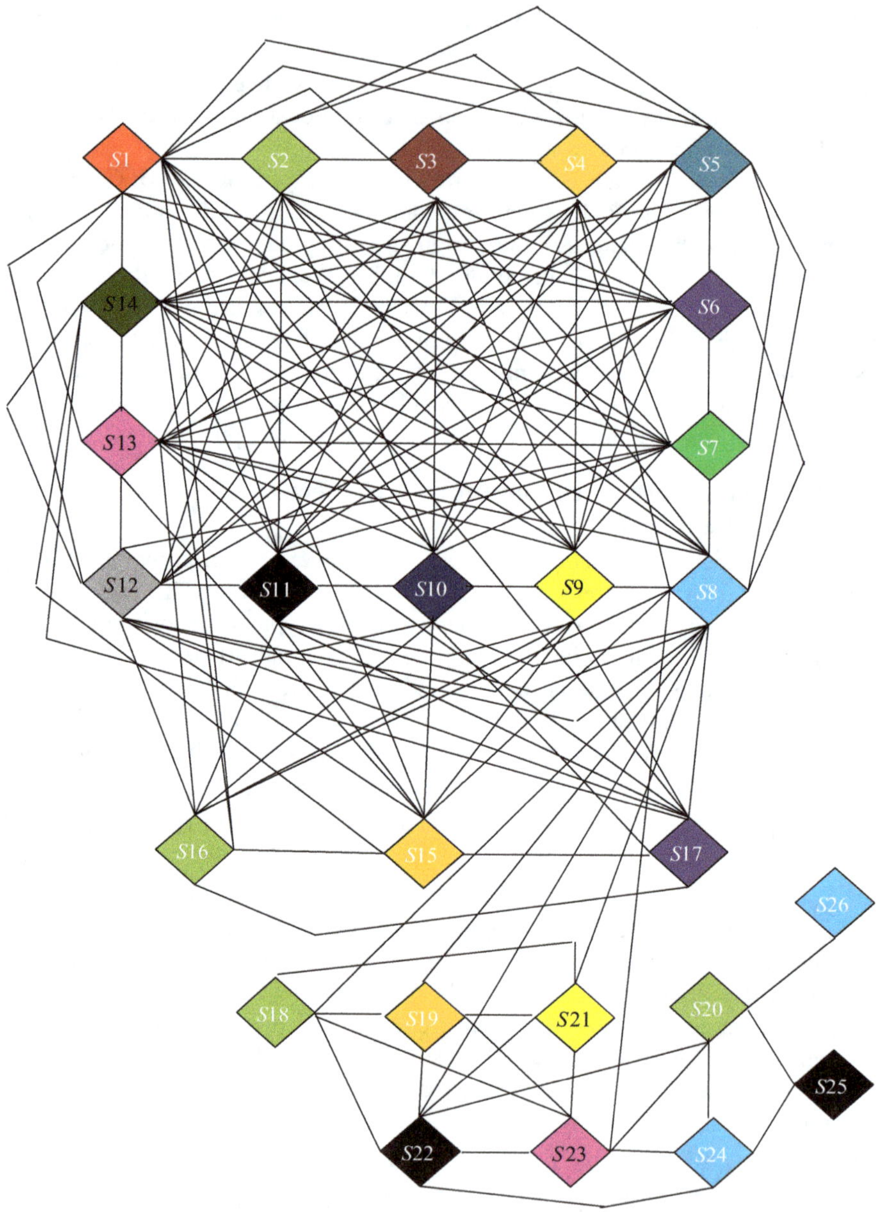

Figure 8.5 CIG of 4-point DFT: pre-embedding stego-constraints

3. Further, the secret design data is converted into a **state matrix** based on secret stego-key1. The detailed discussion on 'state matrix formation using secret stego-key1 has been provided in Chapter 2. For stego-key1 = '001' (mode-2:

Table 8.1 Register allocation (CIG of 4-point DFT) of obfuscated design pre-embedding *steganography*

Q	Red	Lime	Brown	Orange	Blue	Purple	Green	Cyan	Yellow	Navy	Black	Grey	Magenta	Olive
0	S1	S2	S3	S4	S5	S6	S7	S8	S9	S10	S11	S12	S13	S14
1	S1	S16	–	S15	–	S17	–	S8	S9	S10	S11	S12	S13	S14
2		S18	–	S19	–	–	–	S8	S21	–	S22	–	S23	–
3		S20	–	–	–	–	–	S24	–	–	S22	–	S23	–
4		S20	–	–	–	–	–	S24	–	–	S25	–	–	–
5		S20	–	–	–	–	–	S26	–	–	–	–	–	–

choose four elements and skip four elements), the state matrix M_S is given as follows:

$$M_S = \begin{bmatrix} 21 & 23 & 25 & 13 \\ F4 & 62 & 89 & 8B \end{bmatrix} \tag{8.6}$$

4. The next step in crypto-steganography approach is **byte substitution**. The matrix M_B post-byte substitution using forward S-box is as follows:

$$M_B = \begin{bmatrix} FD & 26 & 3F & 7D \\ BF & AA & A7 & 3D \end{bmatrix} \tag{8.7}$$

5. Further, **row diffusion** is performed based on secret stego-key2. The detailed discussion on row diffusion using stego-key2 has been provided in Chapter 2. For stego-key2 = '01 00' (for row-1, the chosen mode is circular right shift by two elements; for row-2, the chosen mode is circular right shift by one element), the matrix M_{Rd} is given as follows:

$$M_{Rd} = \begin{bmatrix} 3F & 7D & FD & 26 \\ 3D & BF & AA & A7 \end{bmatrix} \tag{8.8}$$

6. Next, **Trifid-cipher-based encryption** is performed based on secret stego-key3. The detailed discussion on Trifid-cipher-based encryption using stego-key3 has been provided in Chapter 2. There are four unique alphabets A, B, D and F in the matrix, which are encrypted based on the following encryption keys (the stego-key3):

   Encryption key for A: V # Q A W S E D R F T G Y H U J I K O L P Z M X N C B

   Encryption key for B: Q A W S E D R F T G Y H U J I K # O L P Z M X N C B V

   Encryption key for D: G Y H U J I K # O L P Z M X N C B V Q A W S E D R F T

   Encryption key for F: L P Z M X N C B V Q A W S E D R F T G Y H U J I K # O

Table 8.2 Details of alphabet substitution step of crypto-steganography

Alphabets	Encrypted value	Stego-key4	Selected mathematical expression	Computed equivalent value
A	211	001	$a+b+c$	4
B	323	001	$a+b+c$	8
D	233	100	$\lvert(c+b)/a\rvert$	3
F	322	100	$\lvert(c+b)/a\rvert$	1

Based on the encryption keys, the encrypted values (*abc*) of alphabets *A*, *B*, *D* and *F* are '211', '323', '233' and '322' respectively.

7. Next, **alphabet substitution** is performed in matrix M_{Rd} based on secret stego-key4. The detailed discussion on alphabet substitution using stego-key4 has been provided in Chapter 2. For stego-key4 = '001 001 100 100', modes for computing equivalent values of alphabets *A*, *B*, *D* and *F* are as follows, respectively: $a+b+c$, $a+b+c$, $\lvert(c+b)/a\rvert$ and $\lvert(c+b)/a\rvert$. Table 8.2 highlights the encrypted values of alphabets *A*, *B*, *D* and *F* and their respective equivalent values based on stego-key4 and selected mathematical expressions. Thus obtained equivalent values are used to substitute respective alphabets in the matrix $M_{Rd.}$ The matrix M_{AS} post alphabet substitution is as follows:

$$M_{Rd} = \begin{bmatrix} 31 & 73 & 13 & 26 \\ 33 & 81 & 44 & 47 \end{bmatrix} \tag{8.9}$$

8. Further, **matrix transposition** is performed. The transposed matrix M_T is as follows:

$$M_T = \begin{bmatrix} 31 & 33 \\ 73 & 81 \\ 13 & 44 \\ 26 & 47 \end{bmatrix} \tag{8.10}$$

9. Further, **mix column diffusion** is performed by using a circulant MDS (maximum distance separable) matrix. The detailed process of mix column diffusion has been discussed in Chapter 2. Post-performing mix column diffusion, the matrix M_{Cd} is given as follows:

$$M_{Cd} = \begin{bmatrix} C2 & FD \\ C4 & A1 \\ 0E & F3 \\ 7F & 1E \end{bmatrix} \tag{8.11}$$

10. Next, **byte concatenation** is performed in matrix M_{Cd} based on secret stego-key5. The detailed discussion on byte concatenation using stego-key5 has been provided in Chapter 2. For stego-key5 = '001 000', the following modes are used for column-1 and column-2, respectively: B0B1B3B2, B0B1B2B3.

Hence the concatenated byte stream is as follows: 'C2C47F0EFDA1F31E'. The corresponding bitstream is as follows:

'1100001011000100011111110000111011111101101000011111001100011110'

11. Thus, obtained encrypted bitstream is **truncated** based on designer's specified size of stego-constraints. For stego-constraints size = 27, the truncated bitstream is as follows: '110000101100010001111111000'.

    The truncated bitstream contains fourteen 0s and thirteen 1s.
12. Further, the 0s and 1s in the truncated bitstream are mapped to stego-constraints based on the following mapping rules:
    (i) For each appearance of '0' in the bitstream, embed an artificial constraint edge between node pair (even, even) into the CIG during register allocation phase of HLS.

        Based on this mapping rule of '0' bit, the corresponding stego-constraints to be embedded into the CIG are as follows: $\langle S2,S4\rangle$, $\langle S2,S6\rangle$, $\langle S2,S8\rangle$, $\langle S2,S10\rangle$, $\langle S2,S12\rangle$, $\langle S2,S14\rangle$, $\langle S2,S16\rangle$, $\langle S2,S18\rangle$, $\langle S2,S20\rangle$, $\langle S2,S22\rangle$, $\langle S2,S24\rangle$, $\langle S2,S26\rangle$, $\langle S4,S6\rangle$, $\langle S4,S8\rangle$.
    (ii) For each appearance of '1' in the bitstream, either an odd operation is assigned to FUs of V1 or even operations to FU of vendor V2 during resource allocation phase of HLS.

        Based on this mapping rule of '1' bit, the corresponding stego-constraints to be embedded in the scheduled DFG in the form of FU vendor reallocation to the 12 operations (O1–O12) of the 4-point DFT design are as follows:

    O1→V1, O2→V2, O3→V1, O4→V2, O5→V1, O6→V2, O7→V1, O8→V2, O9→V1, O10→V2, O11→V1, O12→V2

        It is worth noting that there are total thirteen 1s in the truncated bitstream; however, only twelve 1s can be mapped to stego-constraints because of the availability of only 12 operations in the 4-point DFT design.

13. Stego-constraints obtained from mapping of '0' bits into constraints edges are embedded into the CIG in the form of artificial edges. Post-embedding constraint edges, the CIG of 4-point DFT design is shown in Figure 8.6. As shown in the CIG, storage variables (nodes) $S16$, $S18$ and $S20$ are subjected to register (colour) reallocation in order to enable embedding of all constraint edges into the CIG. The tabular representation of register reallocation of storage variables into different control steps is shown in Table 8.3.

Further, stego-constraints obtained from mapping of '1' bits are embedded by performing FU vendor reallocation as follows:

O1→M_1^1, O2→M_1^2, O3→M_2^1, O4→M_1^2, O5→M_2^1, O6→M_1^1, O7→A_1^1, O8→A_1^2, O9→A_1^2, O10→A_1^2, O11→A_1^1, O12→A_1^2

As evident from the FU resource reallocation shown earlier, *operations 6 and 9 have not been allocated to even and odd vendor, respectively*, as per the mapping

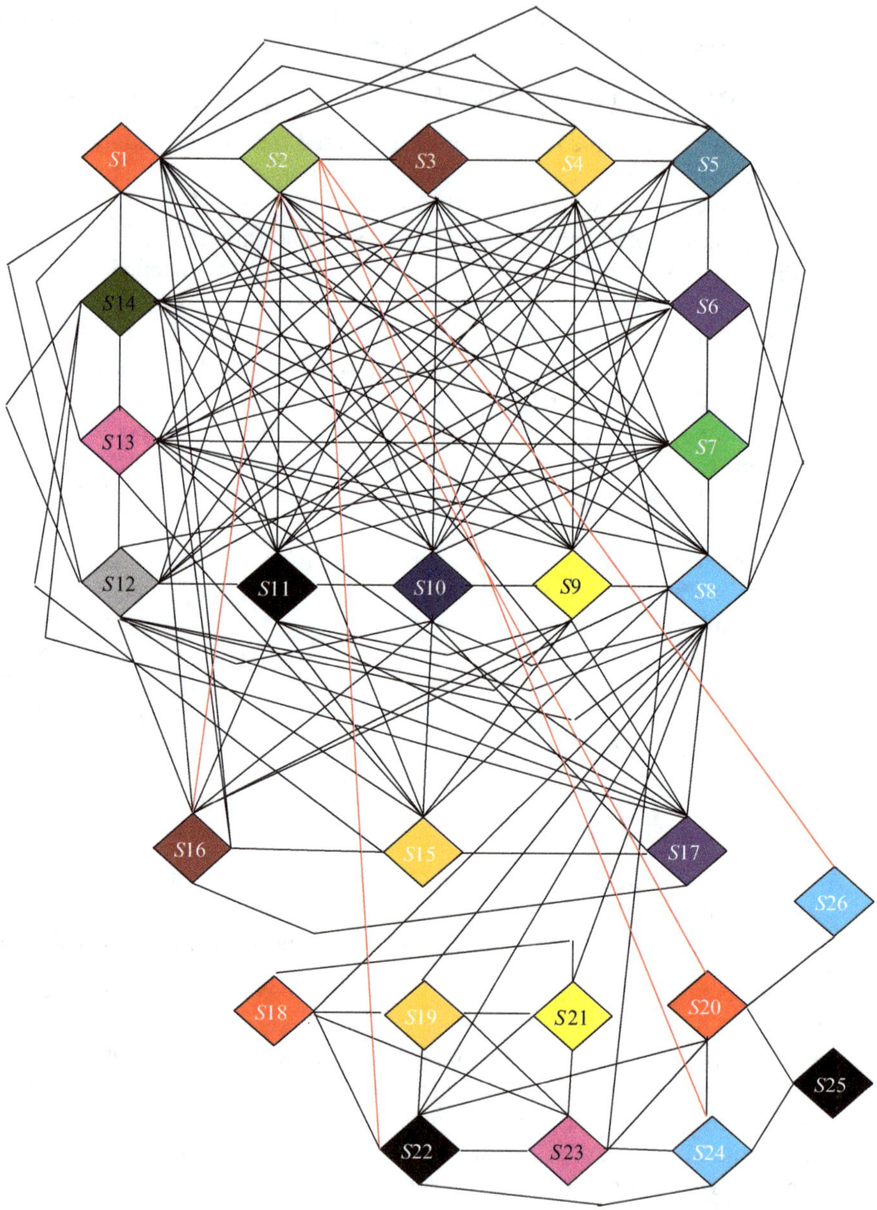

Figure 8.6 CIG of 4-point DFT: post-embedding stego-constraints

rule. Instead, operations 6 and 9 have been allocated to odd and even vendor, respectively. This is because the intended vendor allocation is not possible for operations 6 and 9 in the respective control step.

Table 8.3 Register allocation (CIG of 4-point DFT) of obfuscated design post-embedding steganography

Q	Red	Lime	Brown	Ora-nge	Blue	Purple	Green	Cyan	Yel-low	Navy	Black	Grey	Mag-enta	Olive
0	S1	S2	S3	S4	S5	S6	S7	S8	S9	S10	S11	S12	S13	S14
1	S1	–	S16	S15	–	S17	–	S8	S9	S10	S11	S12	S13	S14
2	S18	–	–	S19	–	–	–	S8	S21	–	S22	–	S23	–
3	S20	–	–	–	–	–	–	S24	–	–	S22	–	S23	–
4	S20	–	–	–	–	–	–	S24	–	–	S25	–	–	–
5	S20	–	–	–	–	–	–	S26	–	–	–	–	–	–

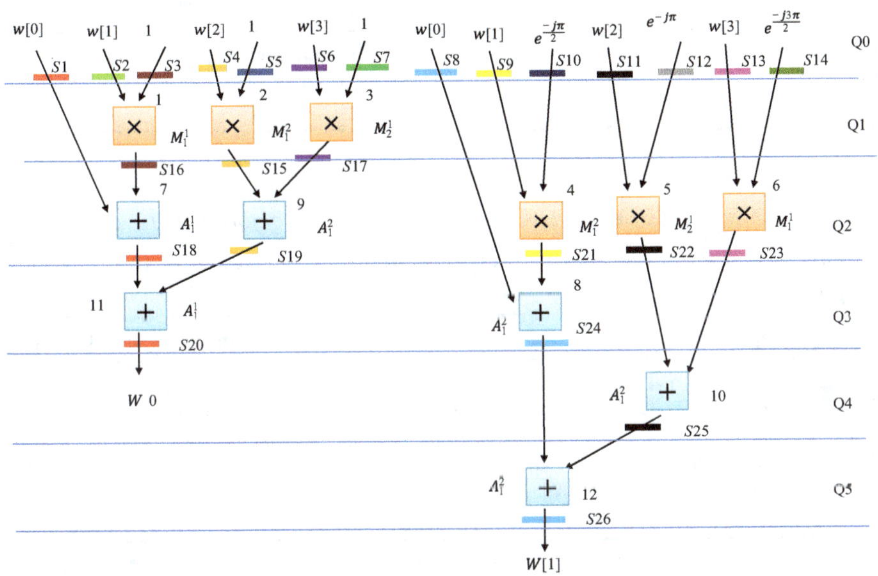

Figure 8.7 Scheduled DFG of obfuscated 4-point DFT post-embedding stego-constraints (Rathor and Sengupta, 2020)

The scheduled and allocated DFG of structurally obfuscated 4-point DFT design, post-embedding stego-constraints, is shown in Figure 8.7. As shown in the figure, the register reallocation of storage variables and FU vendor reallocation to operations highlight the impact of embedding stego-constraints. Thus a stego-embedded structurally obfuscated scheduled and resource-allocated DFG of 4-point DFT processor is obtained.

Formulation of multiplexing and de-multiplexing for registers and FU resources: Tables 8.4(a)–(e) show the multiplexing and de-multiplexing of

Table 8.4(a) Multiplexing tables for multipliers M_1^1 and M_2^1 before steganography

Control steps	M_1^1			M_2^1		
	Input1	**Input2**	**Output**	**Input1**	**Input2**	**Output**
Q0	Lime_out0	Brown_out	–	Orange_out0	Blue_out	–
Q1	Yellow_out0	Navy_out	Lime_in1	Black_out0	Grey_out	Orange_in1
Q2	–	–	Yellow_in1	–	–	Black_in1

Table 8.4(b) Multiplexing tables for multiplier M_1^2 before steganography

Control steps	M_1^2		
	Input1	**Input2**	**Output**
Q0	Purple_out0	Green_out	–
Q1	Magenta_out0	Olive_out	Purple_in1
Q2	–	–	Magenta_in1

Table 8.4(c) Multiplexing tables for adders A_1^1 and A_1^2 before steganography

Control steps	A_1^1			A_1^2		
	Input1	**Input2**	**Output**	**Input1**	**Input2**	**Output**
Q1	Red_out0	Lime_out1	–	Orange_out1	Purple_out1	–
Q2	Lime_out2	Orange_out2	Lime_in2	Cyan_out0	Yellow_out1	Orange_in2
Q3	Black_out1	Magenta_out1	Lime_in3	–	–	Cyan_in1
Q4	Cyan_out1	Black_out2	Black_in2	–	–	–
Q5	–	–	Cyan_in2	–	–	–

multiplier resources, adder resources and registers before embedding stego-constraints. This multiplexing and de-multiplexing are derived from scheduled and resource-allocated DFG (shown in Figure 8.4) of structurally obfuscated 4-point DFT design. The multiplexing and de-multiplexing shown in Tables 8.4(a)–(e) are exploited to synthesize RTL data path and controller of 4-point DFT processor. Since registers lime, orange, purple, cyan, yellow, black and magenta are used to store multiple storage variables throughout the control steps Q0–Q5, they require multiplexing and de-multiplexing. In the multiplexing of registers shown in Tables 8.4(d) and (e), L and R indicate left and right multiplexers associated with FU (M and A) resources. Further, in0–in3 indicates multiplexers input in order and out0–out3 indicates de-multiplexer outputs in order.

Table 8.4(d) Multiplexing tables for registers lime, orange and purple before steganography

Control steps	Lime		Orange		Purple	
	Input	**Output**	**Input**	**Output**	**Input**	**Output**
Q0	$w[1]$	M_1^1_in0_L	$w[2]$	M_2^1_in0_L	$w[3]$	M_1^2_in0_L
Q1	M_1^1_out0	A_1^1_in0_R	M_2^1_out0	A_1^2_in0_L	M_1^2_out0	A_1^2_in0_R
Q2	A_1^1_out0	A_1^1_in1_L	A_1^2_out0	A_1^1_in1_R	–	–
Q3	A_1^1_out1	$W[0]$	–	–	–	–

Table 8.4(e) Multiplexing tables for registers cyan, yellow, black and magenta before steganography

Control steps	Cyan		Yellow		Black		Magenta	
	Input	**Output**	**Input**	**Output**	**Input**	**Output**	**Input**	**Output**
Q0	$w[0]$	A_1^2_in1_L	$w[1]$	M_1^1_in1_L	$w[2]$	M_2^1_in1_L	$w[3]$	M_1^2_in 1_L
Q1	–	–	–	–	–	–	–	–
Q2	–	–	M_1^1_out1	A_1^2_in1_R	M_2^1_ out1	A_1^1_in2_L	M_1^2_ out1	A_1^1_in 2_R
Q3	A_1^2_out1	A_1^1_in3_L	–	–			–	–
Q4		–	–	–	A_1^1_ out2	A_1^1_in3_R	–	–
Q5	A_1^1_out3	$W[1]$	–	–	–	–	–	–

Further, Tables 8.5(a)–(e) show the multiplexing and de-multiplexing of multiplier resources, adder resources and registers post-embedding stego-constraints. This multiplexing and de-multiplexing in Tables 8.5(a)–(e) have been derived from stego-embedded scheduled and resource-allocated DFG (shown in Figure 8.7) of structurally obfuscated 4-point DFT design.

Data path and controller synthesis: Post formulating multiplexing and de-multiplexing of multiplier resources, adder resources and registers, data path and controller are synthesized to obtain RTL data path. Figure 8.8 shows the structurally obfuscated RTL data path of 4-point DFT processor before embedding stego-constraints. This data path shown in Figure 8.8 is based on the scheduling shown in Figure 8.4 and multiplexing–de-multiplexing shown in Tables 8.4(a)–(e). Figure 8.9 shows the stego-embedded structurally obfuscated RTL data path of the 4-point DFT processor. This data path shown in Figure 8.9 is based on the scheduling shown in Figure 8.7 and multiplexing–de-multiplexing shown in Tables 8.5(a)–(e). In both Figures 8.8 and 8.9, multiplexing and de-multiplexing of registers, multiplier resources and adder resources have

Table 8.5(a) Multiplexing tables for multipliers M_1^1 and M_2^1 *after steganography*

Control steps	M_1^1			M_2^1		
	Input1	**Input2**	**Output**	**Input1**	**Input2**	**Output**
Q0	Lime_out	Brown_out0	–	Purple_out0	Green_out	–
Q1	Magenta_out0	Olive_out	Brown_in1	Black_out0	Grey_out	Purple_in1
Q2	–	–	Magenta_in1	–	–	Black_in1

Table 8.5(b) Multiplexing tables for multiplier M_1^2 *after steganography*

Control steps	M_1^2		
	Input1	**Input2**	**Output**
Q0	Orange_out0	Blue_out	–
Q1	Yellow_out0	Navy_out	Orange _in1
Q2	–	–	Yellow_in1

Table 8.5(c) Multiplexing tables for adders A_1^1 and A_1^2 *after steganography*

Control steps	A_1^1			A_1^2		
	Input1	**Input2**	**Output**	**Input1**	**Input2**	**Output**
Q1	Red_out0	Brown_out1	–	Orange_out1	Purple_out1	–
Q2	Red_out1	Orange_out2	Red_in1	Cyan_out0	Yellow_out1	Orange_in2
Q3	–	–	Red_in2	Black_out1	Magenta_out1	Cyan_in1
Q4	–	–	–	Cyan_out1	Black_out2	Black_in2
Q5	–	–	–			Cyan_in2

been highlighted. Further, in Figure 8.9, the impact of embedding stego-constraints on RTL structure has been encircled using dotted red ovals.

8.3 Analysis of case study

The *N*-point DFT hardware accelerator design has been secured using structural obfuscation (security mechanism-1) and crypto-steganography (security mechanism-2). The case study in terms of security analysis and its impact on design cost has been discussed in this section. The following subsections discuss security analysis of structural obfuscation, security analysis of steganography and design

Table 8.5(d) Multiplexing tables for registers red, brown, orange and purple after steganography

Control steps	Red		Brown		Orange		Purple	
	Input	**Output**	**Input**	**Output**	**Input**	**Output**	**Input**	**Output**
Q0	$w[0]$	A_1^1_in 0_L	1	M_1^1_in 0_R	$w[2]$	M_1^2_in0_L	$w[3]$	M_2^1_in 0_L
Q1	–	–	M_1^1_out0	A_1^1_in 0_R	M_1^2_out0	A_1^2_in0_L	M_2^1_out0	A_1^2_in 0_R
Q2	A_1^1_out0	A_1^1_in 1_L	–	–	A_1^2_out0	A_1^1_in1_R	–	–
Q3	A_1^1_out1	$W[0]$	–	–	–	–	–	–

Table 8.5(e) Multiplexing tables for registers cyan, yellow, black and magenta after steganography

Control steps	Cyan		Yellow		Black		Magenta	
	Input	**Output**	**Input**	**Output**	**Input**	**Output**	**Input**	**Output**
Q0	$w[0]$	A_1^2_in 1_L	$w[1]$	M_1^2_in 1_L	$w[2]$	M_2^1_in 1_L	$w[3]$	M_1^1_in 1_L
Q1	–	–	–	–	–	–	–	–
Q2	–	–	M_1^2_out1	A_1^2_in 1_R	M_2^1_out1	A_1^2_in 2_L	M_1^1_out1	A_1^2_in 2_R
Q3	A_1^2_out1	A_1^2_in 3_L	–	–	–	–	–	–
Q4	–	–	–	–	A_1^2_out2	A_1^2_in3_R	–	–
Q5	A_1^2_out3	$W[1]$	–	–	–	–	–	–

cost analysis of *N*-point DFT hardware accelerator design (Rathor and Sengupta, 2020).

8.3.1 Security analysis of structural obfuscation

The THT-based structural obfuscation incurs significant obscurity in the RTL structure of *N*-point DFT design. The obfuscated RTL design further leads to significant obscurity in the gate-level netlist obtained post-RTL synthesis. Therefore, the strength of structural obfuscation has been measured in terms of % gates affected due to obfuscation with respect to baseline (un-obfuscated/unsecured) version. Figure 8.10 shows the difference in gate count (*NGx*), number of gates modified (*NGy*) and total gates affected (*NGx*+*NGy*), post-structural obfuscation. Further, security achieved due to structural obfuscation has been measured using the following formula (Rathor and Sengupta, 2020):

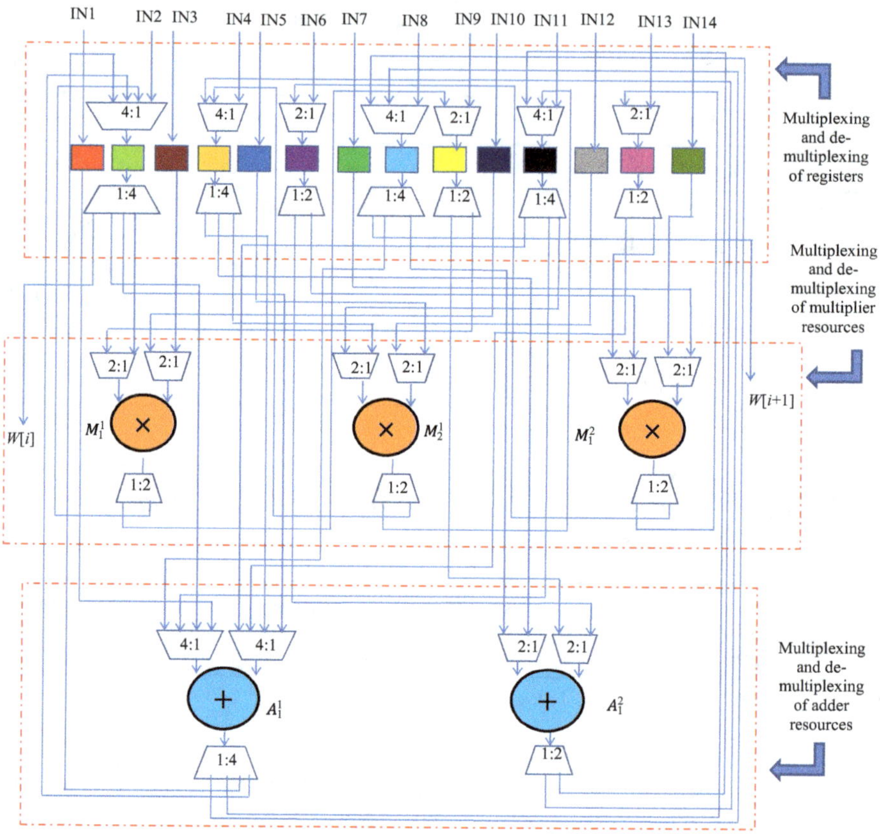

Figure 8.8 Structurally obfuscated RTL data path of 4-point DFT design before embedding steganography

$$\text{Strength of obfuscation w.r.t. baseline}(\%)$$

$$= \frac{\text{total gates affected due to obfuscation}}{\text{total gates in baseline}} \times 100 \tag{8.12}$$

$$\text{Strength of obfuscation w.r.t. baseline}(\%) = \frac{4,336}{5,760} \times 100 = 75.28\%$$

As evident, the strength of obfuscation for the obfuscated 4-point DFT design is obtained to be 75.28%. High value of strength of obfuscation indicates that the structurally obfuscated DFT design is harder to be interpreted through RE by an attacker, thus thwarting Trojan insertion.

8.3.2 Security analysis of steganography

A stego-mark (or stego-constraints) embedded into DFT hardware design ensures security against the false claim of ownership and piracy threats. The robustness of

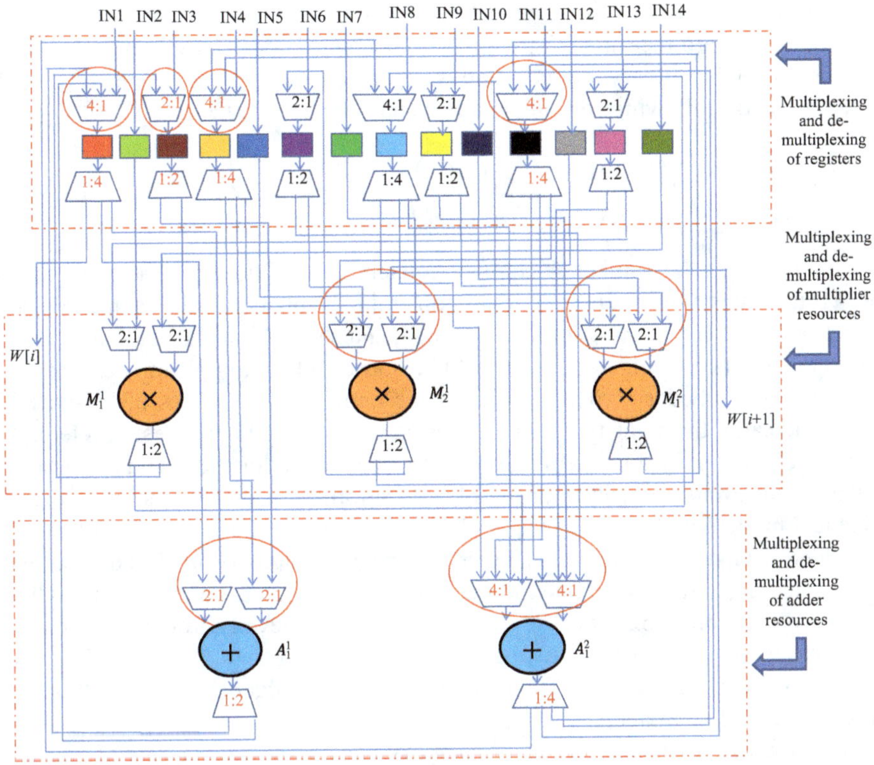

Figure 8.9 Secured 4-point DFT hardware accelerator at RTL (post-structural obfuscation and steganography) (Rathor and Sengupta, 2020)

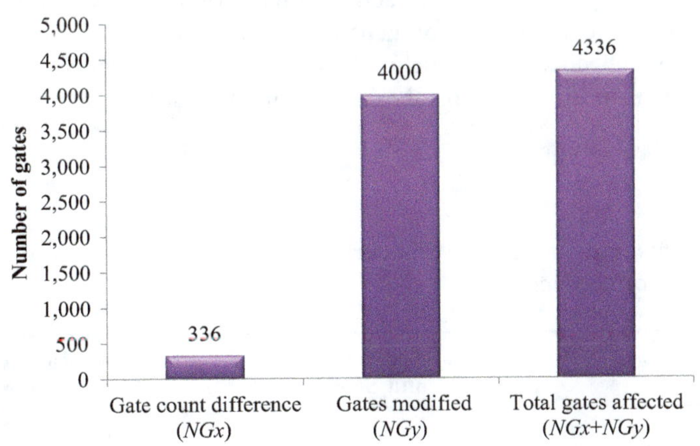

Figure 8.10 Structural obfuscation analysis with respect to baseline (Rathor and Sengupta, 2020)

stego-mark is measured using probability of coincidence (Pc) metric. The Pc metric is a standard measure of strength of ownership proof. To achieve a stronger proof of ownership, a very low Pc value is expected to be achieved. The Pc value is evaluated using following formula (Rathor and Sengupta, 2020):

$$Pc = \left(1 - \frac{1}{h}\right)^{k1} \times \left(1 - \frac{1}{\pi_{j=1}^{m} N(U_j)}\right)^{k2} \tag{8.13}$$

where h indicates the number of colours/registers in the CIG before steganography and $k1$ indicates the number of stego-constraints embedded during the register allocation phase (i.e. number of 0s embedded). Further, $k2$ indicates the number of stego-constraints embedded during the FU resource allocation phase (i.e. effective number of 1s embedded), $N(U_j)$ indicates the number of resources of FU-type U_j and m indicates the total types of FU resources present in the design. Here, $k1$ (number of stego-constraints embedded during register allocation) and $k2$ (number of stego-constraints embedded during FU allocation) indicate the amount of digital evidence hidden within the design. Table 8.6 shows the Pc value of stego-embedded DFT design obtained using crypto-steganography (Rathor and Sengupta, 2020) and compares with a contemporary approach (Sengupta and Rathor, 2019a). As shown in Table 8.6, lower Pc (as desirable) is achieved through (Rathor and Sengupta, 2020) with respect to the contemporary approach. This is because of embedding of more number of constraints in crypto-steganography (Rathor and Sengupta, 2020) approach.

Further, the robustness of stego-mark has been assessed in terms of key size required to produce stego-constraints. Table 8.7 highlights the total stego-key size required in crypto-steganography (Rathor and Sengupta, 2020) approach and compares with the contemporary approach (Sengupta and Rathor, 2019a). As evident, the crypto-steganography approach requires a very large size key (401 bits) which enhances the robustness of generated stego-constraints. Thus, a highly secured stego-embedded and structurally obfuscated DFT processor design is achieved, which is resilient against the false claim of ownership and piracy threats.

Table 8.6 Comparative analysis of security of N*-point DFT in terms of probability of coincidence (Pc)*

Approaches	**Approach (Rathor and Sengupta, 2020)**	**Approach (Sengupta and Rathor, 2019a)**
Number of registers (colours)	14	14
Number of constraints $k1$	14	14
Number of constraints $k2$	10	0
Probability of coincidence	5.72E−2	3.54E−01

Table 8.7 Comparative analysis of security of N-*point DFT in terms of key size*

Approaches	**Approach (Rathor and Sengupta, 2020)**						**Approach (Sengupta and Rathor, 2019a)**
	Stego-key1	**Stego-key2**	**Stego-key3**	**Stego-key4**	**Stego-key5**	**Total key size**	
Key size (in bits)	3	4	376	12	6	401	2

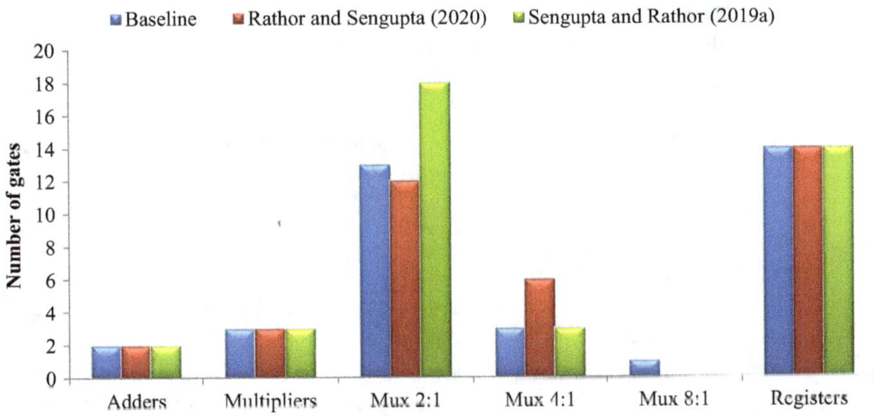

Figure 8.11 RTL components analysis between contemporary approaches and baseline (Rathor and Sengupta, 2020)

8.3.3 Design cost analysis

This subsection discusses the impact of employing structural obfuscation and steganography-based security on design cost. The following equation is used to evaluate the design cost (Sengupta and Rathor, 2019b):

$$C_d(U_i) = \rho_1 \frac{L_d}{L_m} + \rho_2 \frac{A_d}{A_m} \tag{8.14}$$

where $C_d(U_i)$ is the design cost of DFT processor for resource constraints U_i, further L_d and L_m are the design latency at specified resource constraints and maximum design latency, respectively, A_d and A_m are the design area at specified resource constraints and maximum area, respectively, and ρ_1, ρ_2 are the weights which are fixed at 0.5.

Because of employing structural obfuscation and steganography based security, the impact on RTL components with respect to baseline (unsecured) version is highlighted in Figure 8.11. Further, Figure 8.11 also compares the RTL

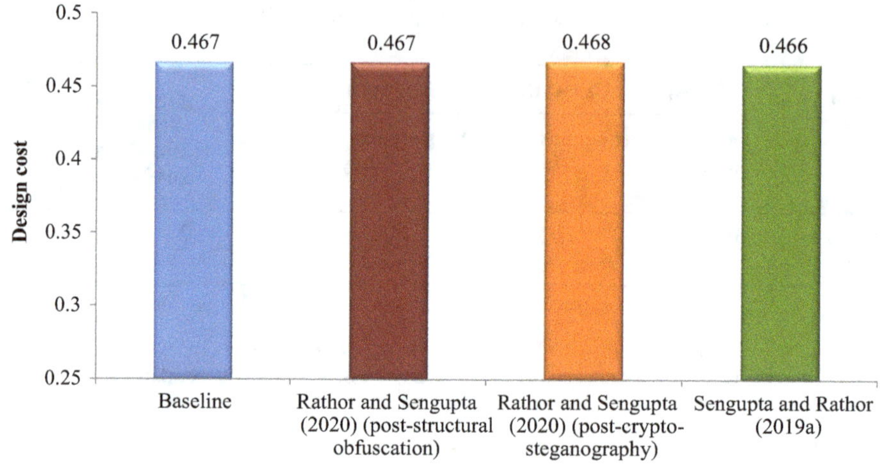

Figure 8.12 Design cost analysis with respect to two security approaches and baseline (Rathor and Sengupta, 2020)

components between the two security approaches made by Sengupta and Rathor (2019a) and Rathor and Sengupta (2020) as well as baseline. Furthermore, the impact on design cost due to employing structural obfuscation and steganography is highlighted in Figure 8.12. As shown in the figure, the design cost overhead post-employing structural obfuscation and steganography is negligible.

8.4 Conclusion

N-Point DFT is an important DSP algorithm which finds numerous applications in electronics. A hardware accelerator design of DFT is vital to improve system performance. However, the design process of a DFT hardware accelerator poses security risks due to growing hardware threats such as Trojan insertion, false claim of ownership and piracy. This entails enabling security of DFT hardware accelerator designs, by the designer/vendor. This chapter discusses a secured design flow of *N*-point DFT hardware accelerator using HLS framework. The robust security of DFT hardware accelerator design has been ensured using two security mechanisms, structural obfuscation and crypto-steganography. Steganography-based security mechanism employed on the top of structural obfuscation enables detective control along with preventive control. The case study of 4-point DFT hardware accelerator in terms of security and design cost analysis shows that robust security has been achieved at the cost of negligible design overhead.

A summary of important concepts that this chapter delivers to prospective readers is as follows:

- need of securing DFT hardware accelerator;
- a secured design flow of *N*-point DFT hardware accelerator;
- structural-obfuscation-based security mechanism-1 for securing DFT hardware accelerator;
- crypto-steganography-based security mechanism-2 for enhancing security of DFT hardware accelerator;
- process of designing a secured 4-point DFT hardware accelerator using integration of structural obfuscation and crypto-steganography;
- multiplexing and de-multiplexing of FU resources and registers to synthesize RTL data path of secured 4-point DFT processor and
- a comparative study in terms of security and design cost analysis between different security approaches.

8.5 Questions and exercise

1. Discuss the design flow of secured *N*-point DFT hardware accelerator.
2. What is the role of high-level synthesis framework in the secured design flow?
3. What are the major phases in crypto-based dual-phase hardware steganography?
4. What is the role of security mechanism-1 in the design flow of secured *N*-point DFT hardware accelerator?
5. What is the role of security mechanism-2 in the design flow of a secured *N*-point DFT hardware accelerator?
6. How is multiplexing scheme performed?
7. Security mechanism-1 safeguards against which threat?
8. Security mechanism-2 safeguards against which threat?
9. What is the generic equation of 4-point DFT?
10. Explain the process of secret design data extraction from the CIG of an *N*-point DFT.
11. What is the role of byte substitution in the design flow of a secured *N*-point DFT hardware accelerator?
12. What is the role of matrix transposition in the design flow of a secured *N*-point DFT hardware accelerator?
13. What is the key size of stego-key 3?
14. Derive the register allocation (CIG of 4-point DFT) of obfuscated design post-embedding steganography.
15. What are the structural differences achieved in an *N*-point DFT hardware accelerator post-structural obfuscation and steganography?
16. How is strength of obfuscation measured?
17. What is the probability of coincidence (*Pc*) metric used for measuring robustness of stego-mark in a 4-point DFT hardware accelerator?
18. What is the total key size used for crypto-steganography approach in a 4-point DFT hardware accelerator?
19. How is stego-constraints derived from crypto-steganography approach?
20. Analyse the design overhead of a 4-point DFT hardware accelerator design.

References

Y. Lao and K. K. Parhi (2015), 'Obfuscating DSP circuits via high-level transformations,' *IEEE Trans. Very Large Scale Integr. VLSI Syst.*, vol. 23(5), pp. 819–830.

C. Pilato, S. Garg, K. Wu, R. Karri and F. Regazzoni (2018), 'Securing hardware accelerators: a new challenge for high-level synthesis,' *IEEE Embedded Syst. Lett.*, vol. 10(3), pp. 77–80.

M. Rathor and A. Sengupta (2020), 'Design flow of secured N-point DFT application specific processor using obfuscation and steganography,' *Lett. IEEE Comput. Soc.*, vol. 3(1), pp. 13–16.

A. Sengupta and S. Bhadauria (2016), 'Exploring low cost optimal watermark for reusable IP cores during high level synthesis,' *IEEE Access*, vol. 4, pp. 2198–2215.

A. Sengupta, S. Bhadauria and S. P. Mohanty (2017b), 'Low-cost security aware HLS methodology,' *IET Comput. Digital Tech.*, vol. 11(2), pp. 68–79.

A. Sengupta, E. R. Kumar and N. P. Chandra (2019), 'Embedding digital signature using encrypted-hashing for protection of DSP cores in CE,' *IEEE Trans. Consum. Electron.*, (3), pp. 398–407.

A. Sengupta and S. P. Mohanty (2019), 'IP core protection and hardware-assisted security for consumer electronics', The Institute of Engineering and Technology (IET), Book ISBN: 978-1-78561-799-7, e-ISBN: 978-1-78561-800-0.

A. Sengupta and M. Rathor (2019a), 'IP core steganography for protecting DSP kernels used in CE systems,' *IEEE Trans. Consum. Electron.*, vol. 65(4), pp. 506–515.

A. Sengupta and M. Rathor (2019b), 'Crypto-based dual-phase hardware steganography for securing IP cores,' *Lett. IEEE Comput. Soc.*, vol. 2(4), pp. 32–35.

A. Sengupta, D. Roy, S. P. Mohanty and P. Corcoran (2017a), 'DSP design protection in CE through algorithmic transformation based structural obfuscation,' *IEEE Trans. Consum. Electron.*, vol. 63(4), pp. 467–476.

A. Sengupta, D. Roy, S. P. Mohanty and P. Corcoran (2018), 'Low-cost obfuscated JPEG CODEC IP core for secure CE hardware,' *IEEE Trans. Consum. Electron.*, vol. 64(3), pp. 365–374.

A. Sengupta (2017), 'Hardware security of CE devices [hardware matters],' *IEEE Consum. Electron. Mag.*, vol. 6(1), pp. 130–133.

A. Sengupta (2020), 'Frontiers in securing IP cores – forensic detective control and obfuscation techniques', The Institute of Engineering and Technology (IET), ISBN-10: 1-83953-031-6, ISBN-13: 978-1-83953-031-9.

X. Zhang and M. Tehranipoor (2011), 'Case study: detecting hardware Trojans in third-party digital IP cores,' *IEEE International Symposium on Hardware-Oriented Security and Trust*, San Diego, CA, pp. 67–70.

Chapter 9

Structural transformation-based obfuscation using pseudo-operation mixing for securing data-intensive IP cores

Anirban Sengupta[1] *and Mahendra Rathor*[1]

The chapter describes a structural transformation-based obfuscation approach using pseudo-operation mixing for securing data-intensive cores or hardware accelerators. The presented approach is based on pseudo-operation mixing algorithm that attains significant structural obscurity in the design to enable unobviousness without affecting the correct functionality.

The chapter is organized as follows: Section 9.1 discusses about the introduction of the chapter; Section 9.2 describes the structural transformation-based obfuscation methodology; Section 9.3 presents pseudo operation mixing based structural obfuscation (POM-SO) tool that is capable of performing pseudo-operation mixing-based structural obfuscation (SOB); Section 9.4 presents analysis of case studies in terms of security and design cost, especially focusing on digital signal processing (DSP) hardware accelerators; Section 9.5 presents conclusion; Section 9.6 provides some exercise and questions for readers.

9.1 Introduction

DSP algorithms such as discrete wavelet transform (DWT) and finite impulse response (FIR) filters are highly computational or data intensive (Schneiderman, 2010; Sengupta, 2020). Therefore, it is highly efficient to integrate such data-intensive intellectual property (IP) cores as hardware accelerators in a system-on-chip (SoC). However, participation of offshore foundries in the very large scale integration design process makes the DSP IP cores or SoC designs vulnerable to Trojan insertion threat posed by an adversary present in an untrusted foundry (Zhang and Tehranipoor, 2011; Sengupta, 2016, 2017; Sengupta and Mohanty 2019; Sengupta *et al.*, 2017; Sengupta and Rathor; 2019; Chakraborty and Bhunia, 2009). In order to address this hardware threat, Rathor and Sengupta (2020) proposed an SOB methodology which is based on structural transformation of the design during

[1]Computer Science and Engineering, Indian Institute of Technology Indore, Indore, India

high-level synthesis (HLS) process. The structural transformation is performed by mixing pseudo-operations during the HLS design process. Mixing of pseudo-operations in the design misguides/deludes a potential attacker who targets to reverse engineer the design to understand its true functionality and structure in order to insert Trojan. Thus, pseudo-operations mixing-based SOB approach (Rathor and Sengupta, 2020) prevents Trojan horse insertion attack by making the reverse engineering (RE) arduous, hence ensuring trust in hardware. The pseudo-operations mixing-based SOB can widely be applied for all kinds of DSP IP cores, regardless of the nature of the DSP application (i.e. operation count and data dependency of operations).

9.2 Structural transformation-based obfuscation methodology

To ensure the security of data-intensive DSP cores against hardware Trojan threat, the structural transformation technique discussed in this chapter is based on the mixing of pseudo-operations in the intended DSP application during HLS process. Let us discuss the high-level perspective followed by in-depth discussion of pseudo-operations mixing-based SOB approach (Rathor and Sengupta, 2020).

9.2.1 High-level perspective

Figure 9.1 shows the pseudo-operations mixing-based SOB approach to generate a structurally obfuscated design of DSP cores. As shown in the figure, the security aware HLS process using pseudo-operations mixing-based SOB takes in data-flow graph (DFG) of a target DSP application as input and produces a secure register transfer level (RTL) design at output. Further, scheduled and resource-allocated DFG is constructed on the basis of resource constraints and module library. Thus, obtained scheduled DFG is fed as an input to the pseudo-operations mixing-based SOB approach (Rathor and Sengupta, 2020). The SOB approach executes the following steps sequentially in order to produce a structurally obfuscate design: (i) determination of fake or pseudo-nodes/operations to be mixed in the scheduled and resource-allocated DFG of DSP application, (ii) insertion of pseudo-nodes/operations into the scheduled and resource-allocated DFG based on mixing rules, (iii) binding of pseudo/fake operations (multiplications, additions, etc.) to the existing functional unit (FU) (multipliers, adders, etc.) resources of the respective type based on binding rules. Once pseudo-operations are mixed and allocated to the existing FU resources, the modified (structurally transformed) scheduled and resource-allocated DFG of the DSP application is subjected to the data path and controller synthesis phase of HLS process. This results into a structurally obfuscated RTL design of a DSP hardware accelerator.

9.2.2 Pseudo-operations mixing-based structural obfuscation

An in-depth discussion of pseudo-operations mixing-based SOB approach for securing data-intensive DSP cores is presented in this subsection (Rathor and Sengupta, 2020).

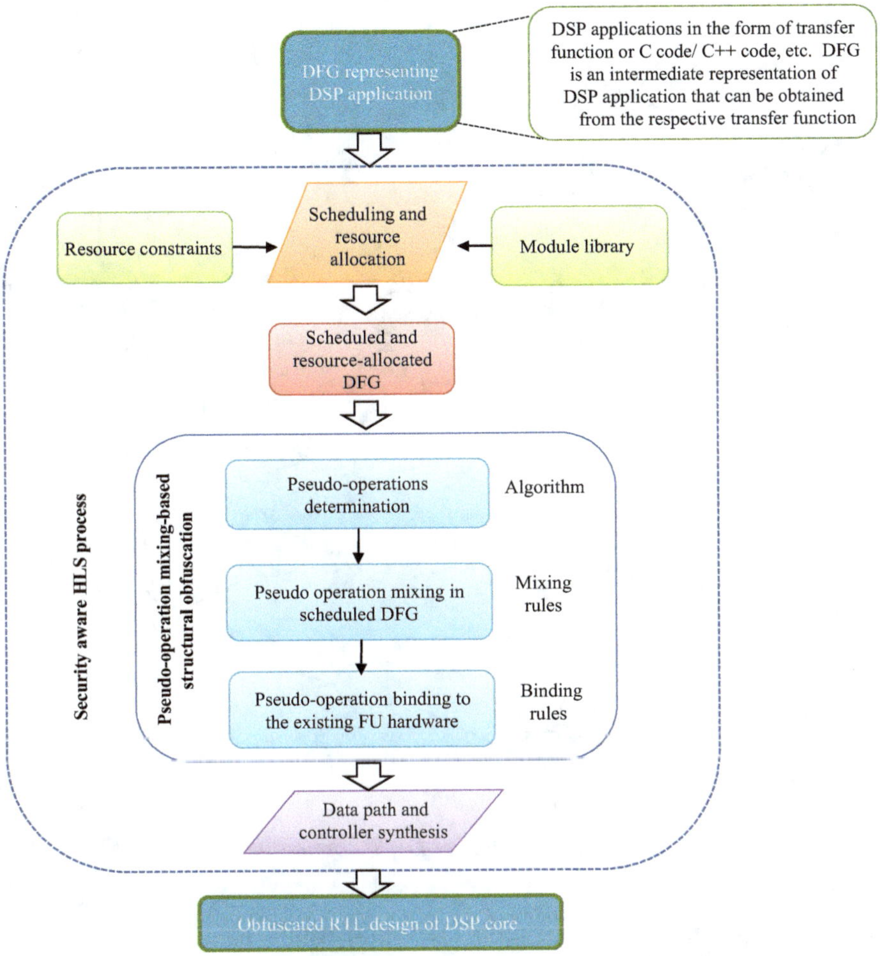

Figure 9.1 Structural transformation-based obfuscation approach for securing data-intensive DSP core (Rathor and Sengupta, 2020)

Additionally, the pseudo-operations mixing-based SOB approach has been demonstrated on DWT application. The process of designing structurally obfuscated DWT core using the SOB approach starts with its DFG representation shown in Figure 9.2. Further, the DFG of DWT is scheduled and resource allocated on the basis of module library and resource constraints of *two multipliers (*) and one adder (+)*. The scheduled and resource-allocated DFG of DWT core is shown in Figure 9.3. Once it is obtained, it is subjected to pseudo-operations mixing-based SOB approach which is executed in the following steps (Rathor and Sengupta, 2020).

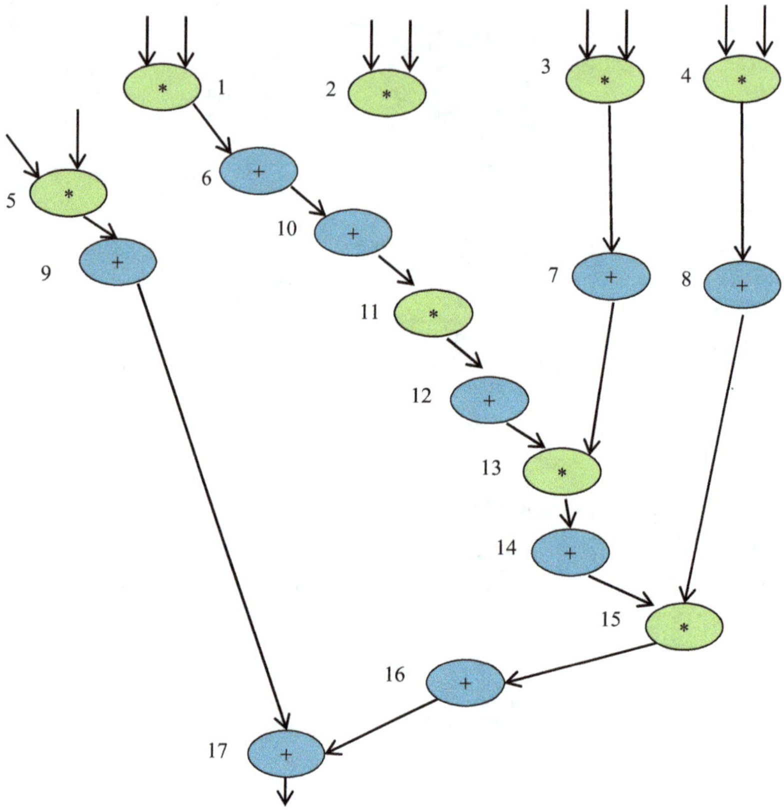

Figure 9.2 DFG of DWT application

9.2.2.1 Determination of pseudo/fake operations

The scheduled and resource-allocated DFG of intended DSP application is the input to this step. The resource constraints adopted during the scheduling and resource allocation govern the process of pseudo-nodes insertion in each control step (Q) of scheduled and resource-allocated DFG. While determining the pseudo-nodes to be mixed in the scheduled and resource-allocated DFG, it is ensured that the mixing of pseudo-nodes does not result into the designer's resource constraints violation. The flow chart/algorithm for determining pseudo-nodes/operations to be mixed in the scheduled and resource-allocated DFG is shown in Figure 9.4. As shown in the flow chart, each control step is checked to determine whether the insertion of pseudo-multiplication operation or pseudo-addition operation in the control step is possible or not. The algorithm runs for all the control steps. *The insertion of pseudo-operations can only be possible when all FU resources are not utilized to their maximum capacity (constraints) in each control step*. There are some control steps where either all instances of an FU resource are free or only some instances are utilized. These unutilized FU instances in a control step are leveraged for

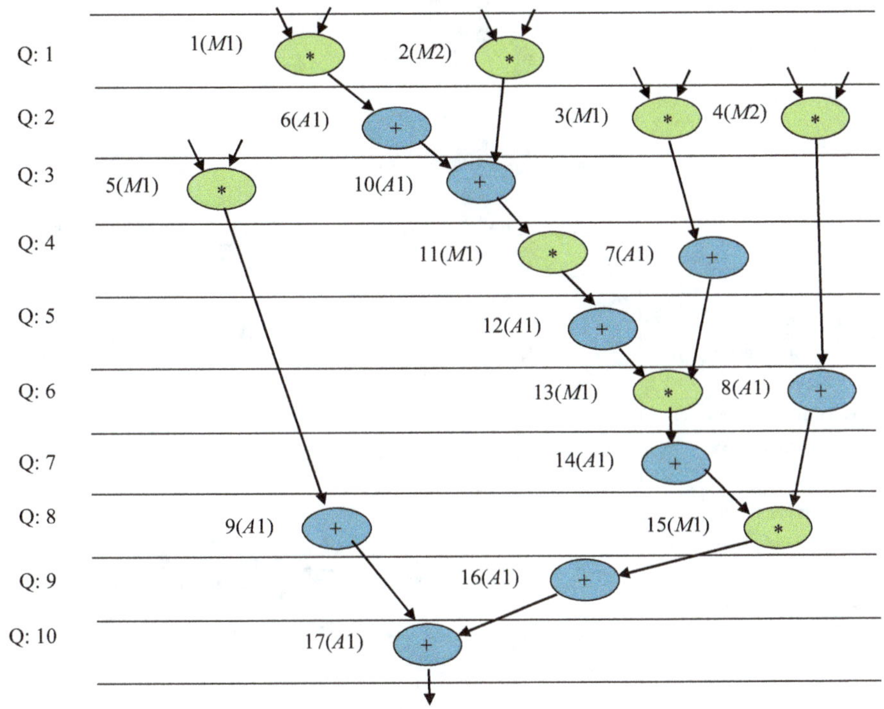

Figure 9.3 Scheduled and resource-allocated DFG of DWT application based on 2(*) and 1(+)

executing pseudo-operation of the respective type. Although multiple pseudo-operations of a specific type can be inserted in a potential control step, the algorithm presented in the flow chart ensures that maximum one addition/one multiplication is inserted. The output of this algorithm of determining pseudo-operations is a list *W* comprising the pseudo-operations to be mixed in the scheduled and resource-allocated DFG and corresponding control step number.

Upon applying the algorithm of pseudo-operations determination on scheduled and resource-allocated DFG of DWT, the resultant list *W* of pseudo-operations and corresponding control step number is shown in Table 9.1.

9.2.2.2 Mixing of pseudo-operations into the scheduled and resource-allocated DFG

The list of pseudo-operations and corresponding Q number obtained from previous step is exploited to insert pseudo-multiplication and additions operation in the scheduled and resource-allocated DFG. The mixing of pseudo-operations with original operations of scheduled and resource-allocated DFG is performed on the basis of the following rules (Rathor and Sengupta, 2020):

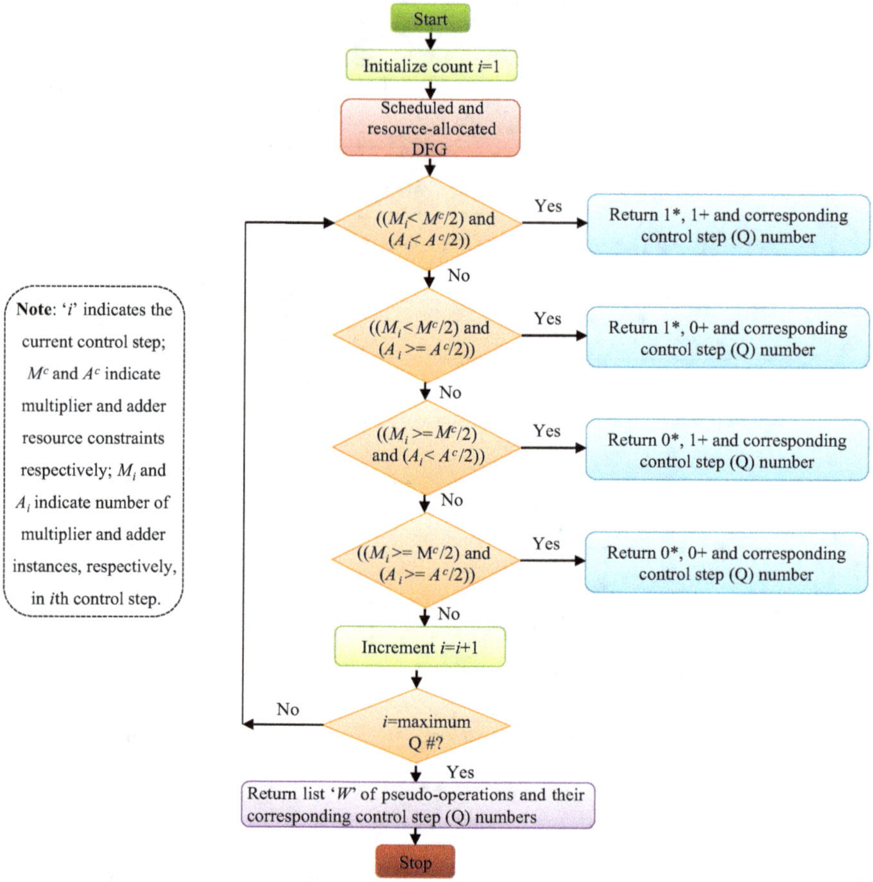

Figure 9.4 Flow chart of determining pseudo-nodes/operations to be mixed in scheduled DFG (Rathor and Sengupta, 2020)

1. Pseudo-operations corresponding to the first control step in the list W use any inputs of those original operations which do not have predecessor operations (i.e. primary inputs).
2. Pseudo-operations corresponding to the remaining control step in the list W use one input from the pseudo-operation and another input from any original operation located in the preceding control step.

Based on the above-mentioned rules, the mixing of pseudo-addition operations and pseudo-multiplication operations (highlighted in Table 9.1) in the scheduled and resource-allocated DFG of DWT core is shown in Figure 9.5. As shown in the figure, operation numbers 18, 19, 20, 21 and 22 (highlighted in red) are the pseudo-operations which have been mixed among the original operations. The mixing has been performed in such a manner that the adversary cannot distinguish the gates

Table 9.1 List W of pseudo-operations and corresponding control step number

Pseudo addition operations	1 '+'	0 '+'	0 '+'	0 '+'	0 '+'	0 '+'	0 '+'	0 '+'	0 '+'	0 '+'
Pseudo-multiplication operations	0 '*'	0 '*'	0 '*'	0 '*'	1 '*'	0 '*'	1 '*'	0 '*'	1 '*'	1 '*'
Control step (Q) number	1	2	3	4	5	6	7	8	9	10

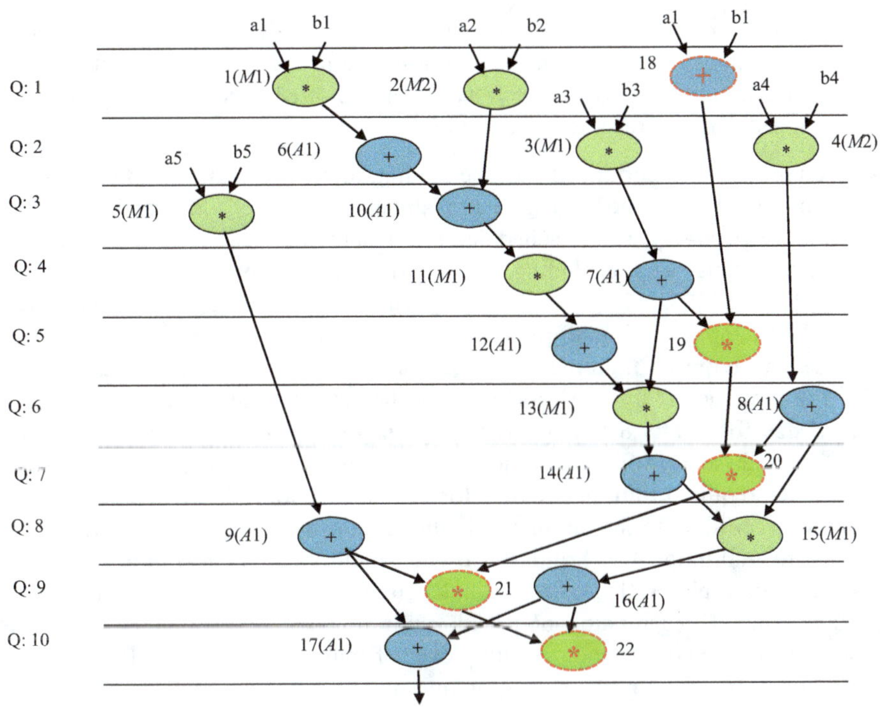

Figure 9.5 Scheduled and resource-allocated DFG of DWT application post mixing pseudo-operations

corresponding to the pseudo-operations among gates corresponding to the original operations during reverse engineering of the design.

9.2.2.3 Binding of pseudo-operations to the existing functional unit (FU) resources

So far, we have seen how the pseudo-operations are mixed into scheduled and resource-allocated DFG. This step shows how the binding of pseudo-operations, to the already available FUs of respective type (multiplier/adder) in scheduled and resource-allocated DFG, is performed. Here, it needs to ensure that the binding of pseudo-operations to the available FUs of respective type should lead to minimal interconnect hardware overhead post synthesizing the RTL data path. The overhead aware binding rules of pseudo-operations, to bind them with the already available FUs of respective type, are as follows (Rathor and Sengupta, 2020):

1. The corresponding multiplexer/demultiplexer size for all the instances of each FU resource type (used for FU sharing while synthesizing RTL data path) is determined in advance.
2. A pseudo-operation is assigned to that instance of corresponding FU type respective multiplexer/demultiplexer of which has the maximum number of free (unused) inputs/outputs.
3. If unused inputs/outputs are not available in the multiplexer/demultiplexer of corresponding FU type, then the pseudo-operation is assigned to that instance of corresponding FU type which has been exploited least number of times in the scheduled and resource-allocated DFG.

Based on the above-mentioned rules, the binding of pseudo-addition operations and pseudo-multiplication operations, to the already available adder and multiplier resources, respectively, in the scheduled and resource-allocated DFG of DWT, is performed. Let us see the binding of pseudo-multiplication operations based on binding rules. While binding pseudo-multiplication operation 19 in control step Q5, the available choices are multiplier $M1$ and $M2$. Before binding, the number of instances and required multiplexer size for $M1$ and $M2$ are determined, as highlighted in the blue column of Table 9.2. As shown in the table, the required multiplexer size for $M1$ is 8:1. However, only six inputs are currently in use. Two inputs are still free to offer two times more sharing of multiplier resource $M1$. In contrast, the required multiplexer size for $M2$ is 2:1 and both inputs are engaged. Therefore, according to the binding rule, the pseudo-multiplication operation 19 is assigned to multiplier $M1$. Similarly, pseudo-multiplication operation 20 is also assigned to multiplier $M1$ because of the availability of unused input in the corresponding Mux. Once binding of pseudo-multiplication operations 19 and 20 is accomplished, the corresponding multiplexer of multiplier $M1$ is not left with any unused input. In addition, the corresponding multiplexer of multiplier $M2$ has also not unused inputs. Therefore, the next pseudo-multiplication operation 21 is assigned to that multiplier instance which has been used least number of times (minimum multiplexer size) in the scheduled and allocated DFG. Therefore, multiplication operation 21 is assigned to multiplier $M2$ according to the binding rules. Similarly, binding of remaining pseudo-operations is performed. Post binding of pseudo-operations with the respective existing FU resources, the structurally transformed scheduled and resource-allocated DFG of DWT is shown in Figure 9.6. Table 9.2 highlights the number of instances and required multiplexer size for FU resources, pre and post performing pseudo-operations mixing-based SOB. As shown in the table, binding of pseudo-operations to the resource $M1$ and $A1$ does not increase their respective multiplexer size. However, binding of pseudo-operations to the resource $M2$ increases its size from 2:1 to 4:1. This results into a slight design cost overhead due to extra interconnect hardware. However, in many cases, the available interconnect hardware is capable of associating pseudo-

Table 9.2 Multiplexer size determination pre and post performing pseudo-operations mixing-based structural obfuscation

FU resources	Before mixing and binding pseudo-operations		Post mixing and binding pseudo-operations	
	Number of instances	Required multi-plexer size	Number of instances	Required multi-plexer size
Adder (*A*1)	9	16:1	10	16:1
Multiplier (*M*1)	6	8:1	8	8:1
Multiplier (*M*2)	2	2:1	4	4:1

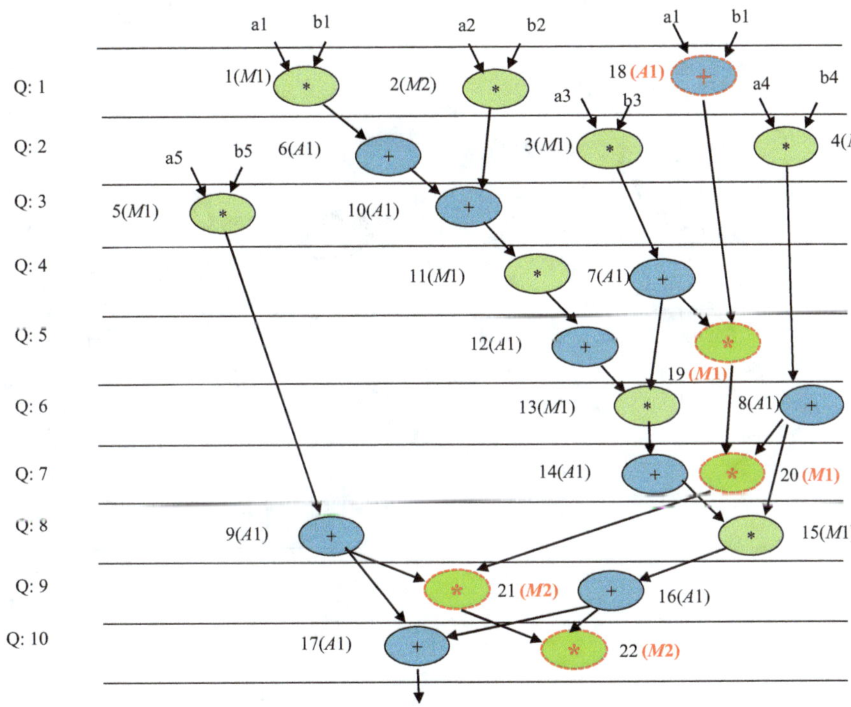

Figure 9.6 Structurally transformed scheduled and resource-allocated DFG of DWT application post mixing and binding of pseudo-operations

operations with the existing respective FU resource without incurring interconnect hardware overhead (or without increasing multiplexer size). Hence, this approach offers the SOB security at minimal area overhead.

RTL data path synthesis: The RTL data path post performing pseudo-operations mixing-based SOB is synthesized from structurally transformed scheduled and allocated DFG. For the sake of comparison, the RTL data path of DWT application pre and post-SOB is shown in Figures 9.7 and 9.8, respectively. The unsecured RTL data path (pre-obfuscation) is obtained from scheduled and allocated DFG shown in Figure 9.3. The secured (structurally obfuscated) RTL data path is obtained from structurally transformed scheduled and allocated DFG, as shown in Figure 9.6. In the structurally obfuscated RTL data path shown in Figure 9.8, the changes due to pseudo-operations mixing-based obfuscation are highlighted in red colour.

9.3 Pseudo-operations mixing-based structural obfuscation tool

Authors have developed a *POM-SO tool* (pseudo-operation mixing-based SOB tool) to simulate and analyse the pseudo-operation mixing-based obfuscation

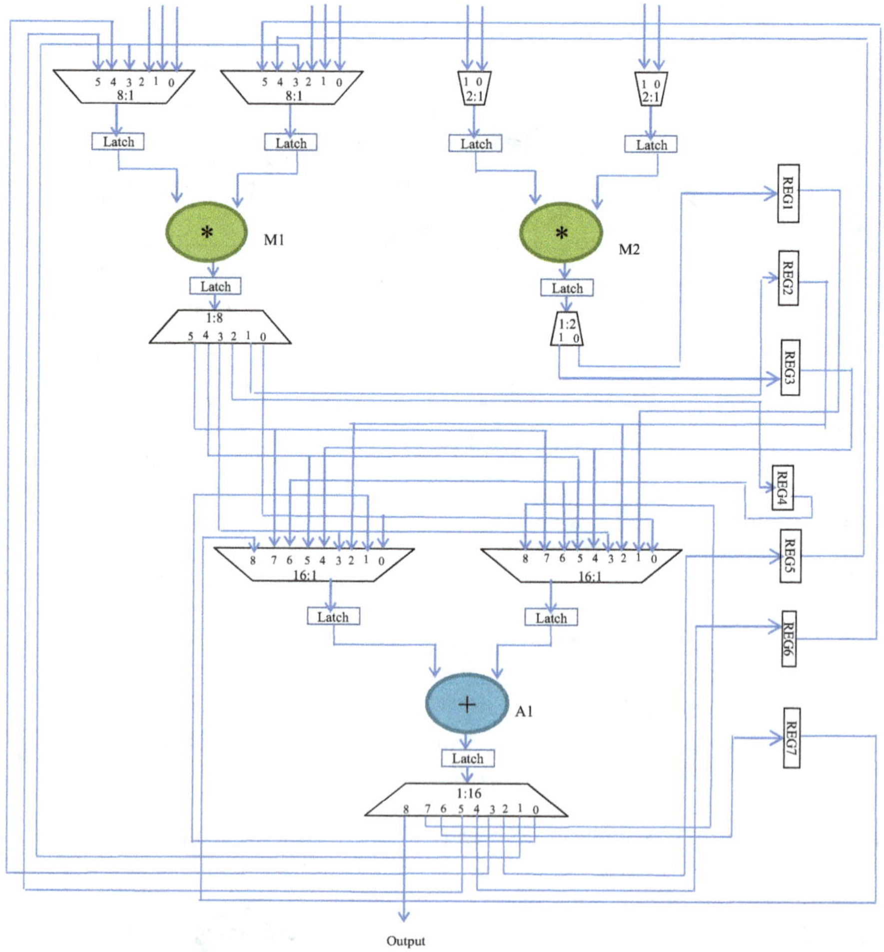

Figure 9.7 RTL data path before structural transformation-based obfuscation

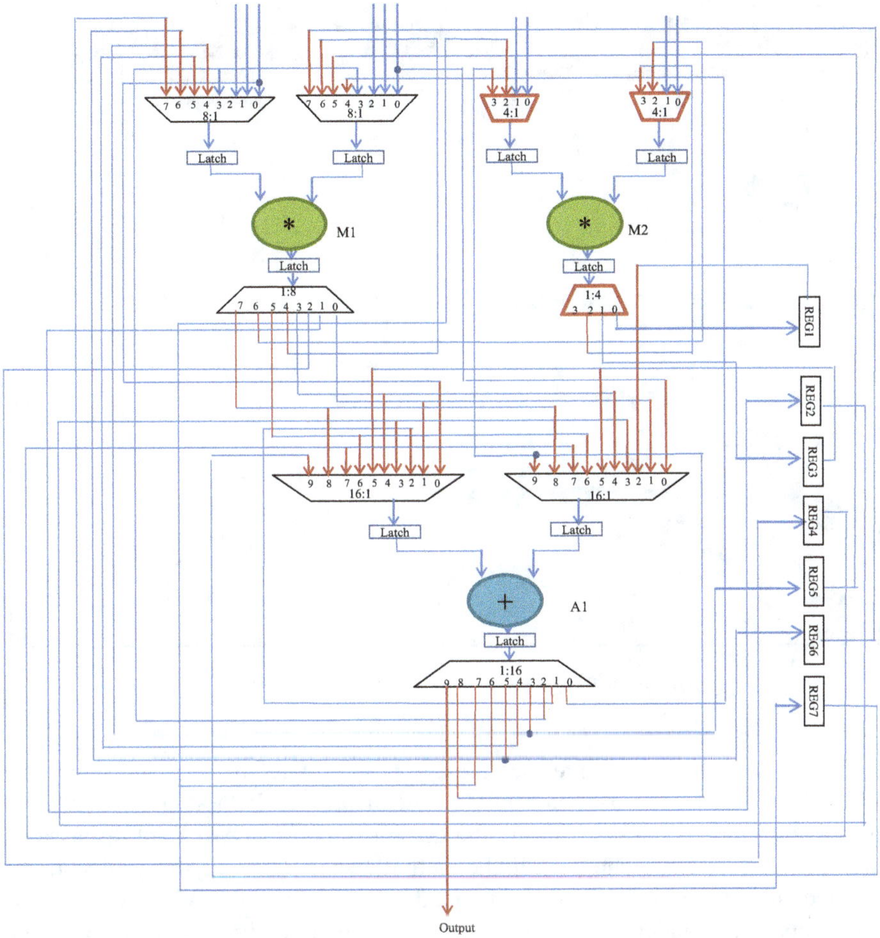

Figure 9.8 Structurally obfuscated RTL data path using pseudo-operations mixing-based structural transformation

approach for securing DSP hardware accelerators. *This tool provides a friendly graphical interface to users and available for free download publicly at:* http://www.anirban-sengupta.com/Hardware_Security_Tools.php.

A snapshot of the graphical user interface (GUI) of the tool is shown in Figure 9.9. The left portion of the tool shows the panel for providing required inputs to the tool, whereas the right portion shows the panel to see the intermediate and final outputs of the pseudo-operation mixing-based SOB approach. The POM-SO tool accepts the DSP application input in the form CDFG along with module library and resource constraints. The tool shows intermediate steps of pseudo-operation mixing and finally generated structurally transformed scheduled and resource-allocated DFG at the output.

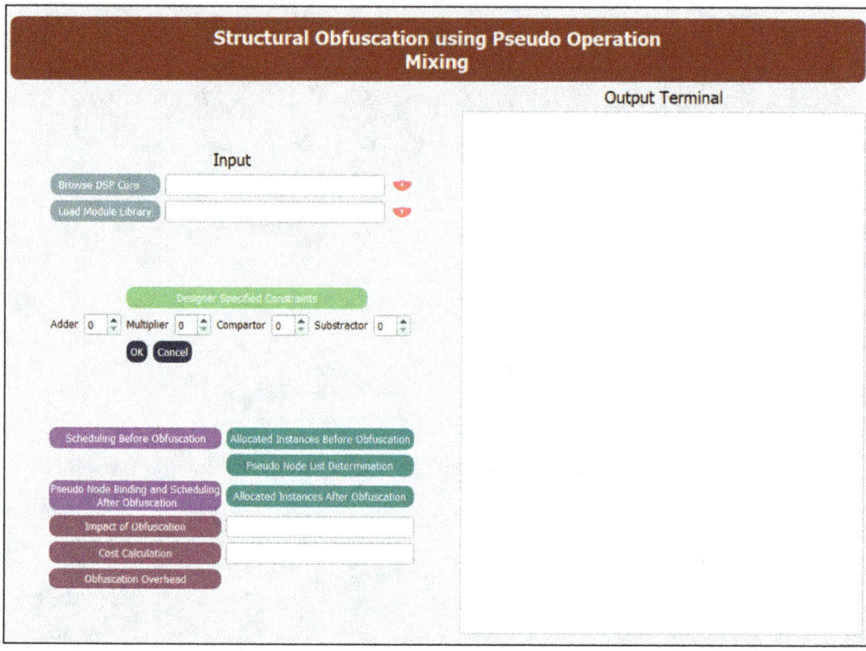

Figure 9.9 Snapshot of GUI of POM-SO tool

Let us generate all the intermediate and final outputs of the pseudo-operation mixing-based SOB approach for DWT core using the POM-SO tool. We will provide the same inputs used during the demonstration of DWT core discussed in Section 9.2. Here, we can match the output generated with the tool and that obtained in the demonstration. First of all, input DFG of DWT core, resource constraints of one adder and two multipliers and module library are fed to the tool as shown in Figure 9.10. Upon clicking on the button 'Scheduling Before Obfuscation', the scheduled and resource-allocated DFG becomes available on to the output terminal. Here, only an excerpt of the graph (up to six control steps) is shown in Figure 9.10. Further, upon clicking on the button 'Pseudo Node List Determination', the list of pseudo-operations to be mixed and corresponding control step number becomes available onto the output terminal as shown in Figure 9.11. The structurally transformed scheduled and resource-allocated DFG post mixing pseudo-operations can also be seen at output terminal by clicking on the respective button. Figure 9.12 shows the structurally transformed scheduled and resource-allocated DFG of DWT core.

Further, Figure 9.13 shows the number of instances of FU resources pre and post performing pseudo-operations mixing-based SOB. The tool produces the desired outputs that match with the demonstration on DWT core discussed in Section 9.2. This tool is useful for case studies of various kinds of DSP hardware accelerator applications such as FIR filter, infinite impulse response (IIR) filter and

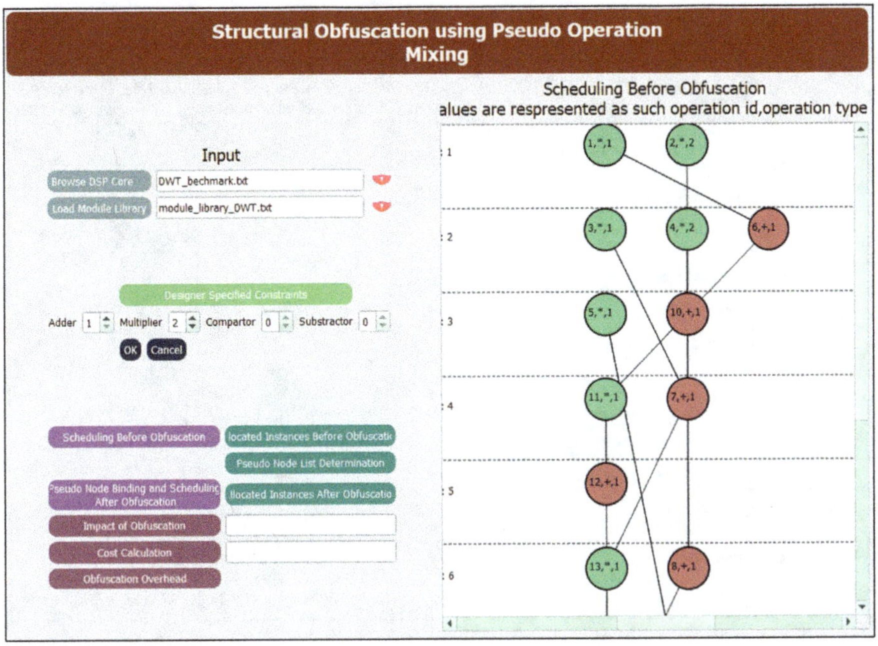

Figure 9.10 Snapshot of scheduled and resource-allocated DWT application

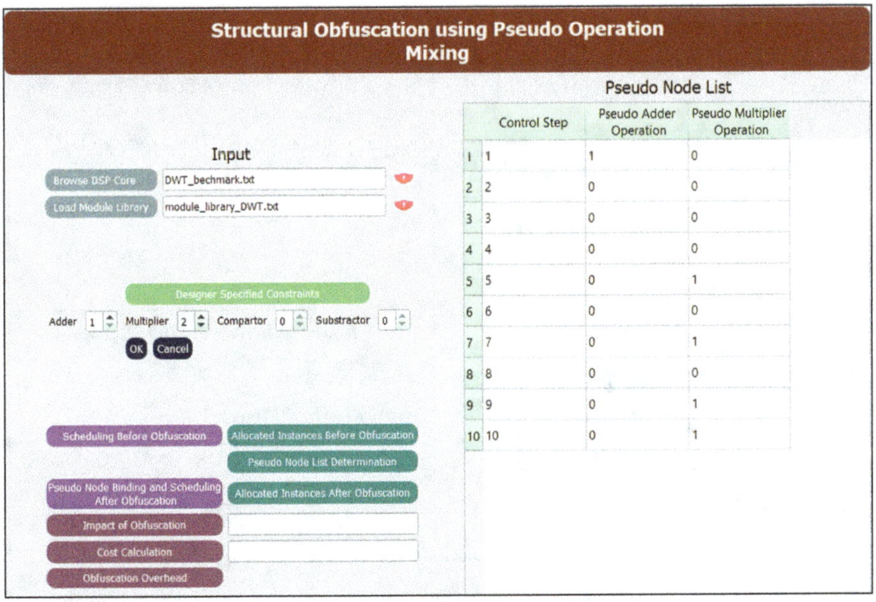

	Control Step	Pseudo Adder Operation	Pseudo Multiplier Operation
1	1	1	0
2	2	0	0
3	3	0	0
4	4	0	0
5	5	0	1
6	6	0	0
7	7	0	1
8	8	0	0
9	9	0	1
10	10	0	1

Figure 9.11 Snapshot post determining the list of pseudo-operations

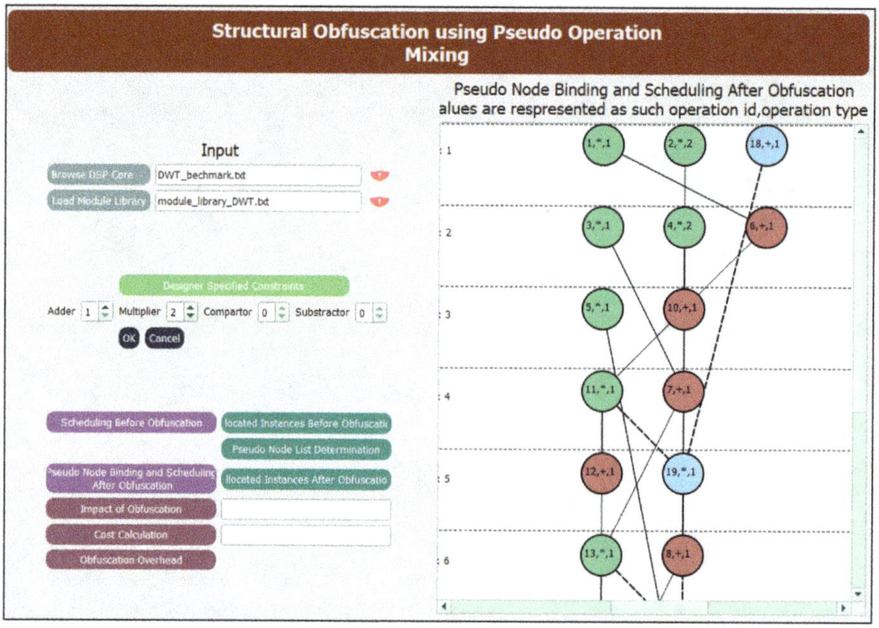

Figure 9.12 Snapshot of structurally transformed scheduled and resource-allocated DWT application

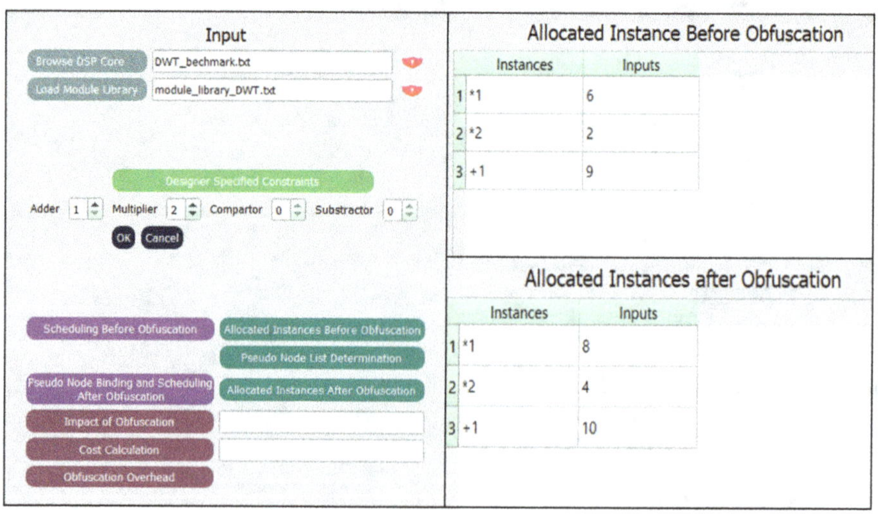

Figure 9.13 Snapshot showing impact on resource instances post structural obfuscation

Table 9.3 Security analysis in terms of strength of structural obfuscation and comparison with (Sengupta and Rathor, 2019)

DSP applications	IIR	Mesa Horner	JPEG	MPEG	DWT
Gate count (baseline)	3,648	4,192	29,856	8,272	6,288
Affected gate count (Rathor and Sengupta, 2020)	3,168	3,168	8,256	4,752	5,472
Strength of obfuscation (Rathor and Sengupta, 2020)	86.84%	75.57%	27.65%	57.44%	93.89%
Strength of obfuscation (Sengupta and Rathor, 2019)	26.3%	NA	NA	NA	NA

Note: NA indicates that the obfuscation approach is 'not applicable'.

DWT. In addition, the tool evaluates and shows the strength of obfuscation and design cost overhead post performing the SOB.

9.4 Analysis on case studies

This section analyses the security due to pseudo-operations mixing-based SOB and its impact on design cost. The case study in terms of security and design cost analysis has been performed on various DSP applications. Further, the security due to pseudo-operations mixing-based SOB (Rathor and Sengupta, 2020) has been compared with a contemporary SOB approach (Sengupta and Rathor, 2019).

9.4.1 Security analysis (Rathor and Sengupta, 2020)

The pseudo-operation mixing-based SOB approach deludes an adversary using pseudo-operations mixed into the design therefore renders the design architecture arduous to be reverse engineered by the adversary. Thus, the pseudo-operation mixing-based obfuscation approach is very useful in preventing Trojan insertion by the adversary. The RTL data path of DSP applications is affected due to pseudo-operation mixing-based SOB in terms of more number of times sharing of available FU resources, among original and pseudo-operations, using multiplexers and demultiplexers. Further, the mixing of pseudo-operations hugely obscures the gate-level netlist obtained post logic synthesis. This is because the mixing of pseudo-operations affects the large percentage of gates. Hence, the per cent gate count affected due to obfuscation is the measure of strength of SOB. The formula for evaluating the strength of SOB in terms of per cent gate count affected is as follows:

$$\%SOB = \frac{AG}{BG} \times 100 \tag{9.1}$$

where AG indicates total affected gate count (with respect to baseline) post SOB, whereas BG indicates total gate count of baseline (un-obfuscated) design. Here, the total AG is calculated as follows:

$$\text{AG} = G^{AR} + G^{C} \tag{9.2}$$

where G^{AR} indicates gate count of affected resources (such as affected multiplexers and demultiplexers) post obfuscation, and G^{C} indicates change in gate count post obfuscation (i.e. difference of gate count of baseline and obfuscated design). Table 9.3 shows the gate count of baseline design, total AG (calculated using (6.2)) due to obfuscation and strength of obfuscation (calculated using (6.1)) in terms of per cent gates affected with respect to baseline. As evident from the table, high value of strength of obfuscation is achieved using pseudo-operations mixing-based SOB. Further, the strength of obfuscation using pseudo-operations mixing-based SOB (Rathor and Sengupta, 2020) has been analysed in terms of comparison with the contemporary approach (Sengupta and Rathor, 2019) as shown in Table 9.3. Since the contemporary approach (Sengupta and Rathor, 2019) is based on the integration of two RTL data path of such DSP applications which have some similarity in their algorithmic description, it is applicable to a limited number of DSP applications. Thus, it is not widely applicable obfuscation approach. However, the pseudo-operations mixing-based SOB (Rathor and Sengupta, 2020) can be applied extensively.

9.4.2 Design cost analysis (Rathor and Sengupta, 2020)

This subsection discusses the impact of employing pseudo-operation mixing-based SOB on design cost. The following equation is used to evaluate the design cost:

$$C_d(U_i) = \rho_1 \frac{L_d}{L_m} + \rho_2 \frac{A_d}{A_m} \tag{9.3}$$

where $C_d(U_i)$ is the design cost towards resource constraints U_i, further L_d and L_m are the design latency at specified resource constraints and maximum design latency, respectively, A_d and A_m are the design area at specified resource constraints and maximum area, respectively, and ρ_1, ρ_2 are the weights which are fixed at 0.5.

The design cost analysis of pseudo-operation mixing-based obfuscation approach has been performed in terms of design cost comparison with the baseline/un-obfuscated version. Figure 9.14 shows the design cost analysis. It is observed from the figure that the design cost overhead due to pseudo-operation mixing is zero for most of the DSP application. The underlying reason is that the binding of pseudo-operations with the available respective FU resources does not result into additional interconnect hardware (multiplexers and demultiplexers). However, in some cases, the sizes of multiplexers and demultiplexers may need to be augmented to accommodate pseudo-operations. Nonetheless, the binding rules discussed in this chapter aim to minimize the interconnect hardware overhead. Hence, the pseudo-operation mixing-based obfuscation approach leads to minimal design cost overhead.

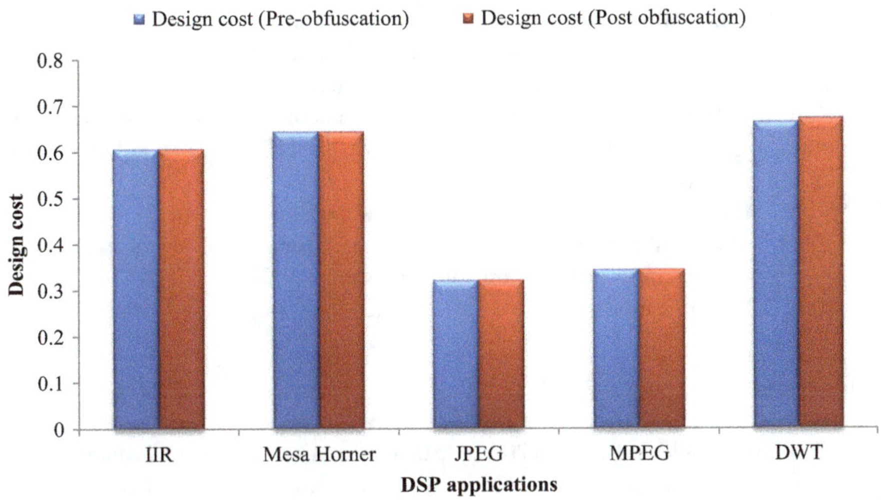

Figure 9.14 Design cost analysis with respect to baseline or un-obfuscated version (Rathor and Sengupta, 2020)

9.5 Conclusion

Employing security during the design process of data-intensive IP cores is required to ensure trust in hardware. This chapter discussed a pseudo-operations mixing-based structural transformation approach which obfuscates the designs of data-intensive DSP cores in order to prevent against Trojan (malicious logic) insertion threat. The robustness of this approach lies in the facts that the applicability of pseudo-operations mixing-based structural transformation approach has extensive coverage of target DSP applications, regardless of the nature of the application. Additionally, the pseudo-operations mixing-based SOB approach provides high security at minimal overhead.

At the end of this chapter, the following concepts are communicated to the readers:

- algorithm of determining pseudo-operations to be inserted in a scheduled and resource-allocated DFG;
- mixing rules of pseudo-operations into the scheduled and resource-allocated DFG;
- binding rules of pseudo-operations with the existing FU resources in the scheduled and resource-allocated DFG;
- demonstration of pseudo-operations mixing-based SOB on DWT core and
- case studies in terms of security and design cost analysis.

9.6 Questions and exercise

1. What is pseudo-operation determination algorithm?
2. Describe the security aware HLS follows.

3. State the algorithm of pseudo-operation mixing in scheduled DFG.
4. State the binding rules of pseudo-operation binding.
5. What is the role of list *W* of pseudo-operations?
6. How does pseudo-operation mixing algorithm achieve structural obfuscation?
7. Perform pseudo-operation mixing algorithm for DCT core to achieve structural obfuscation.
8. How is the multiplexer size determined after resource sharing?
9. What is the input/output of POM-SO tool used for structural obfuscation?
10. What is the strength of structural obfuscation?
11. How is the design cost analysed for a structurally obfuscated design?

References

R. S. Chakraborty and S. Bhunia (2009), 'Security against hardware Trojan through a novel application of design obfuscation,' *Proc. International Conference on Computer-Aided Design*, ACM, pp. 113–116.

M. Rathor and A. Sengupta (2020), 'Obfuscating DSP hardware accelerators in CE systems using pseudo operations mixing,' *Proceedings of 4th IEEE International Conference on Zooming Innovation in Consumer Electronics 2020 (ZINC 2020)*, Serbia, pp. 218-221, doi: 10.1109/ZINC50678.2020.9161775

R. Schneiderman (2010), 'DSPs evolving in consumer electronics applications,' *IEEE Signal Process. Mag.*, vol. 27(3), pp. 6–10.

A. Sengupta (2016), 'Intellectual property cores: protection designs for CE products,' *IEEE Consum. Electron. Mag.*, vol. 5(1), pp. 83–88.

A. Sengupta (2017), 'Hardware security of CE devices [hardware matters],' *IEEE Consum. Electron. Mag.*, vol. 6(1), pp. 130–133.

A. Sengupta, D. Roy, S. P. Mohanty and P. Corcoran (2017), 'DSP design protection in CE through algorithmic transformation based structural obfuscation,' *IEEE Trans. Consum. Electron.*, vol. 63(4), pp. 467–476.

A. Sengupta and S. P. Mohanty (2019), 'IP core and integrated circuit protection using robust watermarking,' *IP Core Protection and Hardware-Assisted Security for Consumer Electronics*, e-ISBN: 9781785618000, pp. 123–170.

A. Sengupta and M. Rathor (2019), 'Protecting DSP kernels using robust hologram-based obfuscation,' *IEEE Trans. Consum. Electron.*, vol. 65(1), pp. 99–108.

A. Sengupta (2020), 'Frontiers in securing IP cores – forensic detective control and obfuscation techniques', The Institute of Engineering and Technology (IET), ISBN-10: 1-83953-031-6, ISBN-13: 978-1-83953-031-9.

X. Zhang and M. Tehranipoor (2011), 'Case study: detecting hardware Trojans in third-party digital IP cores,' *IEEE International Symposium on Hardware-Oriented Security and Trust*, San Diego, CA, pp. 67–70.

Index

www.ingramcontent.com/pod-product-compliance
Lightning Source LLC
LaVergne TN
LVHW011833200926
848484LV00002B/353